Watershed Hydrology

Peter E. Black

State University of New York
College of Environmental Science and Forestry
Syracuse, New York

Prentice Hall
Englewood Cliffs, New Jersey 07632

Library of Congress Cataloging-in-Publication Data

Black, Peter E.
   Watershed Hydrology / by Peter E. Black.
      p.   cm.
   Includes bibliographical references.
   ISBN 0-13-946591-X
   1. Watersheds.   I. Title.
GB980.B57   1990
551.48--dc20   90-33324
                                                                       CIP

Editorial/production supervision: *Jacqueline A. Jeglinski*
Cover design: *Wanda Lubelska Design*
Manufacturing buyer: *Kelly Behr/Susan Brunke*

Prentice Hall Advanced Reference Series
Physical and Life Sciences

© 1991 by Prentice-Hall, Inc.
A division of Simon & Schuster
Englewood Cliffs, New Jersey 07632

This manuscript was prepared, camera-ready, on an IBM compatible 80286 AT computer, using desktop publishing features of WORD®, version 5.0, of the Microsoft Corporation. The scalable fonts used were Tymes Roman, Helvenica, Exchequer Script, and special characters by GLYPHIX®, SWFTE International, Ltd. Certain analyses are the product of STATGRAPHICS® and APL*PLUS/SYSTEM® of the Scientific Time Sharing Corportion. Illustrations were prepared with the aid of TOPO® and SURF® by Golden Software; QUATTRO®PRO by Borland, International; PC PAINT® by Mouse Systems Corporation; and TOUCH UP by Migraph, Inc. Scanned illustrations were handled on a PC SCAN 1000 tabletop scanner, with software by New DEST Corporation and Microsoft WINDOWS®. Printing was performed on a LaserJet III® printer of the Hewlett-Packard Company.

The publisher offers discounts on this book when ordered in bulk quantities. For more information, write:

     Special Sales/College Marketing
     Prentice-Hall, Inc.
     College Technical and Reference Division
     Englewood Cliffs, New Jersey 07632

All rights reserved. No part of this book may be
reproduced, in any form or by any means,
without permission in writing from the publisher.

Printed in the United States of America
10 9 8 7 6 5 4 3 2

ISBN 0-13-946591-X

Prentice-Hall International (UK) Limited, *London*
Prentice-Hall of Australia Pty. Limited, *Sydney*
Prentice-Hall Canada Inc., *Toronto*
Prentice-Hall Hispanoamericana, S.A., *Mexico*
Prentice-Hall of India Private Limited, *New Delhi*
Prentice-Hall of Japan, Inc., *Tokyo*
Simon & Schuster Asia Pte. Ltd., *Singapore*
Editora Prentice-Hall do Brasil, Ltda., *Rio de Janeiro*

To my students

Contents

List of Figures x
List of Tables xiii
Preface xv
Acknowledgments xxiii

1 Introduction

    THE HYDROLOGIC CYCLE 1

    STORAGE 5

        Types of Storage 5

        Concepts of Storage 7

    THE WATERSHED 11

    SUMMARY 14

2 Water and Energy

    ENERGY MOVEMENT 15

        The Source 15

            The Amount of Energy 15

            The Quality of Energy 16

            Energy at the Evaporating/Transpiring Surface 18

        The Geometry of Energy 20

        The Energy Budget 22

        Instruments and Limitations 27

    THE HYDROLOGIC CYCLE 29

        The Amount of Water 29

        The Quality of the Water 30

        The Role of Water in the Energy Sphere 31

    SUMMARY 32

    PROBLEMS 33

3 Water in the Atmosphere

STORAGE 34
- Characteristics 34
  - *Pressure 34*
  - *Temperature 36*
  - *Humidity 42*

PROCESSES 47
- Global Circulation 47
- Precipitation 50
  - *Sources of Precipitation 50*
  - *Forms of Precipitation 57*
  - *Temporal and Spatial Distribution 64*
- Evaporation 73
  - *The Amount 73*
  - *Factors Affecting Evaporation 73*
  - *Dalton's Law 74*
  - *Instruments, Limitations, and Measurement 75*

CLIMATE 79
SUMMARY 80
PROBLEMS 81

4 Water in the Vegetated Zone

PROCESSES 82
- Interception 82
  - *Mechanical Effects of Interception 83*
  - *Quantitative Effects of Interception 83*
  - *Conservational Effects of Interception 91*
- Transpiration 94
  - *The Movement of Water and Nutrients 94*
  - *Measuring Transpiration 96*
- Evapotranspiration 96
  - *Measuring Evapotranspiration 98*
- Using the Water Budget 102
  - *Plots 102*
  - *Lysimeters 102*
  - *Experimental Watersheds 103*
- Evapotranspiration Research 107
  - *General Studies 107*
  - *Vegetation Manipulation 109*

Contents   vii

                Field Investigations of Hydrologic Impacts 119
                Regional Managment Potential for Water Yield Manipulation 121
                Summary 124
        Models   124
        Phreatophytes   127
        Summary   129
    STORAGE   129
        Water Stored in Vegetation   129
        Water Stored on Vegetation   131
        Water Stored in Hydrospheric Systems   132
            Wetland Definition 132
            Wetland Values 132
            Wetland Classification 133
            An Hydrology and Water Quality-Based Classification 135
        Summary   137
    PROBLEMS   138

5 Water in the Terrasphere

    STORAGE   139
        Geologically-Bound Water   140
        Ground Water   140
        Soil Water   143
            Porosity 143
            Soil-Forming Factors 151
            Types of Soil Water Storage 166
        Measurement-Instruments-Limitations   170
            Gravimetric Methods 170
            Tensiometric Methods 172
            Electrical Resistivity Methods 172
            Neutron Scattering Methods 173
            Remote Sensing Methods 174
            Miscellaneous Methods 174
        Linking Storage/Movement from Soil to Stream   175
    PROCESSES   176
        Infiltration   176
            Measurement-Instruments-Limitations 177
        Percolation   179
            Measurement-Instruments-Limitations 180
        Transmission   181
            Measurement-Instruments-Limitations 181

SOIL/VEGETATION/WATER CLASSIFICATION AND HYDROLOGY   181
SUMMARY   182
PROBLEMS   182

6 Water in the Hydrosphere

WATER YIELD   183
    Storage in Lakes and Ponds   184
    Runoff in Streams and Rivers   185
        Measurement-Instruments-Limitations   186
        Patterns of Flow   193
        Stream Behavior   210
    Precipitation-runoff Relations   224
    Models and Modeling   227
        Physical Models   229
        Deterministic Models   229
        Stochastic Models   231
        Electronic Models   232
WATER QUALITY   233
    Properties of Water   234
        Molecular Structure   234
        Density   234
        Viscosity   235
        Specific Heat   235
    Water Quality Characteristics   236
        Physical Characteristics   236
        Chemical Characteristics   237
        Biological Characteristics   240
        Monitoring Water Quality   241
    Using Water Quality to Help Understand Hydrology   246
    Summary   246
PROBLEMS   247

7 Water on the Watershed

THE WATERSHED   248
    Watershed Morphology   249
    Storage on the Watershed   250
    The "Small" Watershed   250
    Characterizing the Watershed   251
        Size   254
        Elevation and Slope   257

        Aspect and Orientation   260
        Watershed Shape   261
        Drainage Network   266
    Summary   270
  WATERSHED MANAGEMENT   270
    Watershed Equilibrium   270
    Inverse Influence   272
    Disproportionate Percentages   273
    Summary   274
  PERSPECTIVES ON WATERSHED MANAGEMENT   274
    Methodologies for Modifying the Water Resources Environment   274
    Watershed Management and Large-Scale Changes   276
    Summary   279
  PRACTICE OF WATERSHED MANAGEMENT   279
    Rehabilitation   280
    Protection   285
    Enhancement   289
    Summary   305
  NONPOINT SOURCES OF POLLUTION   305
    The Legal Basis   305
    The Process of Nonpoint Source Pollution Control   306
    Best Management Practices Principles   308
    Best Management Practices on Wildlands   308
  PROBLEMS   309

Epilogue   310

References   313

Appendices   365
  A Conversion Tables   366
  B Map Scales   370
  C Watershed Eccentricity   371
  D Glossary   372

Author Index   375

Subject Index   385

List of Figures

Plate 1 Realms of water xviii

1-1   The hydrologic cycle 3
1-2   Storage and the hydrologic cycle 6
1-3   Distribution of water on Earth 11
1-4   A typical storm hydrograph 13

2-1   Wavelength of solar and terrestrial and spectral bands 18
2-2   Geometry of radiation relationships between the Sun and the Earth 21
2-3   Radiation relationships with aspect and latitude 22
2-4   Average annual energy budget 24
2-5   Two types of pyroheliometers 28

3-1   Basics of the Coriolis force 36
3-2   The Aneroid Barometer 37
3-3   Standard Weather Bureau Shelter 38
3-4   Standard Thermometer and Sling Psychrometer 39
3-5   Standard Hygrothermograph 41
3-6   Saturation vapor pressure and temperature 44
3-7   The atmosphere on the nonrotating Earth 41
3-8   Atmospheric circulation of the northern hemisphere 49
3-9   Life history of a typical cyclonic storm 51
3-10 Vertical profiles through cold and warm fronts, and occluded fronts 52
3-11 Orographic precipitation patterns between Los Angeles and Denver 54
3-12 Nonrecording Rain and Snow Gage 59
3-13 Universal Recording Rain and Snow Gage 60
3-14 Sample recording rain gage chart 61
3-15 Mount Rose snow sampler 63
3-16 Area-depth curves 67
3-17 Sample of depth-frequency-occurrence in the United States 68
3-18 Methods for calculating areal precipitation 70

List of Figures xi

3-19 Standard Class A Evaporation Pan 75
3-20 Hook Gage and Stilling Well 76

4-1 The interception process in the forest 83
4-2 Stemflow gage 88
4-3 "Misting" in the Colorado Rockies 92
4-4 Climatogram 105
4-5 Moisture index and vegetation along 41°N latitude 108
4-6 Relation between typical hydrograph and pollutograph 116
4-7 Nutrient loss after forest distrubance 117
4-8 Data summary of Coast Redwood study 131

5-1 Water in the terrasphere 142
5-2 Comparison of various systems of particle-size classes 144
5-3 Soil Texture Triangle 145
5-4 Volume Composition of an Average Soil 146
5-5 Capillary conductivity, tension, and soil particle size 150
5-6 Capillary rise in soils 151
5-7 Soil Profile Showing Theoretical Horizonation 153
5-8 Great Soil Group distribution 156
5-9 Vegetation distribution 157
5-10 Distribution of Great Soil Groups in the United States 158
5-11 Distribution of broad vegetative types in teh United States 159
5-12 Distribution of soils under the Comprehensive Soil Classification System 160-1
5-13 A Typical Catena of the High Plains 164
5-14 Mull and Mor Humus Types 165
5-15 Humus type under an eastern white pine stand 167
5-16 Schematic diagrams of soil-moisture relationships 168
5-17 Soil storage water relations 169
5-18 Water relations and soil texture 169
5-19 Infiltration rate and precipitation excess 180

6-1 Open channel streamflow relations: the cubic foot per second 186
6-2 Theoretical horizontal and vertical distribution of flow velocities in a uniform channel 187
6-3 Typical year-by-days streamflow data page 188
6-4 Typical peak runoff data page 189
6-5 Current Meter 190
6-6 Stream divided into trapezoids for measurement of discharge 191
6-7 Details of research weir 193
6-8 Schematic diagram of 120° V-notch weir 194
6-9 Schematic diagram of Venturi flume 195
6-10 Typical streamflow recorder 196
6-11 Average annual runoff in the United States 197
6-12 Normal distribution of runoff by months 197

6-13   A theoretical example illustrating the variable source area concept   199
6-14   A Typical storm hydrograph   202
6-15   Values of the runoff coefficient, c_{max} 206
6-16   Map of n values for durations between 5 min and 60 min   207
6-17   Map of n values for durations between 60 min and 1440 min   207
6-18   Map of k values for durations between 5 min and 60 min   208
6-19   Map of k values for durations between 60 min and 1440 min   208
6-20   Map of x values   209
6-21   Flow duration curve for data shown in Table 6-3   213
6-22   Flow duration curves for four California north coastal streams   214
6-23   Flow duration curves for four Mohawk River streams in New York   215
6-24   Annual flood frequency curve for the 3113-Square-Mile Eel River Watershed at Scotia, California, 1915-1965   219
6-25   Flood frequency curves for data shown in Table 6-5   221
6-26   Runoff per unit precipitation in the Mohawk River watershed   225
6-27   The water molecule   234

7-1   Millswitch Creek Watershed   249
7-2   Topographic and phreatic divides   252
7-3   Topographic map of the Millswitch Creek watershed   253
7-4   Hypsometric curve   258
7-5   The topographic sampler   262
7-6   Watershed shape and time of concentration   263
7-7   Hydrographs from watershed models where storm moved up and down main axis of model   264
7-8   Horton's stream numbers   267
7-9   Examples of drainage patterns   268
7-10   Geologic classification of streams   269
7-11   Water walking strucutre system   283
7-12   Check dam for gully control   284
7-13   Longitudinal profiles of gullies   285
7-14   Forest types in the united States   292
7-15   Diagrammatic representation of wall and step forest in area rotation   298
7-16   Example of stream bank stabilization system   304

List of Tables

1-1 Distribution of Earth's water and residence times   9

2-1 Range of Values of Albedo for Some Natural Objects   17
2-2 Comparison of Energy Budget Components for Forest and Field Crops on Cloudy and Clear Days   26

3-1 Principal Constituents of the Atmosphere   43
3-2 Methods of Calculating Mean Precipitation Shown in Figure 3-18   71
3-3 Climatic Region and Vegetation Type   80

4-1 Summary of Interception Storage on Vegetation   84
4-2 Number of Gages Needed for Throughfall Measurement to Keep the Standard Error under 5 Percent of Mean Throughfall   86
4-3 Evapotranspiration Factors   90
4-4 Sources, Formulas, and Explanations for the Computer Calculation of the Thornthwaite Water Budget   106
4-5 Water Budget for Poughkeepsie and Wappingers Falls, NY   107
4-6 Summary of Changes in Streamflow under Different Conditions of Timber Cutting at the Fernow Experimental Forest, West Virginia   113
4-7 Hydrologic and Water Quality Characteristics of Wetlands   137

5-1 Energy Levels of Common Soil Moisture Constants   146
5-2 Size Limits of Soil Separates   147
5-3 Effect of Particle Size on Space and Surface Area   147
5-4 Textural Classes of Soils and Permeability   148
5-5 Comparison of old and new soil classification systems   162

6-1 Values of Manning's Roughness Coefficient, n 187
6-2 Runoff Coefficients for Use in the Rational Formula   204
6-3 Daily Flow Duration Analysis for 3113-Square-Mile Eel River Watershed at Scotia, California, 1911-1955   212

6-4   Annual Flood Frequency Analysis for the 3113-Square-Mile Eel River Watershed at Scotia, California, 1915-1965   218
6-5   Background Data for Drainages Used for Flood Frequency Analysis in Figure 6-14   220
6-6   Classification of Stream Rises   226
6-7   Relationship Between Particle Diameter and Stream Velocity   237
6-8   Summary of Rock Composition   239

7-1   Calculation of Hypsometric Curve   258
7-2   Calculation of Elevation-Weighted Mean Annual Precipitation   259
7-3   Hydrologic Characteristics of Major Forest Types in the United States   293

Preface

The purpose of this book is twofold. First, I wish to present my approach to and understanding of hydrology. That approach is best designated as "Watershed Hydrology." Second, I hope to provide students of hydrology with a wildland-oriented text with a focus on storage. My understanding is one born of courses in both engineering and ecological hydrology, by practicing field work and research, and through my consulting experience. I have seen engineers who have paid little attention to the natural world around them, and environmentalist protectionists who have paid little attention to the realities of water as a physical entity in the environment. My organization is based on scientifically founded knowledge of the hydrologic environment in its natural state. This posture is essential in order to comprehend the interactions of natural hydrologic events and the impacts of our inadvertent and planned activities on that environment.

Generally, this book is about watershed hydrology. A watershed is a natural unit of land from which the surface, subsurface, and ground water runoff drain to a common outlet. Hydrology is the study of water in the natural environment; the study of the laws, principles, processes, and quantities of water movement and storage. In this book, the emphasis is on wildlands. Watershed hydrology, then, is simply the study of the movement and storage of water on and in the context of the natural land unit of the hydrosphere, the watershed.

Why another book on hydrology? Obviously, none of them fill the bill. I have about a dozen on my bookshelf with the word "hydrology" in the title, and another dozen that deal with hydrologic processes in one way or another, without actually saying so in the title. The beginning of an answer lies in either the inappropriateness or unavailability of any text truly suitable for the Forest Hydrology course taught at the SUNY College of Environmental Science and Forestry and, no doubt, elsewhere. Dunne and Leopold's 1978 book, for instance, entitled "Water in Environmental Planning," comes close, but was unavailable for some time and still presents only minimum attention to wildland hydrology. At the other extreme is Colman's monumental literature review, "Vegetation and Watershed Management" which, in addition to being published in 1953 and, therefore, not being quite up to date, omits useful engineering aspects of hydrology. (The book is normally used for a second course in a wildland hydrology sequence, such as "Practice of Watershed Management," so, presumably, the basics would have been presented in the introductory course.)

Other published hydrology books pay little attention to the vast storage potential of the soil, or to the wildland hydrologic processes sometimes referred to as "Forest Influences":[1] most are arranged by processes.

The processes, of course, are important and obvious. They are in striking evidence as billowing clouds, falling water, or flowing streams. Perhaps the excessive attention to processes, then, should not surprise us. Even the vast storage capacity of oceans appear to us visually as waves, water in motion: it is difficult to comprehend the volume of water involved as evidenced by the fact that we think nothing of traversing — or "seeing" — a mile horizontally on the ocean's surface, but cannot comprehend the same distance in depth. And the primary reason we know as little about and do so little to protect or develop ground water is simply that we cannot see it, much less comprehend its complex physical and chemical characteristics and relationships with its surroundings.

By and large, hydrology texts do discuss storage of water. To my knowledge, however, none of the books have storage as a focus. In fact, storage is a topic that has been seriously ignored in the texts and, as a consequence, in some rather basic hydrologic model research. A rare exception is B. J. Knapp's 1979 booklet entitled "Elements of Geographical Hydrology": in the first figure of the book, the hydrologic cycle is depicted as a series of processes moving from one type of storage reservoir to another. Slight attention to vegetation limits the utility to the wildland manager of this otherwise fine presentation on runoff processes and the hydrograph.

The processes that move water in and out of storage sustain major changes as undisturbed or rural lands undergo transition to urban and suburban uses, profoundly affecting storage. It is here that major floods are generated, that water quality begins to deteriorate, and that other problems such as subsidence begin. It is to a balanced consideration of both process and storage, then, that we must turn our attention. Indeed, it is essential if we are to develop our society without ruining the water resource on which we depend for domestic and residential use, commerce, industry, power, aquatic resources, and recreation.

Major exceptions to the lack of attention to storage in the available hydrology books are the chapters on snow, soil moisture, and ground water. Even these generally focus on movement through the porous media in which the water is found, not how the storage affects runoff. A major interest, of course, is the outflows from snow, soil, and ground water, the three major storage locations in the terrestrial portion of the hydrologic cycle. They deserve major consideration. However, water is also stored in the atmosphere and biosphere. By considering the "spheres" first as types of storage each with characteristics influenced by basic principles and water properties — some patterns of substance and presentation emerge. How is the bulk of water in each realm characterized? How does water get into and out of that segment of our environment in light of its characteristics? And what is the perspective on water's

---

[1] The title of Kittredge's 1948 book in the American Forestry Series. (That book, like Wisler and Brater's "Hydrology," Linsley, Kohler and Paulhus' "Applied Hydrology," and others do have two extensive chapters on snow and soil water storage.) John Manning's 1987 "Applied Principles of Hydrology" presents excellent figures and explanations of both storage and processes, but presumes for the reader a rather low level of scientific background. Many of his plain-language descriptions are works of art, but he omits discussion of vegetative influence. David Miller's "Water at the Surface of the Earth" has a greater discussion about storage than any other text, but it, like several other texts, is arranged by processes ("movement"), and storage is not a theme. So is the venerable "Handbook of Applied Hydrology" edited by Ven Te Chow, which focuses on — is arranged by — processes.

characteristics *between* the several realms of the hydrologic cycle? Out of these questions, an arrangement for studying and presenting watershed hydrology becomes apparent. Thus, the basic organization of this book, by realms or spheres of water.

The chapters are arranged so that each examines the primary characteristics of the parts of the environment, also referred to as "spheres": the energy sphere, the atmosphere, the biosphere, the terrasphere,[2] the hydrosphere, and the cultural sphere, without always using those specific names. In order to do this effectively, it is necessary first to present a brief overview — the terminology and broad relationships — of the hydrologic cycle (Chapter 1). This enables meaningful discussion of the effects of storage and process in one sphere on storage or processes in another to take place without having to skip around from one chapter to another.

This organization did not drop on me out of the blue. About twenty years ago, upon changing the title of the course to "Principles of Watershed Management," I reworked my lecture notes. In the process, I asked myself the question likely to be raised by students enrolling in the course: what *are* the principles? I came up with seven broad paradigms on which I felt I could elaborate and which embrace all of the basic physical (and related biological and chemical) rules of water movement and storage pertaining to the hydrologic cycle. The statements follow (the seventh one is the product of collaboration with my colleague Dr. Arthur R. Eschner):

1. Although it often appears otherwise, there is plenty of water on the Earth: its quality, and temporal and spatial distribution, are often undesirable.

2. The sun is the ultimate source of energy for the movement and storage of water and for its changes of state throughout the hydrologic cycle.

3. In the atmosphere, the movement of water is in response to the general global circulation and to local vapor pressure gradients.

4. In the vegetated zone, movement of water between the atmosphere and the soil plays a diversified role in the storage capacity of and the complex relationships between all three regimes.

5. In the soil, water moves in response to gravity when it is not responding to tension gradients.

6. In the stream, runoff is the integrator of all of the factors which affect its quantity, quality, and regimen, and is one of the factors of the ecology of the watershed itself.

7. Watershed management is the planned manipulation of one or more of the factors of the environment of a natural drainage so as to effect a desired change in or maintain a desired condition of the water resource.

The middle five statements are basic principles of how water moves and is stored within each realm or sphere. I have used these as sub-chapter axioms in the book. I also often use these concise sentences as the basis for essays on a final examination, asking the student to

---

[2] This term is used in preference to "lithosphere" in that the latter term embraces the solid rock portion of the Earth: while water contained therein — ground water — is certainly an important part of the hydrologic cycle, it is well covered in several modern texts. Ground water is dealt with insofar as it is an important part of the hydrologic cycle, but a greater amount of space is devoted to the moisture in the soil's aerated zone. The term was suggested to me by Dr. Robert E. Dils, my former Major Professor, on a recreational outing at about 10,000 feet in the Colorado Rockies, in August of 1989.

elaborate on the statement, explain what it means, how it is ramified, and what evidence there is to support it. Upon a subsequent re-instatement of the title of the course as "Forest Hydrology," I kept the organization and accompanying basic principles. They leave room for continued input of new information and understanding.

For example, during the work on the third draft of this book, I inserted an idea that had occurred to me during a lecture: I had earlier toyed with the idea that infiltration was such an elusive process that it might as well be a "fiction" and, during the class, the thought occurred to me that antecedent moisture conditions and time of concentration were in the same category. I wrote these up, briefly, and then added to them, deciding to discuss them in the Preface. I dramatized the importance of these and two other concepts, return period and variable source area, by referring to them as "synapses" without really being aware of what I was saying. Upon reflection, it further occurred to me that that is exactly what they are, connections ("processes" or "concepts") between spheres. Thus, the following several paragraphs offer another dimension of wildland hydrology, complementing the basic *within*-sphere principles, with *between*-sphere concepts. All are illustrated in Plate 1.

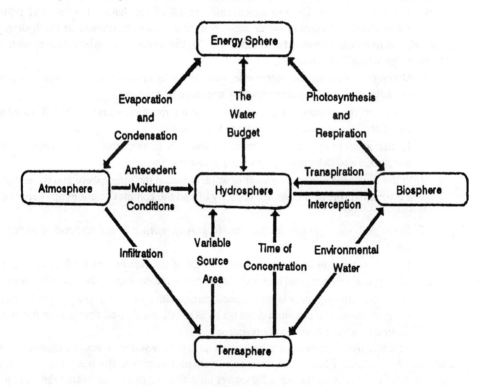

Plate 1 Realms of Water

I am of the opinion that there are eleven critically important concepts underlying the art and science of hydrology that are relatively easy to define abstractly, difficult to illustrate comprehensively, often impossible to measure or predict accurately, and yet are all absolutely essential to the thorough understanding of watershed hydrology and, therefore, to the success of watershed management practices. They are the complex, nonlinear processes that collectively describe the movement of water between spheres.[3] They are:

    1. Evaporation and **Condensation**, the names given to the changes of state between gas and liquid.

These two processes representing a bridge between the *energy sphere* and *atmosphere* play an enormously important role in the energy balance of the Earth, as shown in Chapter 2. They are the principal mechanisms by which the water and energy cycles interlock, driving our weather and climate, and individual storms such as hurricanes.

    2. The Water Budget, the balance of inputs and outputs in a hydrologic system over some specified time period, along with the factors that influence movement and storage within the system, are intimately associated with the energy sphere.

Joint consideration of the *energy sphere* and *hydrosphere* on a worldwide scope (or even over a watershed) reveals the complex interrelationships between the two realms. At the surface of the Earth the principal mobile consituent, water, plays a major role in the energy budget and vice versa. The combined system is so complex that it is chaotic,[4] which explains why evidence for that new science was first detected there (Gleick 1987) and, perhaps, why we cannot forecast global warming any more than we can with confidence predict weather for more than about five days.

    3. Infiltration, the rate of movement of water from the atmosphere into the soil, that is, across the air-soil interface.

This is the most obvious of the synapses: the transfer of water from the *atmosphere* to the *terrasphere*. From the standpoint of the practicing hydrologist, this is a critical concept, for it represents the initial regulation of the process whereby the sharp input pulse of precipitation is attenuated into a runoff hydrograph.

    4. Antecedent Moisture Conditions, the amount of water in storage at the start of an hydrologic event the behavior of which is the subject of interest.

This oft-used and ill-defined term is frequently a consideration between the *atmosphere* and *hydrosphere*, but it is in fact an evaluation of any process (movement) of water from a site where the storage condition at the start of the process is a strong influencing factor.

    5. Variable Source Area, the changeable watershed zone that contributes runoff water to the stream, storm flow, or to between-storm periods.

This is the critical ecological synapse between *terrasphere* and the *hydrosphere*. It actually is a special case of antecedent moisture conditions, extended to include the changes that the near-stream storage goes through in the process of precipitation input. It embraces an additional concept:

    6. Time of Concentration, the length of time for water to travel from some specified point on or zone of the watershed to the outlet.

---

[3] Actually, primarily between the spheres indicated: being complex, there is nothing that simple about most of them!

[4] After Gleick, J. 1987. *Chaos: Making a New Science*.

This concept also links *terrasphere* and *hydrosphere*: the clearest definition is the time between the centroid of precipitation and the occurrence of the peak flow that results from that precipitation, but it is not always easy to identify the centroid of precipitation over the watershed (in time or space). Alternative definitions help understand the specific characteristics of the watershed, explain the effect of watershed shape on the hydrograph, or are related to other time-attributes such as time of rise, lag time, and so on.

  7. Return Period, the time interval between stream discharges of some specified magnitude or frequency.

This is the synapse between the *hydrosphere* and our *cultural sphere* (not illustrated in the figure). The fact that it derives from all of the complex interactions that precede it helps in our understanding of its uncertainty and complexity, if not our frustration with it. It is, of course, tied in with virtually all of the methods involved in evaluating stream regimen, especially flood frequency and flow duration analyses, analyses that provide civilization with some numbers that are useful for living on the water planet called "Earth."

  8. Environmental water, the often large amount of water that exists and moves between soil, root, stem, and leaf, the area that is largely hidden from our probing view; the continuum of water that exists from the tip of the leaves down through twigs and stem to the roots and soil.

This synapse is between the *terrasphere* and the *biosphere*. Forces work in both directions at both ends of the stream of water that is best considered as part of a dynamic continuum from soil to near-atmosphere, an idea inspired by Bates (1960). It embraces how the sap gets up the tree, influencing and influenced by how nutrients move between soil and vegetation. It is part and parcel of

  9. Transpiration, water removed from the soil for the change-of-state of liquid water from the stomata of the leaf to the water vapor of the atmosphere.

Transpiration is the movement of water from *biosphere* to *hydrosphere*. Like the other concepts described here, it is almost impossible to measure, for implantation of any device in a living plant interferes with the process itself, and monitoring the environment of the plant affects the vapor pressure gradients that limit and/or drive the process.

  10. Interception, the interruption of the downward movement of precipitation by vegetation and its consequent redistribution as evaporation, throughfall, and stemflow.[5]

The interception arrow in the figure is in the opposite direction from that of the transpiration arrow, indicating motion from *hydrosphere* to *biosphere*; the two separate arrows highlight the fact the two processes are not the reverse of one another, like evaporation and condensation, or photosynthesis and respiration. Inteception is a (relatively simple) physical process (or amount of water), whereas transpiration, of course, involves a change of state.

  11. Photosynthesis and **Respiration**, the processes that link energy, water chemistry, and organic compounds.

Here is the vital connection between *energy sphere*, *terrasphere*, *hydrosphere*, and *biosphere* involving the critical gases and nutrients that sustain us; the mechanisms that worldwide

---

[5] As noted in the chapter on water in the biosphere, interception is also the amount of water lost by the process.

provide buffer and means to interlock carbonate, oxygen, pH, and energy. Without an understanding of the complex interactions that proceed in the aquatic environment, there can be no understanding of the acid rain problem, nor of managing lakes or wetlands for water quality objectives, nor of the reason or means for controlling nonpoint sources of pollution through Best Management Practices and land management.

These linkages between the spheres are far more complex than the principles that govern the movement and storage of water within the spheres. Furthermore, they are generally more amenable to man's management efforts. Those that are most accessible include infiltration, time of concentration, variable source area, return period, and antecedent moisture conditions: they have received the most attention in the textbooks and management schemes on the ground. In addition, interception and transpiration are directly controllable at the local level by vegetation management. These seven concepts or processes constitute watershed hydrology's most important concepts, yet remain almost myths.[6] Despite their often nebulous nature, they are in fact inextricably linked in what must be considered as the highly dynamic (all are time-dependent) system that is watershed hydrology. The concepts may be used separately or together to explain the behavior of practically any hydrologic system; they are the synapses of the watershed, the decision-points in the complex hydrologic system that embraces the movement and storage of water throughout the natural environment. They are what watershed hydrology is all about.

My approach to problems for the practicing hydrologist is between the often broad-brush view of the geographer (which tends to be dominated by climatic influences on the hydrologic cycle) and the often myopic view of the engineer (which tends to be dominated by numbers that oversimplify and overspecify natural processes). Thus, watershed hydrology is the basis for a practical, middle-of-the-road approach to wildland water resource management. As a consequence, watershed hydrology provides the basic foundation necessary for practical solutions to hydrologic (and water quality, and institutional) problems on wildland and on modified watersheds. "Hydrology ... must deal increasingly with man-induced alterations in the natural order of things" (Ackerman 1969). As more and more notice of how the natural environment is disrupted, the observing public seeks to take more of a role in decisions concerning their surroundings. Today, more and more decisions that affect hydrology are being made at the local level, where the public has access to the decision-making process. In order to have constructive citizen participation, the lay person may need more explanation of the hydrologic cycle than will be found in an encyclopedia or a scout manual. An added purpose of this book would be to meet this need.

Hydrology is an interdisciplinary field. A professional who goes by the label "hydrologist" may be an engineer, oceanographer, or aquatic biologist, a forester, soil scientist, meteorologist, geochemist, or limnologist. The hydrologist must be ready to communicate with the public and members of other disciplines: "an understanding of water in relation to earth processes requires the collaboration of many disciplines" (Hendricks 1962). The public must

---

[6] "An idea that forms part of the beliefs [theorems] of a group [hydrologists] or class but is not founded on fact [measurement]" (Webster).

be ready to communicate with specialists from a wide variety of disciplines. The professional and the public often meet via the planner.

The planner may present decision makers with a proposal for a shopping mall in the form of a plat map or finished drawing, with landscaped lawns, trees, colorful building materials, and tasteful development. The seemingly endless months of bare soil during construction, the period of time for the trees to grow to the size shown, the piles of plowed snow not melting until June, are all absent. *What is their effect on runoff quantity and quality?* The engineer's drawings are just as misleading, often hiding behind incomprehensible formulas, reams of data, and unrealistic, simplifying assumptions. Presented at a public hearing on granting a zone change request, on a Draft Environmental Impact Statement, or some other permit process, the mass of material and scientific jargon confuses the public. More importantly, the information tends to intimidate citizens, forcing them either to accept the experts' collective bamboozling, or to spend good money hiring their own experts who interpret other experts on or off the witness stand at no small expense, or to resort to tactics which are counterproductive. *Are the hydrologic issues really that complex?* Some of them *are* complex. None are so difficult, however, that a competent citizen cannot understand the basics and begin to know which are the appropriate questions that must be asked.

I have resisted utilizing traditional types of problems, that is, ones which set up some numerical example that is a replica of one presented in the text (or that simply involve plugging values into a formula). I have opted, instead, for some brain-ticklers that require some in-depth analysis of some of the assumptions underlying several of the illustrations, or application of principles to the reader's particular location. For most of these, there are no set answers, thus there is no list of "correct" responses. It is my intent to have the problems foster discussion of the topic to enhance understanding of the principle or some particular hydrologic condition. The one problem that has a numeric answer also tells where to find the answer.

While most texts on hydrology identify the instruments used to monitor hydrometeorological phenomena and characteristics, none discusses the limitations of those instruments in achieving a true spatial, temporal, or quantitative measurement. This is of great importance in properly assessing the significance of the movement and storage of water. Furthermore, the units in which hydrologic phenomena are reported vary, depending on the instrument manufacturer, equipment upgrades, and existing data files.

The worldwide trend to convert to the metric system has left hydrology (and hydrologists) behind (hydrologists might consider it the other way around!). A major reason is the tremendous body of data already collected in and comprehension of the "English" system, especially the very "neat" conversion of cubic feet per second to acre-feet per day. Comprehension of quantities of water by experts in the field is very much unit-based, and this is a major reason for resistance to change. Even the American Society of Testing Materials maintains a combination of units in its recent publications. In the text, both systems are used, with the unit in which the research results were originally reported cited first. Given a choice, I have used the English system of units. A table of conversion factors is presented in the Appendix.

I collected, over the years, a large number of references, mostly articles from *Science, Water Resources Research, Water Resources Bulletin, Journal of Hydrology,* conference and symposium *proceedings,* and reports from Forest Service Experiment Stations. I have used over

750 of these extensively throughout the book. Yet, when I felt I had finished, I was humbled by discovering that there were at least 700 more that I had not used. There are, undoubtedly, *another* 700 or so that would be worthy of reviewing for and citation in the book. Some of the references I did use provide contrasting viewpoints; some support my views; some document recent research results; and others ensure that credit is given where it is due. A lot of what is known in the topics covered by this book has been known — and published — for a long time. I have tried to provide a balance of citations in both the older "classics" and more recent innovations and insights: there is much of value in the older works that is overlooked by eager researchers, and there is much that a current writer can forget to review. I'm sure that I have overlooked some important ones, and hope that readers will both forgive and inform me of any errors of omission or commission.

ACKNOWLEDGMENTS

As an educator, environmental consultant, and project manager for the preparation of environmental impact statements, I have had the opportunity to communicate with a wide variety of experts and lay persons about watershed hydrology. I acknowledge a large debt to them for providing me with the situations in which I had to elaborate or, conversely, succinctly summarize. I did not always feel grateful at the time but, in retrospect, I see that there was tremendous value in being forced to "get to the essence" of a complex topic.

Specifically, I am grateful to my former Instructor and Major Professor at the University of Michigan and Colorado State University, Dr. Robert E. Dils, who provided me with my introduction to the hydrologic cycle on wildlands and watershed management.

I also wish to especially thank my friend and colleague Dr. Arthur R. Eschner who has continually provided me with information in response to questions, insights, and review of drafts and sketches. Drs. Philip J. Craul, John P. Felleman, Lee P. Herrington, and Norman R. Richards have provided much constructive criticism and "outsider" viewpoints as partners in our joint entrepreneurial activity as impact consultants. Dr. Craul, in particular, has provided invaluable assistance in bringing me up to date on the new taxonomy and nomenclature of soils. Dr. Craig J. Davis, Jr. provided much help with software operations, as well as with the section on models and modeling, as did Dr. William M. Stiteler. Dr. Edwin H. White has been a constant source of encouragement and information, especially with regard to questions on soils. I am deeply indebted to all these fine friends and colleagues at the SUNY College of Environmental Science and Forestry (ESF). The Late Dr. J. V. Berglund who, as my immediate superior, continually provided the support and encouragement that ultimately resulted in my being able to write this book, deserves special thanks.

Dr. Richard C. Schultz of Iowa State University and Dr. Donald F. Potts of the University of Montana provided early review and constructive comments on drafts of the manuscript: their frank, constructive, and often humorous input has been both a source of inspiration and challenge. I am deeply indebted to both of them for their assistance, support, and continued interest. Randolph L. Bitely, a student at ESF provided review and a great deal of assistance to the project, not the least of which was inventorying and monitoring references and illustrations. He has my special thanks for a job well done. Nanette S. Rutkowski, a

Graduate Student at ESF provided timely, excellent, and vital drafting expertise. I am deeply indebted to her for stepping in at the last minute to revise, assist, and redraft most of the graphics. I am indebted, too, to ESF Graduate Students H. Chandler Rowell and Philippe A. Thibault who provided ideas and responded creatively to questions regarding presentation of text and graphic material.

I greatly appreciate the support and good humor of the many professionals at Prentice-Hall who were always at the telephone with friendly advice and answers to endless questions.

Finally, I express my deep appreciation to my students over the years: they have forced me to explain and rethink and rework my presentation of the hydrologic cycle. Any errors remain my responsibility, and I will appreciate having them called to my attention.

Peter E. Black

1 Introduction

*Although it often appears otherwise,
there is plenty of water on the earth:
its quality, and temporal and spatial distribution
are often undesirable*

This chapter provides an overview of the hydrologic cycle. Its primary purpose is to present an introduction to the hydrologic cycle in its entirety so that:

1. essential terminology is disclosed, and
2. important interactions of parts[1] of the hydrologic cycle can be discussed without having to turn to various sections of the book for related information.

One runs into this problem throughout the hydrologic literature. For example, how does one classify an article entitled "Rainfall-Runoff-Soil Interactions"? Or "Stormflow Responses to Road Building and Partial Cutting in Small Streams"? Or, even the elegantly simple title "Subsurface Stormflow"? Each of these might otherwise require two or more separate chapters.

A second purpose is to present some basic concepts of storage in the hydrologic cycle, and the third is to be able to comprehensively consider the hydrologic environment.

THE HYDROLOGIC CYCLE

Hydrology is the study of the movement and storage of water in the hydrologic cycle. In scout manuals, elementary science textbooks, and encyclopedias, the hydrologic cycle, sometimes called the "water cycle," is often depicted as shown in Fig. 1-1. The basic premises behind this view are that: (1) the hydrologic cycle is a closed system, that is, there are no new water inputs of any significance, and (2) what you see in the diagram is representative of reality. The first of these, for the purposes of this book which considers the entire hydrologic environment, can be

---

[1] The arrangement of the text is explained in detail in the Preface.

accepted; the second cannot. In fact, it is this very depiction of the hydrologic cycle that generates misinformation, misconception, and, as a result, much of the mismanagement of the water resource.

For example, since not all land has slope like that shown, there is no guarantee that water will run off as the arrow shows. Nor is there universal truth to the idea that water will move vertically through the soil and get back to the atmosphere via evapotranspiration,[2] condensing and falling as precipitation on the land. Most precipitation, in fact, falls on the oceans since they cover 70 percent of the earth's surface. The precipitation which does fall on the land tends to fall in mountainous areas and near the coast.[3] Water will not necessarily follow *any* of the arrows in the diagram at any particular point or any particular time. By oversimplifying the rather complex hydrologic cycle, the idea that it is simple is inadvertently and mistakenly conveyed.

Perhaps the greatest misconception presented in Fig. 1-1 is the process identified as "surface runoff." On a bare-rock or relatively impervious watershed, water will travel over the surface in what is defined as overland flow. On an undisturbed vegetated watershed, however, it is quite difficult to find evidence of surface runoff. The evidence, erosion or small accumulations of vegetative debris on the upslope-side of tree trunks or rocks, is hard to find because on an undisturbed vegetated watershed, there generally is no surface runoff. Thus, infiltration, the rate at which water enters the soil, is normally greater than the precipitation intensity on those watersheds, the rate at which the water is delivered to the surface. There are exceptions to this rule. On very steep slopes, or on areas of thin soils, and where there are factors that lead to very intense rains, surface runoff can and does occur. But, by and large, it doesn't. The importance of this is that when we see water running across a parking lot, playground, or even a lawn, we accept it as a normal part of our environment. It isn't. What's more, we needn't accept it. With knowledge of the hydrologic cycle, we can build features into our man-modified environment that will compensate for the decreased infiltration rate. By planning for expected decreases in infiltration rate, floods can be controlled, even from built-up urban areas.

Another important misconception arises from the obvious uncertainty about the location and extent of the line identifying the water table (often shown by a dashed line). Does the water table terminate by extending horizontally to an impervious layer? Does it turn upward and remain parallel with the land surface? Or does it remain parallel with the impervious surface? Or does it quietly disappear? The answer depends on the circumstances of the specific site. Any of the above possibilities may apply. In contrast, no one of them applies universally. The difficulty lies in the fact that many people think of this visual representation of the extent and nature of the "water table" when the term is used. Correcting that habit is an important goal. Yevjevich (1968) suggested doing this by getting away from descriptive presentations (such as Fig. 1-1) and creating mathematical and statistical models: "Use of

---

[2] The term that stands for all of the evaporative types of movement of water from earth or water surfaces to the atmosphere. (Terms are defined at their first usage in the text, either as a footnote or in context, and also appear in the Glossary.)

[3] Contrary to G. B. Shaw and popular Broadway rhyme, it is not true that "the rain in Spain falls mainly on the plain"; mostly, it falls in the mountains (see Chapter 3).

descriptive instead of numerical variables has delayed the application of modern statistical methods to planning the exploration of the ground-water environment." Unfortunately, numerical and statistical models are often even less comprehensible to the interested citizen. Ultimately, a variety of means of hydrologic cycle representation need to be used: each contains some truth and, consequently, provides a learning opportunity.

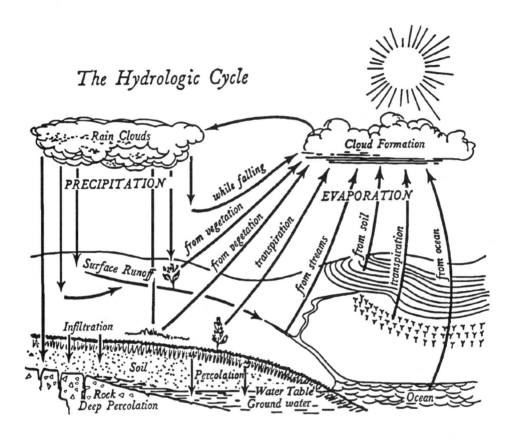

Figure 1-1 The hydrologic cycle. From the 1955 Yearbook of Agriculture, courtesy U.S Department of Agriculture (1955)

Yet another important misconception is implied by representing all the processes and locations of storage on one diagram. Not all the processes occur simultaneously. Also, not all storage locations are shown or are consistently represented; nor are the non-daylight hours represented. Clearly, some discussion of the ramifications of these implications is necessary. They are taken up in each chapter, as the topic is presented.

Still another misconception is an implied assumption of uniformity. Until one is confounded by the diverse readings from several adjacent rain gages exposed to the same storm, looks up at the variegated underside of a forest canopy, or starts to dig a soil pit, one cannot appreciate the variability of the natural hydrologic environment.

There are some true conceptions and some important terminology both presented and suggested in Fig. 1-1 (otherwise it wouldn't be shown at all). And, it *is* useful to start out with a simplistic model of the hydrologic cycle or of any complex system.

Starting in the atmosphere, water condenses to form drops of water (if the temperature at which they form is above 32°F) or ice crystals (if below 32°F). When driven by movement of air (wind) or when they have grown to sufficient size, they fall to earth as **precipitation**. There are a variety of forms of precipitation, including **rain, snow, hail, freezing rain**, and **fog drip**. The type, intensity, amount, and frequency of the precipitation varies with several other factors, including time of year, geography, and climate.

Precipitation intercepted by natural vegetative cover is redistributed to runoff or may evaporate directly back to the atmosphere. Water may move directly to the atmosphere by **sublimation** (vaporization of the solid or frozen phase, bypassing the liquid phase). This process was first recognized as an important part of the hydrologic cycle by Pierre Perrault[4] (Nace 1974).

If and when precipitation finally gets to the leaf, soil, or water surface (where, under natural vegetation, litter layers may present another intercepting layer), it moves from the atmosphere into the soil in one of the most important processes in the hydrologic cycle, **infiltration**. Infiltration is the rate at which water enters the soil profile from the atmosphere. The process of infiltration is best regarded as a concept since it is impossible to see or directly measure without influencing its value. It may, however, be approximated by a variety of instruments and calculations. From a distance, the interface between air and soil is rather clear. Upon close examination, however, it is almost impossible to identify where each begins and ends. There is, indeed, atmosphere in the soil **pores** (spaces between soil particles) when they are not filled with water.

Percolation is the downward movement of water through the soil profile once infiltration has occurred. Percolation may be saturated (the force governing the flow is gravity) or unsaturated (the force governing the movement of the water is capillarity) flow. The word **transmission** is normally used in a general sense to describe movement in *any direction* within the aerated portion of the soil that is above the water table, and **transmissibility** is normally used quite specifically to indicate the *capacity* of a rock strata to have water move through it, a function of the rock's thickness and **permeability**. Permeability is normally expressed as the **coefficient of permeability**, defined as the flow of water through a one-square-foot cross section under an hydraulic gradient of one foot, at 15.6°C.

Evaporation, the movement of water from the liquid to the vapor state and then to the atmosphere, occurs from any wet surface. The process effectively reduces the moisture in the soil, on leaves wet by rain, or in a water body itself. The term **interception** is used dually. It indicates the *process* by which the downward movement of water is interrupted and

---

[4] Perrault is known as the "father of modern hydrology." He made measurements of precipitation and runoff on the watershed of the Seine River, and published *On the Origins of Springs* in 1674.

redistributed by whatever surface gets in the way of the precipitation. And, second, the term indicates the *amount* of water lost through the evaporative process following precipitation. Some of the precipitation intercepted by vegetation may collect and run down the tree trunk as **stemflow**. Or, it may fall directly off the leaf or twig and join with drops that fall directly through the canopy in what is collectively termed **throughfall**. Water that moves through the stomata[5] from the soil and roots via the plant's internal moisture supply system to the atmosphere is **transpiration**, another major evaporative-type process. The combined evaporative processes are termed **evapotranspiration**.

Once in the stream, water is referred to as **streamflow**, **runoff**, or **discharge**. At this point, it is difficult to tell just by looking at it where the runoff originated. Water in the stream may be supplied by **surface runoff**, **subsurface flow**, **storm flow**, and **base flow**, and may come from channel, bank, or ground water storage reservoirs. These different classes of runoff have different characteristics, may occur at different times, and are affected by different factors in the hydrologic environment.

Storage occurs at several locations in the hydrologic cycle, as suggested by Fig. 1-1. Water is stored in the aerated unsaturated portion of the soil mantle and beneath the water table under saturated conditions. A great deal of water can also be stored on top of the soil in the form of snow, especially at high elevations where conditions favor heavy precipitation and low temperatures. A considerable amount of water is also stored in vegetation. On the surface, water is stored in the form of puddles, ponds, lakes, and wetlands of all types. Water is also stored in river and stream channels, and this is a significant part of certain hydrologic environments.

The fact that the water stored in river and stream channels is in motion should not deter reference to it as "storage." Water in the atmosphere, the soil, and the ground water is also in motion, although to widely varying degrees. At times, even the water stored in an apparently inactive snowpack is in motion. This impreciseness of categories exemplifies some of the complexity of the hydrologic cycle and, therefore, the difficulty of describing it. For example, water in the soil is both in storage and represents a process. As with most classification schemes, there comes a point where the class boundaries break down. If some benefit has been obtained by using the classification, then it should be abandoned only when it is no longer useful. At that time, it is appropriate to fall back and regroup, or construct another model. Identifying process and storage in the hydrologic cycle still has considerable utility. Having briefly identified the major components of the hydrologic cycle, it is time to examine concepts of storage.

STORAGE

TYPES OF STORAGE

Figure 1-2 is a representation of the hydrologic cycle which, by vertically aligning the sites of storage, calls attention to them. The symbols used are derived from computer flow

---

[5] The minute openings in leaves, mostly on their undersides, that allow the passage of oxygen, carbon dioxide, water vapor, and other gases.

charts, with parallelograms for process labels, diamonds for decision-points, and triangles for storage. The connection by dotted lines indicates *conditional* flow, that is, flow may not always occur. The horizontal combination of the parallelograms at the top of the diagram is the **water budget**, an equation which will be used more extensively in later chapters. For the moment, it describes the balancing nature of the hydrologic cycle, as well as the fact that it is a closed system.

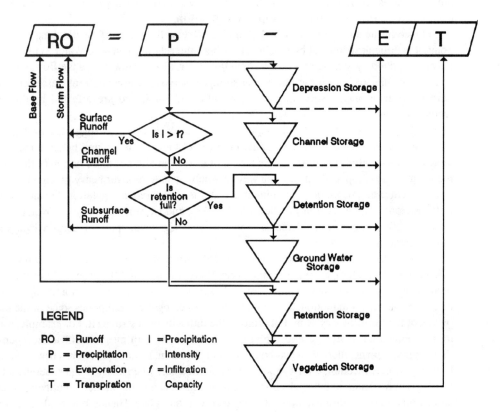

Figure 1-2 Storage and the hydrologic cycle.

By focusing on storage, one is first aware of the several types of storage and their locations. Several of these have already been briefly described without being labeled "storage." Some additional comments are appropriate here. **Depression** storage occurs on any surface where water collects and, because there is no outlet for flow, remains until it either seeps through the surface on which it is located, or evaporates. Puddles and **ponds** (bodies of still surface water) are examples of depression storage, as is the water that collects on some man-made object and remains until it evaporates.

Water in **bank storage** may not actually move, but is detained in the porous media surrounding and immediately adjacent to a stream. This water usually joins runoff waters during a runoff event. Bank storage is rather nebulously conceived, thus the term is often substituted for or equated with channel storage. **Channel storage** occurs in the open portion of the stream or river that is visible, but also occurs in the nonvisible porous banks and bed of a stream. Channel storage may be very large, in places larger than the river's visual water limits. Such waters are usually markedly cooler than the "river water" and, having traversed through sands which filter it, may be cleaner as well. Waters from bank or channel storage are valued for a variety of purposes for which the river water may not be as well suited.

Pore space in the soil occurs in a distribution of sizes, from microscopic interstices to readily visible cavities. These are classified as capillary and noncapillary pores. In the zone of aeration above the **water table** (below which is the zone of saturation), these pores may be filled with water or air.

In the capillary pores, the combined forces of **cohesion** (between the water molecules) and **adhesion** (between the water and the soil particle) are strong enough to counteract the force of gravity. In the noncapillary pores created and maintained by a variety of fauna, and by frost action, the water is free to move by the force of gravity. A small amount of soil water is also bound in thin layers to soil particles by strong electrochemical forces, known as **hygroscopic storage**. The water in the capillary pores is called **retention storage**. Water in the noncapillary pores is known as **detention storage**. Retention and detention storage are collectively referred to as **soil water**, which is not separately identified as a unit of storage on the diagram.

If retention and detention storage reservoirs are both filled, then water cannot enter the soil (unless drainage just equals infiltration), and it puddles and/or runs off as **surface runoff**.[6] Following infiltration, water moves into storage dependent upon tension gradients, as explained in Chapter 5. Because it is held at high tensions, water does not move out of retention storage. For 24 hours following recharge of detention storage, that water will move out of storage, becoming water in motion, subsurface flow, a *process*. **Subsurface flow** water moves either laterally to a spring or stream, or vertically to **ground water** storage. Subsurface flow also may occur under undisturbed conditions.

This perception of storage conditions affecting how water flows out of the soil introduces the concept of **antecedent moisture conditions**. Generally, antecedent moisture conditions are the combined amounts and locations of water in storage at the start of the hydrologic event or the time period being studied. This is an essential concern in many considerations of the hydrologic cycle.

CONCEPTS OF STORAGE

Storage might best be defined as "space available." Although obviously, if use is made of that space by putting something in it, it is no longer available, or no longer *as* available. Storage of water in the hydrologic cycle is also a quantity of water temporarily out of circulation. Neither of these definitions is completely satisfactory, nor is either incorrect. But, by combining them, one derives the idea of *capacity*, the property that expresses the storage

---

[6] This term is used here in preference to **overland flow**. In using the latter, the engineer often implies that there is no subsurface movement through the soil. Overland flow may be defined as all surficial runoff on the watershed.

reservoir's ability to temporarily hold water out of its normal circulation. An analogy in flood control is that of a man-made reservoir which, if maintained empty, will be able to accept and temporarily (at least) hold back excess flows, referred to as "flood waters." This is artificial storage, to which a lot of space in engineering hydrology and water management texts is devoted.

A lake or wetland is the easiest to envision. If it is full, it cannot accept more water without yielding some at its outlet. If it is empty, obviously its capacity is equal to its volume, plus whatever spillage can take place during the filling period. If the lake is partially filled, its capacity to store water is somewhere between the two extremes, and the ability of the lake to buffer or attenuate a flood input is moderated over the empty condition.

The lake is easy to visualize because it has definite boundaries. Such boundaries are not always definitive in other types of storage in the hydrologic cycle. But the vagueness of storage site boundaries also provides a high degree of flexibility in definition and, perhaps, greater total capacity, as well. Indeed, some of the capacities are so great that it always seems possible to "add one more unit" of water to the undisturbed soil, ground water, or atmosphere without having a major reaction of some sort.

The relationship between the amount of water in storage and the amount delivered to it in a given time period has some interesting properties. First, if the storage is the same at the end of the period as it was at the beginning of the period, the amount of output must be equal to the amount of input. In simple formula form, this can be expressed as:

$$Output = (S_e - S_b) + Input$$

where S_e is the amount stored at the end of the period, and S_b is the amount stored at the beginning of the period. Thus, if the two storage terms are the same, that is, if their sum equals zero, then

$$Output = Input$$

Since there is a fixed amount of water on earth and long-term storage is neither increasing nor decreasing, output from any of the several global storage types is equal to input. For example, on a global basis, precipitation must equal evapotranspiration.

Furthermore, the amount of this flow (into or out of storage) may be divided into the total amount stored to provide what is termed the **residence time** for storage. The units involved are some measure of *volume per unit time* for the flow, and *volume* for the storage. Thus,

$$volume/volume \ per \ unit \ time = time$$

For example, the storage volume for a reservoir might be 100,000 acre-feet[7] and the annual river flow into the reservoir might be 30,000 acre-feet (AF); in this case the residence time would be 3.33 yr since it takes 3.33 yr of runoff to fill the reservoir. The residence times for the large storage types in the hydrologic cycle are shown in Table 1-1, along with the total volume of water in various types of storage on earth.

---

[7] The amount of water necessary to cover an acre (43,560 sq ft) to a depth of 1 ft, thus, 43,560 cu ft.

Chapter 1 Introduction

Table 1-1 Distribution of Earth's Water and Residence Times[a]

Location of Storage	Total Water (acre-feet)	Percent of Total	Percent of Fresh Water	Residence Time (years)
Total Water on Earth	1.033×10^{15}	100.0		
Oceans	1.0×10^{15}	96.8		
Total Fresh Water	3.314×10^{13}	3.2	100.0	6977.
Ice and Glaciers	2.475×10^{13}		75.0	5210.
Ground water: deep	4.62×10^{12}		14.0	973.
shallow	3.63×10^{12}		11.0	764.
Lakes	9.9×10^{10}		0.3	21.
Biosphere	8.1×10^{10}		0.24	17.1
Soil Moisture	1.98×10^{10}		0.06	4.17
Atmosphere	1.155×10^{10}		0.035	2.43
Rivers	9.9×10^{9}		0.003	2.1

[a] Data from Chow (1964), Freeze and Cherry (1979), and other sources. There was little agreement on some of the figures in different sources, so discrepancies can be expected. Nevertheless, the relative values and overall scale of the distribution of the earth's water are substantially as shown. The Residence Time is calculated, without regard to relative area on the earth's surface, using the total annual volume of precipitation, 4.75×10^{9} acre-feet. This figure is not included in the total, as at any given time it is distributed in the sites identified. The table is a bit misleading in that the residence times are averages. The residence time for ice and glaciers is greater than that for the oceans: the ocean waters are highly mobile, whereas the molecules in the ice and glaciers are not. Also, for example, some individual lakes may have a residence time of a fraction of a year, as little as a week: the 21 years refers to all the lakes collectively.

Note that the oceans comprise 96.8 percent of the total water on earth. Of the total fresh water, slightly less than 75 percent is in the polar ice and glaciers, and slightly less than 25 percent is in ground water. The infinitesimally small percentage that remains is divided between the lakes, biosphere,[8] soil moisture, atmosphere, and rivers. This is the portion we normally see, and which we generally manage in the hydrologic cycle.

A particularly significant observation that can be made based on the information in Table 1-1 concerns man's access to the waters of the hydrosphere. Certainly, the rivers have been themost useful for commerce, water supply, defense, navigation, fisheries, and recreation. They are also the most accessible (see Chapter 6) and, therefore, are inviting objects for water management, including attempts to redirect and control flows. The residence time, however, suggests that the relatively quick turnover of the water in the river means that attempts at control will be futile. In some usually short period of time, our controls will be destroyed by floods. While the same may be said in terms of residence time for the atmosphere, that reservoir of water is neither as accessible nor as deceptively easy to control as are the

---

[8] Those portions of the lithosphere and atmosphere occupied by living matter.

rivers. Controlling soil water will be easier, but a lot less spectacular (not as seductive as building dams, flood walls, and levees) and, being diffuse, not as accessible as rivers.

At the other extreme, efforts of manipulation or control of the oceans or the world's ice are puny and hardly noticeable (muCh less statistically verifiable) in view of the massive volumes of water involved. Thus, weather modification, hurricane control, towing icebergs, diverting major rivers, etc., are impressive engineering feats, but usually do not consider environmental impacts of such bulldozing approaches. In contrast, portions of storage in our local environs, such as vegetative or soil water, are more likely than others to yield large returns for little effort. It is inevitable, too, that manipulation of one regime will result in effects elsewhere in the hydrologic cycle (Achuthan 1974).

For example, Achuthan also notes it is more likely for man to succeed in intervening in the evaporation component of the hydrologic cycle than in the precipitation component, which "is not a phase which is subject to manipulation on a large scale." Between the two, he observes, storage offers a wide variety of options and sites for control.

Wagle (1971) emphasizes the role of the climate in attempts to manipulate storage on the watershed: "In any given region, then, the watershed manager must understand that the primary limitation on his efforts to manipulate vegetation and increase water yield is climate. He must work with climatic phenomena and relate his efforts in every instance to these phenomena.... Water yield is secondarily controlled by soil conditions and their effects are wholly within conditions superimposed by local climate and the vegetation sustained by same." As an example, manipulation of the vegetative cover to alter streamflow behavior is a catalytic type of "control" in that the actual effect is to let a climatic controlled runoff pattern show once the vegetation effect has been removed (see Chapter 6).

The disproportionate distribution of water evident in Table 1-1 is a fundamental characteristic of the hydrosphere. It is emphasized further in Fig. 1-3. As in Table 1-1, the small percentage of fresh water on the earth is shown in (a). The distribution of this small percentage is elaborated on in (b), showing the less than 4 percent that is "in circulation." This term, as suggested above, is somewhat misleading here, for some of the water in each of the other three categories of (b) is also in circulation. However, the amount in circulation is also a very small portion of that which is stored in the ice and ground water. Part (c) of Fig. 1-3 shows the lopsided distribution of water that is in circulation, noting, again, that some of the 4.75×10^9 AF that actively falls as precipitation each year is included in each of the storage categories. Finally, (d) shows the distribution of fresh waters in the world's lakes (The Caspian Sea is sometimes considered a lake, as it is fresh water, but is treated here as part of the "oceans and seas"; thus, it is not included in the "lakes" category).[9] Nearly one-fifth is in one lake, Lake Baikal in the Soviet Union. About the same amount is in the Great Lakes system of North America. Noting that it is as difficult to define lakes at the lower limits of size (as ponds, or wetlands, or wide spots along stream and rivers), as it is when one tries to classify the Caspian Sea, the total should not be construed as an exact figure.

---

[9] The Caspian Sea is sometimes considered a lake, as it is fresh water, but is treated ehre as part of the "oceans and seas"; thus, it is not inlcuded in the "lakes" category.

Chapter 1 Introduction

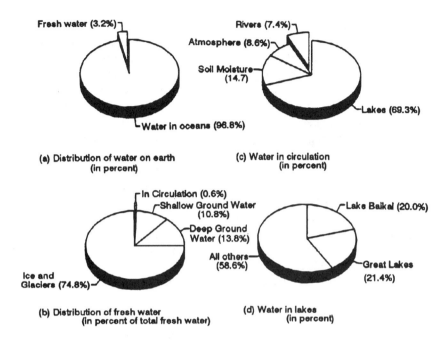

Figure 1-3 Distribution of water on earth

THE WATERSHED

The **watershed** is the natural unit of land upon which water from direct precipitation, snowmelt, and other storage collects in a (usually surface) channel and flows downhill to a common outlet at which the water enters another water body such as a stream, river, wetland, lake, or the ocean.[10] The watershed may be quite small, an acre, for example; or it may be quite large, to hundreds of thousands of square miles embracing much of a continent. At first glance, the watershed would appear to be a rather static unit of our landscape, yet from an hydrologic standpoint it is a dynamic and changeable area.

During the period immediately following a runoff-producing event, the amount of water on the watershed naturally diminishes. As this occurs, the source of the water for streamflow shrinks in size, with the distant, upper slopes drying out earlier than those areas closer to the stream. In the last stages of runoff, if allowed to continue long enough, only the channel will be contributing to streamflow.

Conversely, when a runoff-producing event occurs (under the assumption here that the entire watershed is instantly affected by a rainstorm, for example), the area contributing to runoff gradually grows, in the reverse manner from that described in the previous paragraph. The concept expressing this phenomenon is referred to as the **variable source area**, first

---

[10] Occasionally, the watershed's channel may terminate in an underground passage, providing flow directly to a ground water reservoir.

suggested (but not with that name) by Hewlett (1961). Thus, the zone of saturation in and around a stream expands and contracts in response to water available for runoff. The details of this phenomenon are described in Chapter 6, but at this point the basic concept is important to comprehension of the individual pulse of water during a runoff-causing event (often a storm) itself, as well as during nonstorm periods. The pulse of water has been referred to as "quickflow" or storm flow.

Over time, a graphic plot of the flow of water from the watershed is known as a **hydrograph**. Stream gages typically record the depth of water in a known cross section of the stream (the volume or rate of flow is a function of the depth) and, as a consequence, the hydrograph is often represented simply as a plot of depth of water over time. The term "hydrograph" is occasionally applied to this long-term record of runoff, but it is also used to refer to the short pulse of water that accompanies a runoff-producing event (a rainstorm or snowmelt event), in which case it should be properly referred to as a "storm hydrograph." Probably because the latter application is more frequently used than the former, the storm hydrograph is often simply referred to as a "hydrograph": usually, the meaning is implicit and easily inferred from the context.

A typical hydrograph for a storm period is shown in Fig. 1-4. In general, the flow at any given point in time after the peak of the runoff event has passed, Q_t, is a function of the flow at the start of the interval, Q_0, and a recession constant, K. Thus,

$$Q_t = Q_0 K^t$$

The value of K is a characteristic of the storage unit. For any given type of storage, including the watershed as a whole, K is a constant and, therefore, one of several characteristics that distinguish one watershed from another. A great deal of research has been conducted on the nature of the flow out of storage, especially, ground water storage,[11] and on the nature of the **recession curve**, the downward-sloping portion of the hydrograph that reflects the draining of the watershed's ground water reservoir. The **rising limb** and **falling limb**, both of which reflect the rate of storm flow and the variable source area from which it comes, are also characteristic of the watershed. Together, the rising and falling limbs superimposed on the recession curve make up the storm hydrograph.

Integrating the above equation, and noting that

$$Q_t \, dt = -dS_t$$

the amount of water remaining in storage at any given time, S_t, can be determined by

$$S_t = -Q_t/\ln K$$

A characteristic of this equation is that it plots as a straight line on semi-logarithmic paper (Chow 1964). To whatever extent the depletion curve deviates from a straight line, it is an indication that flow is being contributed by types of storage other than ground water. In fact, working backward from the end of the depletion curve, one can determine where the

---

[11] On large watersheds, storage is dominated by the ground water component. In fact, Chow (1964) defines a small watershed as one on which ground water storage is not a dominant contributor to runoff.

contribution to runoff included flow from other types of storage.[12] Thus, the upper portion of the hydrograph reflects drainage from other types of storage. For example, during a runoff-causing event on a small watershed under dry antecedent moisture conditions, water in the stream comes primarily from storm flow. Since the watershed is made up of many different types of storage (and processes), it follows that each unit of storage has its own characteristic K which, for its unique waterholding capacity, also exhibits straight-line depletion. While these have not been researched as extensively as has the ground water component, the basic characteristic is an important one: it means, among other things, that the several types of storage are different in how each contributes to runoff, especially on small watersheds where channel and ground water storage are not the dominant sources of water to the storm hydrograph.

Typically, during a storm or other runoff-causing event, such as snowmelt runoff, the hydrograph of an undisturbed stream rises fairly rapidly, and after reaching a peak value, falls off rather gradually. A typical storm hydrograph is shown in Fig. 1-4.

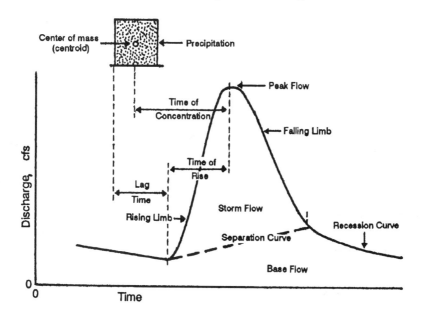

Figure 1-4 A typical storm hydrograph

The storm hydrograph is profoundly influenced by the first four of the five basic principles underlying watershed hydrology identified in the Preface: infiltration, antecedent moisture conditions, time of concentration, and variable source area are critical determinants of the storm or unit hydrograph. The remaining principle, the return period, expresses in part the relationship among several observed peak flows.

---

[12] This practice, known as **hydrograph separation** is a useful tool in hydrology and is well covered in several texts. It is presented here, initially, to convey the important characteristic of storage.

Many other factors, including the timing, nature, and distribution of the precipitation, soil, vegetation, watershed characteristics, antecedent moisture conditions, and land use contribute to the shape of the storm hydrograph. For example, a heavily urbanized watershed, with a high proportion of its land impervious owing to pavement, roofs, and compacted lawns, etc., will cause a rapidly rising limb and a correspondingly steep falling limb, and the recession curve. The peak will be higher than that of the undisturbed watershed, and the time of runoff will be shorter even though the total amount of runoff may be greater.

Early research (Sherman 1932; Horton 1933) into stream behavior led to the concept of the **unit hydrograph**, defined by Wisler and Brater (1949) as the "hydrograph of surface runoff resulting from a relatively short, intense rain, called a unit storm." Since storm flow may be made up, in part, of flow *through* the soil as well as surface runoff (which may not exist on undisturbed vegetated watersheds anyway), the unit hydrograph may be practically thought of as consisting of all storm flow from a given storm. This concept has shortcomings, however, and the complex, ecologically oriented variable source area concept of short-term stream behavior is of considerable interest to researchers, modelers, and managers.

SUMMARY

The amount of water precipitating in any given year (the 4.75×10^9 AF) in the hydrosphere is a scant 0.00046 percent of the total water on earth. It is what makes up that part of the hydrologic cycle with which man most commonly interacts. It is the link between various units of storage. It is the source of floods and droughts. And it includes all the water we have available and which we may manage.

Once on the ground, water may enter any of several types of storage and, on occasion, move among them in a series of processes that are largely governed by the physical properties of the storage medium. Watershed runoff exhibits differing characteristics dependent upon the storage from which it derived. Two theories are set forth to explain, evaluate, and predict the flow from the watershed, the traditional engineering unit hydrograph, and the ecologically oriented variable source area concept. While the latter is a more appropriate conceptualization of the runoff process, the former is a useful tool in hydrologic research and management: both are discussed in detail in Chapter 6.

2 Water and Energy

*The Sun
is the ultimate source of energy
for the movement and storage of water
and for its changes of state
throughout the hydrologic cycle*

The hydrologic cycle and energy movement on the face of planet Earth are intimately related. The Sun provides the energy to evaporate water and drives the ocean and atmospheric currents. And the water stores energy and moves it around the surface of the globe. Both are very complex cycles, and there is a danger in oversimplifying the energy budget much like that described in Chapter 1 when the hydrologic cycle is simplified. Nevertheless, the first consideration must be of basic energy relations at the surface of the Earth.

ENERGY MOVEMENT

THE SOURCE

Characteristics of solar radiation are important for understanding the energy distribution at the surface of the globe.

The Amount of Energy

The tremendous amount of energy emitted by the Sun is spread more and more until, by the time one two-billionth of it is intercepted by the Earth (Reifsnyder and Lull 1965), it amounts to approximately 1.94 cal per sq cm per min. Actually, this is the average annual amounts intercepted on a square centimeter above the atmosphere, and perpendicular to the incoming radiation. The units are usually expressed as Langleys/minute (Ly/min), a Langley being 1 cal per sq cm. Since the distance between Earth and Sun varies with the time of year and with the tilt of the Earth's axis, the value actually varies between 2.068 Ly/min (in winter, when the Earth is closer to the Sun) and 1.932 Ly/min (in summer). With such a small variation, the amount of incoming radiation is often referred to as the Solar Constant, about 2 Ly/min.

To give some perspective to this quantity of energy, recall that it takes approximately 1 calorie to raise 1 gram of water 1 cu cm at standard pressure and temperature 1° Celsius (C), and 540 cal to evaporate that cu cm of water at 100°C. Thus, to evaporate a cubic centimeter of water, energy at the rate of the solar constant irradiating the upper surface of the cubic centimeter (above the atmosphere) would take 270 min, or more than 4 hr. If the cubic centimeter of water were at 0°C, it would require 640 cal for complete evaporation, or about 5-1/3 hr. One gets the impression that the energy-water machine is slow; it is. Water has a high specific heat, therefore, it can accumulate a large amount of energy — number of calories — before changing temperature or state from liquid to vapor. Thus, water *buffers* the energy distribution. If this were not the case, our environment would be much more volatile. So slow is the evaporation process, on the average, that over a 10-hr period, only about 0.41 cm would be evaporated. If this rate applied for one half year of sunlight, however, about 75 cm or 30 in. would evaporate. This figure is, of course, approximately equal to the average annual rainfall in the United States.

On the other hand, the amount of the solar constant radiating a hypothetical square mile above the atmosphere would generate 4.8 million horsepower!

The Quality of the Energy

All objects radiate energy if their temperature is above absolute zero in direct proportion (the Stefan-Boltzmann Constant) to the fourth power of their temperature (Planck's Law). A perfect radiator is referred to as a black body,[1] which is a theoretical, but useful, concept:

> The Stefan-Boltzmann relationship applies rigorously only to perfect black bodies, which do not exist in nature. The radiation emitted by any natural body is always less than black-body radiation because the emissivity of the material is always less than 1 (Reifsnyder and Lull 1965).

The peak wavelength of the emitted radiation from a black body is an inverse function of the fourth power of the absolute temperature of the radiating body (Wien's Law), a relationship that produces dramatic increases in energy emitted with small increases in temperature. Therefore, the hot Sun radiates more energy, and that energy is at shorter wavelengths than those of the cooler earth; the radiation coming from the Sun is in the short-wave range of the electromagnetic spectrum, from about 0.1 microns (μ) to 4 μ, with a peak at about 0.47 μ, in the visible light region, while those emitted by the Earth are in the long-wave region, between 4 μ and 100 μ, at 20°C, in the infrared range and invisible to the human eye (Fig. 2-1). For the purposes of this text, it is appropriate simply to refer to solar energy as "short-wave" and terrestrial energy as "long-wave" radiation.

---

[1] One which absorbs and radiates equally over the entire wavelength spectrum to which the body is exposed. The term is derived from the idea that if the object absorbs all the visible radiation incident upon it, it will reflect no color and therefore appears to be black.

Deviation from the ability of an object to absorb radiation as a black body is expressed by the albedo,[2] the percentage of the incoming (incident) energy reflected, expressed as a decimal fraction. Since natural objects do not, in fact, absorb equally across the spectrum, a substance like new-fallen snow, which appears to have a very high albedo (about 0.90), actually absorbs long-wave radiation, which is why winter precipitation accumulated into deep snowpacks and glaciers appears bluish. In a forest canopy, the albedo of snow has been reported to be as low as 0.20 and decreases to about 0.14 under a low Sun angle (Leonard and Eschner 1968). Moist earth is a good absorber of both long- and short-wave lengths (Reifsnyder and Lull 1965). The range of albedo values for some natural objects is shown in Table 2-1.

Table 2-1 Range of Values of Albedo for Some Natural Objects

Material	Albedo
Fresh, dry snow	0.80 - 0.95
Snow, covered with ice	0.71 - 0.76
Dry, light, sandy soils	0.25 - 0.45
Dry, clay, or grey soils	0.20 - 0.35
Meadows	0.15 - 0.25
Desert	0.10 - 0.35
Rye and wheat fields	0.10 - 0.25
Moist, grey soils	0.10 - 0.20
Deciduous forests, autumn	0.06 - 0.57
Deciduous forests, summer	0.05 - 0.54
Dark soils	0.05 - 0.15
Coniferous forests, summer	0.03 - 0 32
Coniferous forests, autumn	0.02 - 0 19
Water surface, viewed obliquely	0.02 - 0.13

Source: Reifsnyder and Lull (1965). The values are correct in the short wave lengths only.

Incident short-wave radiation is either transmitted, reflected, or absorbed. Thus, short wave radiation can penetrate the atmosphere, but some of it is reflected off the top, and some is absorbed by various substances in the atmosphere. Some of that which is transmitted through the atmosphere and received at the surface of the Earth is also reflected, with transmission occurring at the surface of water, snow, or ice, or even a forest. The remainder is absorbed, the consequence being that the receiving body is heated. That body will, as a result, emit more energy, accompanied by a change from short- to long-wave radiation in the wavelength emitted. Similarly, the clouds and atmospheric gases absorb in the short-wave range of the spectrum and emit long-wave energy. The emission of long-wave radiation from

---

[2] The term albedo is generally used for a broad energy spectrum, whereas the term "reflectivity" is used for a restricted portion of a spectrum (Reifsnyder and Lull 1965).

the surface of the earth is upwards; from the atmosphere it is in all directions. As noted below, there are some other forms by which energy is transferred as well.

Since, for the purposes of this book, the earth in the long run is neither heating up nor cooling off, the amount of incoming and outgoing energy must balance, which fact can be used in an energy budget. Since the incoming energy and outgoing energy are at different wavelengths, the intervening energy has been reflected, transmitted, or absorbed and dissipated by conduction or convection away from the surface being heated. One of the complex and highly variable factors influencing the energy budget, of course, involves the movement of water itself as it changes state from liquid to vapor. Other factors and processes that are involved in or are a part of the hydrologic cycle, such as clouds and their formation, also influence the distribution of energy.

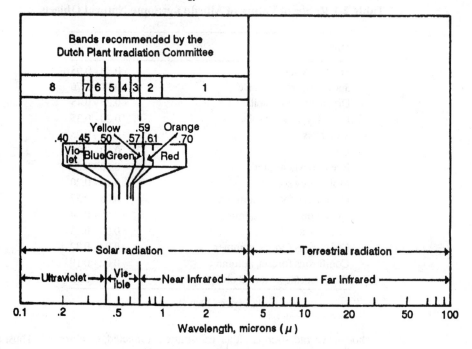

Figure 2-1 Wavelength of solar and terrestrial and spectral bands.
From Reifsnyder and Lull (1965)

Energy at the Evaporating/Transpiring Surface

A long list of factors that affect the amount of energy absorbed/reflected at a vaporizing surface can be identified. They may be grouped into three classes: (1) those that relate to inherent characteristics of the surface (e.g., its color or roughness, or the elevation, aspect, or latitude at which it occurs); (2) those that relate to properties of the radiation, including the angle of incidence (some of which may be characteristic of the incident radiation and some of which may be a peculiarity of how the surface is exposed), and the quality of the energy (e.g., which portion of the spectrum); and (3) current moisture conditions, since water (and snow) exhibits different albedo at different Sun angles, as well as under some other environemntal conditions.

In the case of rough surfaces, such as a forest canopy or stand of brush or herbaceous vegetation, the complexities of the relationships are greatly magnified. Albedo of a plant cover also varies with soil fertility, and season of the year (both of which may affect leaf color, for example). Crown density (affected both by position of the species in ecological succession and by silvicultural activities and past land use), depth of the canopy, branching habit, canopy roughness, and several related factors also play a role. For example, leaf reflectivity increases when the habitat is drier (Reifsnyder and Lull 1965), a beneficial water-conserving reaction for, with greater reflectivity, less energy is available for evaporative-type processes. Even such an apparently smooth surface as that of a bare soil is quite rough when viewed on a microscale level and, as a consequence, energy relations at the air-soil interface are not easily described, as evidenced by the wide range of albedo values in Table 2-1.

According to Reifsnyder and Lull (1965), a "solid forest canopy approximates a black body in the long-wave portion of the spectrum, absorbing and emitting almost all possible radiation." The actual amount of energy radiated varies with crown density, forest condition during seasonal development, stand uniformity, time of day, and season. Dependent primarily upon crown density, some of the incident radiation can be transmitted through the forest canopy to the forest floor, or to understory canopies. The amount of energy in the short wave portion of the spectrum transmitted through a "complete canopy" is approximately 10 percent (80 percent is absorbed and 10 percent is reflected). In the long wave portion of the spectrum, as much as 90 percent of the energy radiated upward from the earth may be absorbed and only 10 percent reflected (down toward the forest floor). Long-wave radiation from the crown is in all directions, but the net flux is upward, with most of the upward-directed component transmitted through the atmosphere on a clear, midsummer day. High moisture and carbon dioxide content cause most of the long-wave radiation to be absorbed in the atmosphere.

Thus, the forest has a marked effect on short-wave radiation (shade) and very little effect on long-wave radiation. The latter is dependent upon its temperature, the result of heating by short-wave radiation. Cloudiness, which absorbs and re-radiates long-wave radiation emitted from the earth, and the lack of short-wave radiation at night have little effect on long-wave radiation from the forest. Because radiation in the long-wave range of the spectrum continues throughout the night and because the forest is such an efficient radiator, temperatures in the forest may actually be reduced beyond that of surrounding openings. However, leaves in the interior of the canopy will not radiate as much to nearby, warmer leaves as will upper leaves to the colder sky (Reifsnyder and Lull 1965). The net result is that the forest tends to be cooler than its surroundings during the day, and warmer at night.

The amount of evapotranspiration from different vegetative cover is directly related to the amount of energy available, which is profoundly affected by both the latent and sensible heat fluxes. These two energy fluxes often compete with one another: "usually the latent-heat flux prevails over that of sensible heat, if the surface is wet [, and] if it is dry, there is of course, little or no evaporation or latent-heat conversion taking place" (Miller 1977). The ratio between the two fluxes, known as the **Bowen Ratio**, may be used to evaluate and predict E_t

values. Three-year averages of evapotranspiration-to-net radiation (E_t/R_n) ratios calculated by the Bowen Ratio equation for growing season forest, clear-cut opening, and natural meadow were found to be 19.4 in., 15.3 in., and 13.2 in., respectively (Thompson 1974). Thus, the forest, with seasonal differences integrated, was most efficient in using the available energy for evapotranspiration processes. Lindroth (1985), however, reported that variation in canopy conductance of energy between forests was small compared to seasonal variation.

The Geometry of Energy

The angle of the Sun above the horizon is of primary importance in evaluating the amount of incoming energy at any given location and time (Reifsnyder and Lull 1965). As the Earth travels around the Sun during the year, it is closer to the Sun at the winter solstice (December 22) than at the summer solstice (June 22). The Earth's axis is tilted away from the Sun in winter, however, thus making the Sun's rays (1) appear to come from closer to the horizon; and (2) travel through more of the atmosphere, thus subjecting the radiation to more reflection. The overall effect is to reduce the radiation received at the Earth's surface since the atmosphere also absorbs more of the energy passing through it obliquely.

With the Earth's axis pointed away from the Sun in the northern hemisphere, the southern hemisphere has the Sun more directly overhead when the Earth is closest to the Sun during the northern hemisphere's winter, and vice versa. The southern hemisphere, therefore, exhibits warmer summers and colder winters than the northern hemisphere, but this is tempered by the scant land masses in the southern hemisphere.

Only at the equinoxes (September 23 and March 21), is the Sun directly overhead at the equator. On those dates, and at **local solar noon** (by definition), the **zenith angle**, the angle between a point directly overhead (the zenith) and the Sun, will be equal to the latitude of the observer. In other words, and as a complement to the opening sentence in this paragraph, the greater the zenith angle, the less intense the incident radiation (Lambert's Cosine Law). These basic relationships are shown in Fig. 2-2.

Aspect and slope of the surface on which an observer is located or on which radiation is incident is identified by both compass direction and degrees (or percent) of slope. Obviously, a slope tilted toward the Sun will have a smaller zenith angle and, therefore, a higher intensity of incident radiation.

All south-facing slopes in the northern hemisphere do not automatically receive greater radiant energy all the time, however. The geometry is quite complex and cyclical. For example, an horizontal surface at 23½° north latitude has the Sun directly overhead at local solar noon at the summer solstice: any horizontal surface between 23½° N and the equator has less radiation on that date and, dependent upon its latitude, has two radiation maxima as the Sun "travels" (as the Earth's axis tips toward and away from the Sun) north and south during the year.

The geometry of a few different aspects at different latitudes is shown in Fig. 2-3. Note that the south-facing 45° slope at 45° north latitude is parallel to an horizontal surface at the equator. Since the Sun's rays will travel a greater distance through the atmosphere to reach the sloping surface, the amount of energy it receives will be reduced. In other words, in the absence of any attenuation of radiation by the atmosphere, both surfaces should receive the same amount of radiant energy. Note that at the summer solstice, the two surfaces will receive the same amount of energy because the declination of the earth's axis is about one half (23½°) the latitude of the sloping surface. The basic relationship between North latitude and equivalent south-facing slopes holds for all latitudes. That is, the sloping surface will always be parallel to the horizontal surface at the equator.

The same holds true for north-facing slopes in the southern hemisphere. Conversely, for a 30° north-facing slope at 60° North latitude (or for a 45° north-facing slope at 45° North latitude), the slope will be parallel to an horizontal surface at the North Pole.

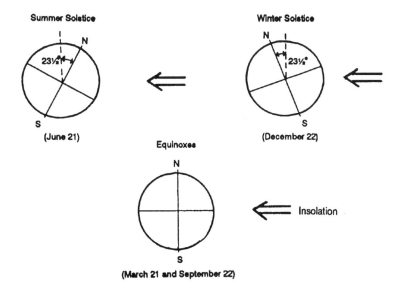

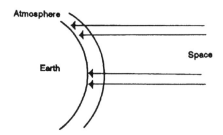

Figure 2-2 Geometry of radiation relationships between the Sun and the Earth. After Reifsnyder and Lull (1965). Note that the Sun is sufficiently far away that the rays may be considered as being parallel upon reaching the Earth: their intensity is further attenuated by the atmosphere (lower diagram)

The amount of radiation received on such sloping surfaces will, in addition, be modified by other local factors, such as surrounding topography, reflection by clouds and neighboring surfaces, and refraction by the atmosphere. The actual temperature of the air at or near the surface will also be modified by the movement of air (and the factors that affect that movement) in the immediate vicinity.

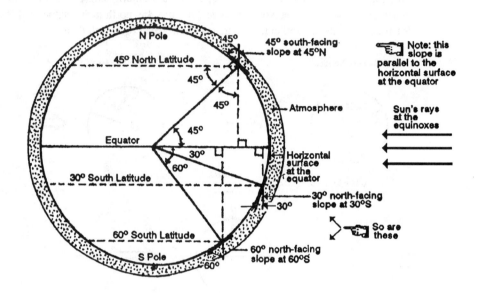

Figure 2-3 Radiation relationships with aspect and latitude

The significance of these observations lies in the fact that watersheds are made up of many different slope/aspect combinations and, as a consequence, may receive radiation that is quite different from what might be expected for a parcel of land at the actual latitude of the watershed. Thus, an east-west river might have a watershed composed primarily of north- and south-sloping sides, one of which may dominate the area. This would, in effect, "move" the watershed to an "equivalent" latitude. Since the radiant energy to which the watershed is exposed limits the amount of water that can be evaporated, it is important to evaluate watershed aspect and slope, often represented collectively by the word **facet**, for both comprehension and management.

THE ENERGY BUDGET

Keeping these potentially confusing factors in mind but temporarily out of the way, a theoretical energy budget itself can be presented. The "view" of the earth "seen" by the sun can be represented by a disc, or circle, the area of which is $A_c = \pi r^2$. The surface area of the earth, however, is in fact a sphere, calculated by $A_s = 4\pi r^2$; thus, the ratio between the two

Chapter 2 Water and Energy

areas is 4. Half of the time, of course, half of the earth is not facing the sun at all, but this cancels the observation that the direct solar radiation received at the equator gives way to zero received at the poles (there is some scattered, or diffuse radiation). Therefore, on the average, the earth actually receives:

$$(1.94 \text{ Ly/min})/4 = 0.485 \text{ Ly/min}$$

This is the figure used as the incoming radiation in Fig. 2-4, which shows the average annual energy budget for the earth.

The incoming radiation of 0.485 Ly/min is presumed to occur at local solar noon with the sun directly overhead.[3] The diagram is arranged horizontally into three zones: space, atmosphere, and earth. Vertically, the diagram is arranged from left to right as follows: (1) the short-wave portion of the spectrum, (2) the long-wave portion of the spectrum, (3) earth-atmosphere transfers of sensible and latent heat, and (4) a net balance column.

At the upper left corner of the diagram, the 0.485 figure is shown as negative, a loss to space. Upon reaching the upper limit of the atmosphere, about 10 percent is reflected back to space, represented by the +0.05 Ly/min. Another 6 percent is reflected off the top of the average cloud cover, shown as +0.03 Ly/min. About 25 percent of the incoming radiation that penetrates the upper regions of the atmosphere diffuses therein: about two fifths of this (10 percent of the incoming) is absorbed in the atmosphere, and three fifths of it (15 percent of the incoming) reaches the surface of the earth. This latter figure of 0.7 Ly/min is combined with 0.01 Ly/min reaching the earth's surface by cloud scattering (in which 0.1 Ly/min is absorbed) and another 0.215 Ly/min that penetrates directly. All three amount to 0.345 Ly/min of direct short-wave radiation reaching the earth's surface, about 14 percent of which is reflected to space (+0.05 Ly/min), leaving 0.295 Ly/min. Note that all the incoming radiation in the short-wave range is accounted for:

- 0.485 Ly/min =	.05	+ .03	+ .05	+ .05	+ .01	+ .295 Ly/min
(from space) =	(	*to space*	)	+ (*to atmos*)		+ (*to earth*)

The net balance in space, at this point, is negative, which will be adjusted by long-wave additions later. The net balance at this stage in the atmosphere, is positive, which will also be balanced shortly.

Starting at the surface of the earth, the positive incoming short-wave radiation energy heats the surface to an average annual temperature of about 20°C. A table of black body radiation indicates that an object at that temperature would radiate 0.6038 Ly/min in the long-wave range. This, like the incoming radiation, must be divided by 4. Thus, 0.151 Ly/min is radiated from the surface of the earth in the long-wave range. About 90 percent of it is absorbed by the atmosphere (+0.136 Ly/min), leaving about 10 percent to penetrate the atmosphere (-0.015 Ly/min to space).

This component, known as **back radiation**, continues all the time, as it is in the long-wave range and dependent upon temperature, not incoming short-wave radiation. Back

---

[3] Under such conditions, the zenith angle is zero, the cosine of which is 1.00, thus there need be no correction for angle of incidence of radiation.

radiation transfers energy from the surface to the atmosphere and beyond, causing the ground and any other radiating surface to cool. The ability of the atmosphere to absorb radiation in the long-wave more efficiently than in the short-wave range is the basis of the so-called "greenhouse effect." It is enhanced by the high absorptive capacity of water vapor (and carbon dioxide). Thus, a heavy cloud cover will prevent even more long-wave back radiation from penetrating the atmosphere, which will warm up accordingly.

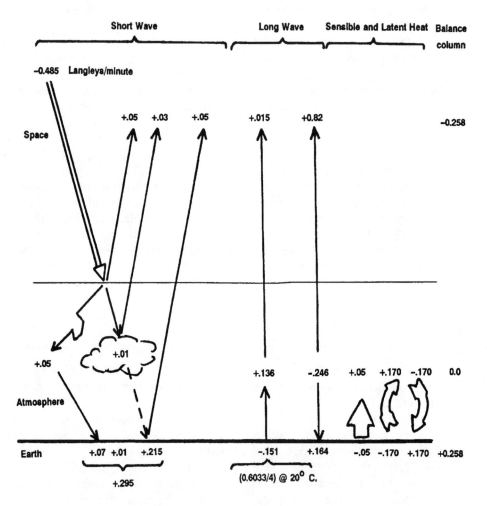

Figure 2-4 Average annual energy budget

The remaining positive input of energy at the surface of the earth is balanced by: (1) sensible heat, energy that can be felt as the warmed surface heats subsurface materials and the adjacent layer of air by conduction which, upon being warmed, is convected (rises) and conveys heat energy upward; and (2) by latent heat, energy that goes into increased kinetic energy of water molecules, evaporating the liquid and changing it to the vapor state. This loss from the surface is an addition to the atmosphere. It must be balanced by an equal loss to the atmosphere and increment to the earth in order for all evaporative type processes to equal precipitation. The balance is accomplished by the heat of condensation, where heat energy is given up as water vapor condenses, although the two would not necessarily be equal for a given day, as shown. The balance, however, provides the core of the hydrologic cycle, and involves, on the average, about 35 percent of the incoming radiation, as shown on the diagram. The amount of energy used by evapotranspiration has been reported as high as 81 percent under favorable conditions by Storr, Tomlain, Cork, and Munn (1970).

If the sensible heat is about 10 percent of the incoming radiation, and the atmosphere is assumed to be neither heating up nor cooling off during the day, then it must lose (adding algebraically within the atmosphere):

$$+.05 +.01 + .05 +.17 +.136 -.17 = 0.246 \text{ Ly/min}$$

by long-wave radiation to show a balance of 0.0 Ly/min in the right margin. If this assumption is correct (and it probably is not, precisely), then the atmosphere, according to the black body function, would show an average temperature of -40°C, which is a typical temperature at about 25,000 ft. Of the 0.246 Ly/min in the long-wave that the atmosphere radiates, about one third radiates to space and two thirds is returned to the earth. This completes the budget.

If the positive values to and from space are now totaled, the balance is -.258 Ly/min, while the balance at the surface of the earth is +0.258 Ly/min. Thus, the entire system balances, even though portions are positive or negative. The imbalance between space and Earth expresses the fact that the diagram intentionally represents the incoming radiation pattern, actually only a portion of the day. To find the situation at night, one must remove all the short-wave portions of the diagram, and adjust the sensible heat. This is not as easy as it sounds, since the contribution to the atmosphere by latent heat may continue at night. Consequently, the nighttime energy budget can be just as complex as that during daylight hours. Calculating it is not important here, as the point in presenting this diagram is in (1) establishing the relative values of the different portions of the typical energy budget, and (2) illustrating the substantial interactions between the hydrologic cycle and the energy budget.

The consequences of increasing the cloud cover on the energy balance are considerable and widely ramified. First, the amount of energy absorbed by the atmosphere would increase in both the short- and long-wave ranges. The amount of short wave reaching the surface of the earth would decrease; therefore, it would not re-radiate as much in the long wave (because it doesn't get as warm) and would not contribute as much to sensible or latent heat transfers. The ultimate consequences of such changes remain a matter of controversy, debated under arguments about burning fossil fuels which increases carbon dioxide, water vapor, and particulate matter in the atmosphere (Council on Environmental Quality 1981), and the extent of the Nuclear Winter (Turco, et al. 1983).

An important perspective on the scale of this energy budget concerns the 35 percent of the incoming radiation of 0.5 Ly/min for evaporative processes. This (0.5 cal/cm^2/min x 35%) amounts to:

$$0.175 \text{ cal/cm}^2/\text{min} \times 12 \text{ hrs/day} \times 60 \text{ min/hr} = 126 \text{ cal/cm}^2/\text{day}$$

If the water to be evaporated is at 20°C (the average temperature of the earth's surface), then approximately 620 cal will be required to heat the water to 100°C and to vaporize it. This would amount to:

$$(126 \text{ cal/cm}^2/\text{day})/(620 \text{ cal/cm}^3) = 0.20 \text{ cm/day}$$

or

$$0.2 \text{ cm/day} \times 365 \text{ days/yr} = 74 \text{ cm/yr} = 29 \text{ in/yr}$$

Based upon the fact that this budget is only for the incoming — daylight — situation, the figure is probably somewhat high. Evaporative processes often continue at a reduced rate in the absence of short-wave radiation at night. Given a favorable vapor pressure deficit (the gradient of water vapor from high or saturated levels near the evaporating water surface to low levels at some altitude above it), evaporation may even be greater at night, especially in deserts (Dake 1972). The order of magnitude, however, is correct.

Just because the energy is available does not mean that evapotranspiration will automatically increase: in a study reported by Tajchman (1971) where soil moisture was not a limiting factor, net radiation over a forest was greater than over fields, yet forest evapotranspiration was less than from one crop, greater than another. With greater canopy roughness and cooler temperatures, long-wave emission from the forest was 15 percent to 22 percent less than from the field crops, while sensible heat loss was 1.6 times to 2.6 times greater from the crops than from the forest. Comparison of radiation components on clear (one with high and one with low humidity) and cloudy days over forest and field crops (average of potato and alfalfa) are shown in Table 2-2. The data clearly show the impact of relative humidity on the latent heat flux, as well as the difference between forest and field crop evapotranspiration.

Table 2-2 Comparison of Energy Budget Components for Forest and Field Crops on Cloudy and Clear Days

Crop	Condition	R_n	B	H	V	V/R_n	K
Forest	Cloudy	92	16	-37	-71	0.77	1304
	Clear, high H_r	353	1	-99	-255	0.72	3924
	Clear, low H_r	318	0	-15	-303	0.95	3402
Field	Cloudy	88	14	-32	-70	0.80	1333
	Clear, high H_r	288	-10	-78	-200	0.70	1540
	Clear, low H_r	26	-5	-28	-228	0.86	1632

After Tjachman (1971)

Symbols and Units:
R_n - net radiation, in langleys
B - energy used to heat the plant, in langleys
H - sensible heat flux, in langleys
V - latent heat flux, in langleys
K - Bowen Ratio, in cm^2/sec^{-1}
H_r - Relative Humidity, in percent

Note, finally, that the many simplifying assumptions mask the budget's extreme complexity. For example, calculating the budget at local solar noon precludes having to acknowledge that albedos and, as a consequence, absorptivity and re-radiation all vary with the angle of incidence for most natural surfaces (Table 2-1). So, too, the degree of cloud cover, relative land and water masses and proximities, aspects, slopes, and (relatively small amount of energy going to) photosynthesis, all affect the energy budget. These and other related factors that also affect the nature of the interactions between the energy budget and the hydrologic cycle are discussed subsequently, as appropriate. The amount of energy available for evaporative processes, however, provides an upper limit to, and thus provides the basis for, an estimate of the potential evapotranspiration.

A similar model for the fate of radiant energy has been described for a water body by Viskanta and Toor (1972), and for snowpacks (E. A. Anderson 1968). In both cases, very different energy budgets may be expected. A thick snow cover insulates the warm soil beneath it, for example, reflecting short-wave and trapping long-wave radiation (Santeford, Alger, and Meier 1972), thus profoundly influencing snow melt and spring-time hydrology (Chapter 3).

Fluctuations in the solar output, belying the concept of a "solar constant," appear to have profound effects on terrestrial weather and climate. The ability to separate signals from a "noisy climatic record"[4] based on high-tech data acquisition and long-term records, and on the use of satellites to observe the Sun above the atmosphere, have contributed to our understanding of drought, floods, and other important hydrologic factors of Earth's climate (Sofia, Demarque, and Endal 1985; WQED 1986).

INSTRUMENTS AND LIMITATIONS

A number of different types of instruments are available for measuring the solar beam, the short-wave component of incoming radiation on any given slope or watershed facet (the homogeneous combination of slope and aspect). They may be organized into three groups. The first group of devices, used at first-order or Class A weather stations, contains the Marvin Sunshine Recorder, Epply Thermoelectric Pyroheliometer (Fig. 2-5), and the Smithsonian Silver-Disk Pyroheliometer. The latter is the standard instrument for measuring short-wave radiation. It does so by capturing energy on a blackened, evacuated glass tube which is connected to a mamometer and alternately exposed to and shielded from radiation.

The Epply sensor is an expensive piece of equipment (in excess of $2000), which exposes two concentric rings of metal, one white and one black, to the solar beam inside an evacuated globe. The different absorptivity of the rings affects their temperatures which, in turn, affects the resistivity of coupled circuits. The change in resistivity during the period of exposure is recorded on a clock-driven paper chart or, more recently, can be input directly into punched or magnetic tape for computer compilation and analysis. The Marvin sunshine Recorder is used for monitoring radiation intensity at the Class A weather stations, and operates on differential air and alcohol vapor expansion that is monitored by an electronic circuit,

---

[4] There are a large number of factors that affect atmospheric behavior. The atmosphere is much more volatile than are the oceans and its patterns are much more erratic, that is, there is more "noise" in the record. As a consequence, extracting information is more difficult.

activated by a mercury column that responds to the changes in pressure (Linsley, Kohler, and Paulhus 1949).

(a) Epply Pyroheliometer sensing unit. Photo courtesy of Science Associates, Princeton, NJ

(b) Pyroheliograph: a less expensive self-contained radiometer. Photo courtesy of Belfort Instrument Co., Baltimore, MD

Figure 2-5 Two types of pyroheliometers

The second group contains a number of less expensive commercial instruments that measure all or part of the electromagnetic spectrum in a variety of innovative ways. Differential thermal expansion of metals exposed to different types and amounts of radiation (Fig. 2-5b), or electrical conductivity/resistivity, such as is employed in the Epply, are used in some, while others, such as the Gunn-Bellani radiometer, utilizes an evaporation/condensation cycle that is driven by incident radiation and cool temperatures in the lower extremity of the device that is below ground. The devices expose different types of surfaces to the solar beam and, in the case of those that measure long-wave radiation, to any radiating body. Some of the less expensive units do not adequately isolate the incoming radiation as the only source of energy measured, and are therefore not reliable. Cost is from $650 to $1000. These measurements may be affected by diffuse radiation and thermal conductivity, as is also the case, albeit to a lesser extent, with the Class A weather station devices.

The third group contains an almost limitless number of instruments constructed for specific research projects. The first-order instruments, and many of those in the second group, are expensive. They, therefore, are not suitable for studying solar beam radiation where a large number of observations are needed, for example, under a forest canopy. Researchers have made use of modern materials to construct inexpensive devices to measure both long- and short-wave radiation, many for under $100. A typical radiometer made for field research consists of two blackened copper plates, each about a square inch in size, separated by about 4 in. of styrofoam mounted in a cylinder which is isolated from air temperatures by plastic covers. The vertically mounted cylinder is positioned to have one plate exposed to the incoming and one to the outgoing radiation. Behind each plate is a thermocouple which operates on the same principle as the Epply pyroheliometer. Employment of these instruments is in the interest of relative rather than absolute values, so interference by scattering or diffuse radiation and by thermal conductivity is either not important or is the same for all units.

Expensive instruments need not be widely used, however, since the maximum amount of radiation is known based on the geometry of the Earth-Sun system, and a few monitoring stations suffice to establish diurnal and annual patterns in each region or state. Cloud cover variation can affect those readings, however, and it is therefore important to have a few first-order instruments scattered around to establish radiation levels and patterns. Thus, there are usually only a few first-order weather stations in each state or region.

The limitations of the existing data on incoming radiation consist of three types. First, with the Class A monitoring stations widely separated, it is almost always difficult to get the radiation for the particular site desired. Second, determination of the incoming radiation by tables of the incoming solar beam radiation, while available (Frank and Lee 1966), is a time-consuming exercise owing to the need for corrections for date, latitude, and position of the observer in the time zone relative to local solar noon: calculation of potential solar beam radiation at a particular site is now readily possible with the aid of a computer. Third, the instruments or tables do not necessarily provide the researcher with the actual amount of energy incident to any given location: a large number of local conditions affect the energy balance, and may need to be taken into account, dependent upon study objectives.

THE HYDROLOGIC CYCLE

Providing average annual figures to the theoretical hydrologic cycle is a much easier task than is the case for the energy budget. Nevertheless, the range of values is quite high, as the many processes vary considerably with proximity to large bodies of water and latitude, and the influence on and by vegetative cover.

THE AMOUNT OF WATER

Reckoning with the actual volume of water in circulation (4.75×10^9 mi^3) is mind-boggling. In addition to being too large to comprehend, it is impractical to refer to the amount of precipitation in terms of volume. It is preferable, and standard practice, to convert the volume to some measure of depth, usually in inches (in.) or millimeters (mm). For many years the literature referred to the precipitation reported in this way as "area-inches," but the term has, by and large, dropped out of common usage. Nevertheless, that is what is referred to, in fact. Area must be considered, of course, when calculating the total volume of water that will run off from a particular rainfall or snowmelt event, for example. For every day reporting, for characterizing a geographic region or climatic province, and for many scientific calculations, the data are given simply in terms of depth.

The depth figures for the highly generalized water balance equation shown in Fig. 1-2, for most temperate zones (between about 30° and 60°), may be quantified as:

$$8" \; Runoff = 30" \; Precipitation - 22" \; E_t$$

or

$$200 \text{ mm } Runoff = 760 \text{ mm } Precipitation - 560 \text{ mm } E_t$$

The driving component of the equation, precipitation, can be as high as 900 in. (22,500 mm) per year in monsoon regions to 0 in. in desert regions. Since, where precipitation is high, the upper limit of evapotranspiration is fixed by the amount of energy available and, with the increased cloud cover associated with the high precipitation, that maximum is even lower. Under such conditions, one can expect annual runoff to approach annual precipitation. That is, both potential and actual evapotranspiration losses are likely to be very small both in quantity and as a percentage of annual precipitation.

The equation is based upon a year so that the change in storage (the difference between storage at the end and beginning of the period) does not affect the balance. In fact, the accounting is done on an annual basis so as to eliminate the effect of storage. The date chosen for this accounting cutoff is October 1, the beginning of the Water Year. At this time, for most of the northern temperate regions of the globe, soil storage is about the same, thus the change from year to year is minimized, and the accounting simplified.

THE QUALITY OF THE WATER

The quality of the water in the three components of the above equation can be characterized in very general terms. When water evaporates, it is, of course, going through a distilling process whereby the materials dissolved in the water are left behind as the energy heats the water to the boiling point and then converts it to vapor. The vapor is pure water molecules consisting of hydrogen and oxygen atoms. This is in contrast to the structurally associated forms of the compound in the liquid and solid states (Hem 1971). In these two forms, the molecule is attracted to nearby molecules owing to the slight polarization that occurs caused by the inherent imbalance of the two hydrogen and the single oxygen atoms with their respective electrical charges. This attraction of the molecules for each other is the basis for the strong cohesive force in liquid water. These characteristics are further discussed in Chapter 6.

Precipitation is not "pure," as is often cited by poets. Upon condensation of water molecules into a droplet, there is the potential and great likelihood for materials to be included. In fact, condensation must take place on some nucleus,[5] either a dust particle (from terrestrial or extraterrestrial sources), or an ice crystal. The materials present in the water include dust from meteor showers entering the atmosphere as the Earth passes through concentrations of materials in space; particulates from the burning of fossil fuels, volcanic explosions, and wind-blown dust; and gases of the atmosphere. A particularly important component of precipitation is carbon dioxide which dissolves readily in water, even while raindrops fall through the atmosphere. Rather than being pure water, then, precipitation is often a weak solution of carbonic acid, typically exhibiting a pH of 5.7, a pH value that is considerably lower, and thus more acid, than the neutral value of 7.0. The water which falls on the land, therefore, is somewhat corrosive and already has some dissolved impurities. As a

---

[5] An exception to this is where spontaneous condensation occurs, that is, condensation without a nucleus. This phenomenon can occur when the ambient temperature is below -40°F or -40°C (where they happen to be the same).

weak acid, it dissolves certain minerals and thus becomes even less pure. Falling water also has considerable capacity to erode soil particles by virtue of physical impact, the result of the water droplets' potential energy, and some of these materials may become suspended in the solution.

Runoff water can be expected to have a greater amount and variety of dissolved materials in it than precipitation. Minerals from the earth's surface and from within the soil are likely to be dissolved. These materials are known collectively as **total dissolved solids** (TDS). As water collects in rivulets and channels, its increased volume and velocity permit carrying larger amounts and larger sizes of particles in suspension, too. The materials thus carried are known as **total suspended solids** (TSS). The deposition of these materials, and the residual dissolved materials, provide the sediments and saltiness of the oceans.

The generalized differences in the water quality of the three broad components of the water budget equation provide a basis for further analysis of the hydrologic cycle itself, and provide background for additional discussion of water quality in Chapter 6. The instruments for measuring both water quantity and quality, and their imposed limitations on data analysis, interpretation, and use are discussed in each chapter.

THE ROLE OF WATER IN THE ENERGY SPHERE

The quality of the energy in each of the three components is also different, as is the nature of the energy needed to dissolve and carry materials, and to drive chemical reactions. The relationship between the energy sphere and the hydrologic cycle varies, as well.

Water's most important quality with regard to the energy sphere is its specific heat.[6] The high specific heat value of water provides the opportunity for the large masses of water in the more than two thirds of the globe's oceans to buffer (absorb and slowly redistribute) heat. Thus, the most massive relationship between water and energy involves the movement of large quantities of energy from the tropics toward the poles. The major circulation patterns of the oceans display patterns of heat transfer. For the United States, for example, the Gulf Stream conveys warm water northward off the East Coast, whereas the California Current brings cooled water from the Gulf of Alaska southward along the West Coast. Similar currents brace the other continents. These currents play a major role in the climate and weather of regions bordering the oceans (Chapter 3).

Similar buffering takes place in the atmosphere. Heated air rises at the equator and sinks at the poles;[7] hurricanes are apparently a mechanism whereby excess concentrations of energy at or near the equator are transported poleward where precipitation, winds, and general dissipation spread the energy, raising local temperatures. Weather modification efforts directed at controlling hurricanes must reckon with the energy involved, not only from the standpoint of our ability to control such storms, but also with the largely unknown effects of interfering with the energy-transfer mechanism itself.

---

[6] Defined in terms of water: the ratio of thermal capacity to that of water at 15°C. The specific heat of water is unity, derived from the number of calories (1) necessary to raise 1 cc of water 1°C.

[7] There are intermediate cycles as well, described in detail in Chapter 3.

Evaporation of water from any surface is a cooling process. In order to bring the activity of the water molecules to the boiling point, heat energy is required. When this energy is taken from the most immediate source, the remaining water or surface from which the water is evaporated is cooled. As noted earlier, this latent heat (of vaporization) is conveyed to and distributed in the atmosphere. Conversely, energy must be given up when water vapor condenses to form precipitation, which may result in warm rain and/or a heated atmosphere in the region of the condensation. Both of these energy relationships figure importantly in the discussions of precipitation and evapotranspiration, which follow in Chapters 3 and 4.

The peak rate of evapotranspiration does not occur simultaneously with the peak radiation. McNaughton and Black (1973) report that a lag of two to three hours after local solar noon was observed for 18 July days in a young stand of Douglasfir. They attribute the delay to "large forest roughness," meaning that the vapor pressure deficit is not uniformly distributed throughout the canopy, and that buffering of the incoming energy is effective in smoothing out the radiation peak.

Water at high elevations on land surfaces has high potential energy which, in accordance with the first law of thermodynamics that expresses the fact that energy can be neither created nor destroyed, is converted to kinetic energy as the water moves toward the ocean. Dissipation of this energy is by creation of heat energy through friction, and erosion, the lifting and transporting of materials such as silt, sand, and gravels comprising the bed of the stream and the formation of the stream network (Yang 1971). It is particularly important to recall this fundamental relationship when considering manipulation of the water in the stream: there is a given amount of energy therein which must be dissipated. The many instream practices of stream improvement and watershed management must be based upon reasoned consideration of this basic concept.

The amount of energy used in natural chemical processes in water bodies is small in comparison with the overall energy budget and the other energy relationships of the hydrologic cycle. What can be more significant, however, is the fact that relatively small changes in the temperature of water can cause very large changes in the rate of chemical reactions (Hem 1970) and in fish metabolism. Such changes may be wrought by intentional and inadvertent tampering with the hydrologic cycle, such as converting vegetated land to a parking lot, where temperatures of runoff waters may be dramatically increased.

SUMMARY

The energy budget is a representation of a complex portion of the hydrologic environment. The magnitude of its role varies in different parts of the hydrologic cycle, as well as at different times of the year and day. The interrelationships of the hydrologic cycle and the energy budget provide a necessary background for further consideration of the hydrologic cycle in terms of water quality, quantity, and distribution in time and space. In terms of accessibility to hydrologic components and fulfillment of management practices, the energy relationships, and the resultant type, nature, and distribution of precipitation and runoff-causing events, influence the nature and magnitude of the role that vegetation plays in the hydrologic cycle.

Chapter 2 Water and Energy

PROBLEMS

1. Working with the average annual energy budget shown in Fig. 2-4 (p. 25):
    a. Why is the net energy figure at the surface of the earth positive?
    b. What would the average annual energy budget look like if the area of cloud cover is doubled?
    c. What would the energy budget look like if the system is considered at night?
2. Create a diagram (network) between the energy and water spheres showing linkage impacts on water quality.
3. What happens to the energy available for evporative type processes at the surface of the earth when a naturally-vegetated surface is turned into a parking lot?

3 Water in the Atmosphere

*In the atmosphere
the movement and storage of water
is in response to the general global circulation
and to local vapor pressure gradients*

The residence time for water in the atmosphere, as shown in Table 1-1, is 2.43 years, meaning that the amount stored is almost two and one half times the amount in process. The first consideration is of this "stored" water. The word "stored" is enclosed in quotation marks since atmospheric water is actually in motion and often is actively and simultaneously involved in one of the several processes that transfer water within the atmosphere or between the atmosphere and the earth.

The primary factors that affect the storage (and movement) of water in the atmosphere are temperature and pressure, as influenced by radiation. The several conditions of the atmosphere (moisture, temperature, pressure, and wind movement) at any given point in time constitute the weather. The summation of weather over time is climate. All are important concerns in the study of hydrology. These topics, and the chapter quotation, provide the substantive basis of this chapter, as well as some concept of its organization.

STORAGE

The water content of the atmosphere is expressed in any of several ways. Some of these are important to the fields of atmospheric physics, research meteorology, and flight dynamics. Others are important to understanding weather and climate. It is the latter on which this chapter will focus.

CHARACTERISTICS

Pressure

The random motion of molecules in the atmosphere is typical of its gaseous nature. Attracted by gravity, the atmosphere has a greater **density** (number of molecules per unit

Chapter 3 Water in the Atmosphere

volume) at lower altitudes near the surface of the earth. The density decreases with increasing altitude. The mix of substances changes, as well, owing in part to gravity, which acts differentially on the different molecular weights of the substances, in part to different factors which influence how these substances are cycled in and out of the atmosphere, and in part to chemical processes which change them under a variety of conditions.

The collective motion of the molecules of all the substances in the atmosphere exerts a pressure in all directions, the result of molecular collisions. The net result of these collisions in the presence of gravity is a positive pressure at the surface of the earth which can support a vertical column of nearly 30 in. of mercury or 14.7 lbs/in.^{2},[1] also cited as 1013.250 dynes/cm^2. The latter figure is frequently expressed in **millibars** (mb, or one one-thousandth of a "bar," defined as one million dynes/cm^2). Standard pressure is referred to as 1013.25 mb.

Pressure in the atmosphere decreases logarithmically with altitude. Thus, while the atmosphere may be considered to be roughly 400 mi deep, the pressure at 200 mi is less than one ten-millionth of a millibar. The halfway-up point (500 mb) in the atmosphere in terms of pressure is only about 3.4 mi, or 18,000 ft (6,000 m), well below the 30,000 ft altitude at which commercial jets fly. The direction of the high-speed wind movement at or near this level, the so-called **jet stream**, is an important factor in continental weather and is related to the provision of moisture for precipitation (described in detail later in this chapter).

Being quite fluid, the atmosphere can pile up and conversely, thin down, causing areas of **high** and **low pressure**. These are caused by tidal influences of the moon and sun, and by solar heating. There are two forces that create winds in the resting atmosphere (Willet and Sanders 1959): the force of **gravity** which acts on differentially heated and, therefore, varying density air, and the **pressure gradient force** (the force due to difference in atmospheric pressure, which results in air being moved from areas of high pressure to those of low pressure). Once the air is in motion, the **Coriolis Force**[2] (Fig. 3-1) influences air movement such that the direction of air around a high-pressure "dome" is to the left around a low-pressure "trough," and is to the right around an area of high pressure in the northern hemisphere. The reversed directions are applicable in the southern hemisphere. This force is shown in Fig. 3-1 as an arrow at right angles to the pressure gradient force; the wind that results is shown as a dotted line. Thus, the deflection of air movement in the northern hemisphere is *clockwise* around a high-pressure system; *counterclockwise* around the low-pressure system.

Instruments and Limitations

The device for measuring atmospheric pressure is the barometer, of which there are two types. The **mercury barometer** is the standard[3] instrument (about $200). It makes use of a

---

[1] One can also think of the weight of the air column above a single square inch: that weight is 14.7 lb.

[2] Named after the French scientist who explained the deflection of objects (like air masses or winds) moving between the axis and outer limb of a rotating disk (which is how the earth appears if viewed from above either Pole).

[3] The term "standard" here denotes a benchmark which can be relied upon to establish accuracy, continuity, consistency, and reliability between successive readings and between different locations and observers. The standard instruments have the least number of moving parts, are simple to operate and read, and generally are *not* the recording instruments, which may have a variety of error sources.

33-in. column of mercury in a noncapillary glass tube, the upper end of which is sealed. The lower, open end is immersed in a pool of mercury, the surface of which is exposed to the atmosphere. The weight of the atmosphere keeps the mercury in the tube, and variations in atmospheric pressure are reflected in different heights of the mercury column. The atmosphere will support a column of mercury about 28 in. in height, the equivalent of (and a great deal more convenient than) the 32 ft of water it could support in a similar instrument. The barometer's primary error source is expansion and contraction of the mercury due to temperature changes. The error can be calculated and compensated for since the mercury is relatively free of impurities, and confined. Thus, a thermometer with all of *its* limitations (see below), and adjustments for temperature and gravity, are functional parts of the standard barometer.

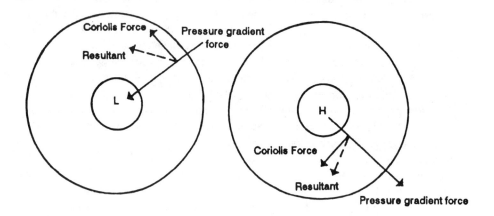

Figure 3-1 Basics of the Coriolis Force (in the northern hemisphere). In the lower atmosphere, friction plays a role in decreasing the Coriolis Force; friction is less of a factor in the upper atmosphere (Miller and Thompson 1970)

The mercury barometer is not very mobile, and this is the principal reason for the widespread use of the more convenient **aneroid barometer** ($50 to $100); see Fig. 3-2. This device merely substitutes an evacuated chamber and a spring for the weight of the column of mercury. It is more portable, less expensive, and less accurate than the mercury barometer. Both are sufficiently accurate, however, for purposes of hydrologic investigation, and need no further mention here. The interested reader is referred to standard meteorological texts such as Willet and Sanders (1959) or Miller and Thompson (1970) for additional information and references.

Temperature

Temperature is an expression of the heat energy of any object, solid, liquid, or gas. Temperature of the atmosphere decreases with altitude, too, but not in the same exponential or even uniform manner as does pressure. Temperature in the atmosphere undulates and the

Chapter 3 Water in the Atmosphere

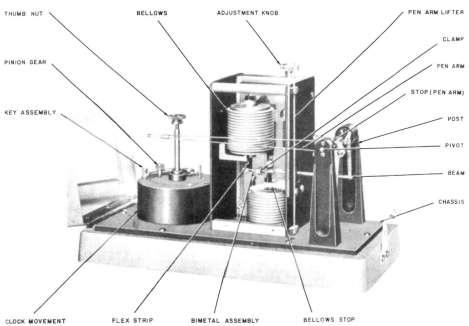

Figure 3-2 The Aneroid Barometer. Photos courtesy of Belfort Instrument Company, Baltimore, MD

Figure 3-3 Standard Weather Bureau Shelter, open, showing normal position of maximum-minimum thermometer assembly and hygrothermograph. Photo courtesy Belfort Instrument Co., Baltimore, MD

pattern changes cyclically. The alternate rising and falling of temperature mark the boundaries between the several identified layers of the lower atmosphere (Miller and Thompson 1970). The resultant characteristics influence (and reflect) distribution of constituents, physical properties, ability to absorb short- and long-wave radiation, and many optical phenomena. For example, moist air will absorb more long-wave radiation from the earth which, in turn, builds up heat energy in the atmosphere, increasing the potential for evaporation. That process eventually transfers enough water from the earth to produce clouds, which reflect incoming short-wave radiation and keep it from reaching the earth, thus countering the effect. In this manner, the atmosphere physically *buffers* the potential extremes of energy trnasfer via long- and short-wave activity, as well as the dependent temperature, water vapor content of the atmosphere, cloud cover, and all related processes.

Standard temperature of the atmosphere at sea level and a pressure of 1013 mb is 15.0°C. At the 18,000-foot (6000 m) level (500 mb) it is -24.0°C, and at 25,000 ft (8,500 m) it is about -40°C. The general **gas law** combines the relationship between pressure (p) and volume (v) at initial conditions ($_0$) and some other time in Boyles' Law:

$$p_v = p_0 v_0$$

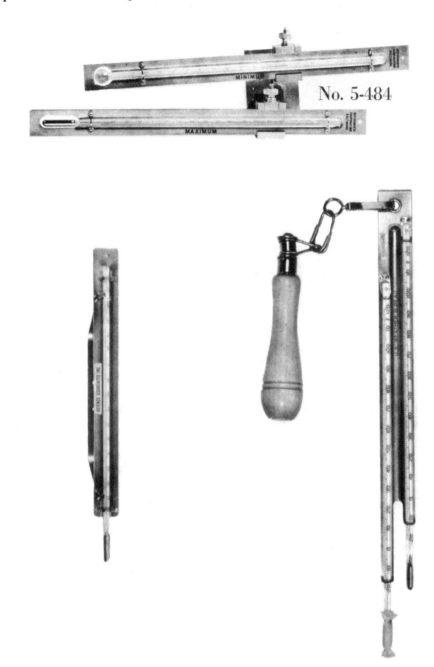

Figure 3-4 Standard Thermometer and Sling Psychrometer and maximum-minimum thermometer assembly. Photos courtesy Science Associates, Princeton, NJ and Belfort Instrument Co., Baltimore, MD, respectively

(at constant temperature the volume of a gas varies inversely as the pressure), and between volume and temperature (t) in Charles' Law:

$$vT_0 = v_0 T$$

(at constant pressure, volume is directly proportional to absolute temperature) into a single formula:

$$pv = (p_0 v_0/T_0)T = RT$$

where R is the **gas constant**, which varies dependent upon whether the air is dry or damp, that is, its moisture content. Since the molecular weight of air that has a high water vapor content is different from that which has a low water vapor content, the rate of change in volume per unit change in temperature, as expressed in the basic gas law, varies over the natural range of atmospheric pressure and temperature. The expression is useful in defining and understanding the adiabatic process, by which condensation and subsequent precipitation occur, discussed below.

Instruments and Limitations

Temperature is measured by several different instruments in a variety of exposures. The U. S. National Weather Service standard instrument is a liquid-in-glass thermometer exposed in a white, wooden, louvered shelter (Fig. 3-3) with a double top "to protect the thermometers from precipitation, condensation, and radiation" (Weather Bureau 1962). The shelter (about $400) is approximately 30 in. x 30 in. x 12 in. in size, and is mounted 4 ft above the ground, with a 45° cone of obstruction-free exposure, if possible, and the door facing north so that sunlight will not affect the thermometers when it is opened. While the resultant reading of the thermometer may or may not truly be the temperature of the atmosphere at that time and location, it does provide a reliable and replicable reading of the temperature inside the 30 in. x 30 in. x 12 in. white, wooden, louvered box: a standard so that readings at different times and places may be reliably compared. There are three different thermometers for measuring current, maximum, and minimum air temperatures (Fig. 3-4).

The **Standard Thermometer** is a mercury-in-glass instrument with which most readers will be familiar. The standard instrument ($20 to $25) is more carefully calibrated than is the run-of-the-mill household thermometer, and is more accurate.

The **Maximum Thermometer** also contains mercury, but is constructed with a slight restriction in the capillary tube, immediately above the reservoir. Upon whirling vigorously, the mercury is forced down past the constriction into the reservoir. Exposed in an horizontal position, the mercury expands and forces its way past the constriction as the temperature rises. When the maximum has been reached and the temperature falls, the mercury column breaks at the constriction and the upper end of the column may be read on a standard scale. The reading must be made before the next maximum and, of course, does not indicate when the event occurred. The Max-Min thermometer assembly costs about $135.

The **Minimum Thermometer** is an alcohol-in-glass thermometer that contains a small dumbbell (like a two-headed pin) in the fluid. This instrument is "set" by tipping it upside down until the dumbbell rests against the inside ("upper") surface of the fluid (meniscus), and then is also exposed (nearly) horizontally. Upon contraction of the fluid, the dumbbell is pulled toward the reservoir and, upon warming and re-expansion, is left behind. Its upper end thus indicates the minimum temperature during the interval since the last reading. Like the maximum thermometer, the instrument does not indicate the time of occurrence.

Recording thermometers are often combined with a device to monitor relative humidity into a single instrument called an **hygrothermograph** (Fig. 3-5) that makes use of differential expansion of metals. A sandwich of two such metals will bend upon undergoing a temperature change, and the resultant deflection is transferred mechanically to a pen which rests against a chart that is wrapped around a drum driven by a daily, weekly, or monthly clock. The typical Class A Weather Station instrument has an 8-day clock and is used to indicate the times at which the maxima and minima occurred. It is recommended that the current temperature (and the other instruments) are read and re-set between 6 a.m. and 8 a.m. and between 5 p.m. and 8 p.m. each day (Weather Bureau 1962). This instrument currently lists for about $365.

Figure 3-5 Standard Hygrothermograph. Photo courtesy Belfort Instrument Co., Baltimore, MD

Temperature is an important and easily obtained indicator of the energy available for evapotranspiration, thus it is important to understand the methods by which average temperatures are determined. Mean daily temperature can be calculated by:
1. averaging the readings taken on the current temperature thermometer in the morning and evening;
2. averaging the maximum and minimum readings; or
3. striking a horizontal line (which will be the mean) through the undulating pen trace on the hygrothermograph chart so that the area above and below it between the line and the trace are equal.

The first two methods will work only if the temperature change during the day is uniform. And the third needs to be adjusted for any error deviations of the mechanical/bimetallic instrument from the "true" temperature as indicated by the maximum and minimum thermometers. Obviously, the average of 24 hourly readings taken at one-hour intervals provides a more accurate estimate of the mean daily temperature, and may be employed to calibrate the other, less time-consuming methods finally used. In the long run, that is, over a year or more, averaging the max and min, or the morning and afternoon readings, will indicate the average temperature of the site.

As is the case with other atmospheric characteristics, it is the conditions near the ground, that is, within 3 m or so, that are particularly important in most of the hydrologic processes. A treatise on this topic was prepared by Geiger (1957). One must exercise care in utilizing temperature and other data in light of hydrologic processes and moisture storage: their accuracy, reproducibility, and applicability should be continually evaluated. Complex interrelationships among temperature, wind, humidity, and radiation and interactions between these factors and aspect, slope, latitude, vegetative type, season of the year, and time of day all affect the rate of evaporative-type processes.

Humidity

Water stored in the atmosphere as a gas is called **water vapor**. At any given time, some water may also occur in the atmosphere in liquid form as drops, either buoyed by winds or too small and too light to fall as precipitation.

The Amount

In the well-mixed lower portion ("homosphere") of the atmosphere, water vapor is the third most plentiful substance, as shown in Table 3-1. The relative amounts of the constituents vary from season to season, and from the Equator toward the Poles. For example, water content of the atmosphere in centimeters (cm) varies from 5.4 cm at the Equator to 2.2 cm at 60°N in the summer or wet season, and from 3.2 cm at 0° to 0.5 cm at 60°N in the winter or dry season. Water content also decreases by about 75 percent at about 3000 m (9000 ft) altitude, as well.

Such wide latitudinal, seasonal, and altitudinal variation makes it difficult to monitor water content of the atmosphere by direct measurements. The amount of water vapor that the atmosphere can hold at any given temperature and pressure is limited. The normal method of expressing potential water content is by the **vapor pressure**. The **saturation vapor**

pressure is defined simply as the amount of water vapor that the air could hold at the specified temperature and pressure. Vapor pressure is measured in millibars (mb) or in. of mercury, and is often expressed as the "**partial pressure** of the atmosphere due to water vapor." The saturation vapor pressure is on the order of 15 mb to 30 mb.

Table 3-1 Principal Constituents of the Lower Atmosphere

Constituent	Percent by volume
Nitrogen (N_2)	78.08
Oxygen (O_2)	20.95
Water Vapor (H_2O)	~4.
Carbon Dioxide (CO_2)	0.033[a]

[a] An average: the value can be as high as 0.1 percent.
Source: Miller and Thompson (1970).

A plot of the saturation vapor pressure relationship with temperature is shown in Fig. 3-6. Theoretically, the air cannot hold more than the indicated water vapor at the corresponding temperature. In the absence of condensation nuclei, however, the air may become supersaturated with water vapor. With the aid of Fig. 3-1, one can consider what happens to air that is less than saturated as it is cooled: the amount of water vapor remains the same, but the cooler temperature means that the air becomes closer and closer to the saturation vapor pressure. That is, as the air is cooled, the actual vapor pressure, e_a, approaches the saturation vapor pressure, e_s. When the two are equal, the air is, indeed, saturated. The temperature at which this occurs is called the **dew point**, specifically defined as *the temperature to which the air must be cooled to reach saturation at constant pressure*. The dew point is important because, since condensation cannot occur until saturation is achieved, it determines whether the condensate will be water (if the ambient temperature is above 32°F) or ice (if the ambient temperature is below 32°F), in which case it is the frost point (Miller and Thompson 1970). The air may be *supercooled*, that is, cooled below the dewpoint without condensation. At -40°C (either C, or F, the cross-over point) spontaneous condensation occurs; if dust particles or ice crystals are present, they act as "templates" for condensation (Dingman 1984).

A particularly important feature of the relationship shown in Fig. 3-6 is highlighted in the inset portion of the graph. Below 0°C, the vapor pressure over ice is less than that over liquid, super-cooled water. As a consequence, ice crystals make good condensation nuclei. Water vapor will migrate toward ice crystals along a vapor pressure gradient caused by the lowered value of e_s over the ice. Weather modification by cloud seeding with dry ice (solid carbon dioxide) is based on this phenomenon (Dennis 1970; Squires 1971). Silver iodide is used instead of dry ice because its crystalline structure behaves in the same manner. Dry ice is bulky to carry aloft, and silver iodide generators may be located on the ground, firing the crystals up into the atmosphere. Furthermore, the phenomenon has been shown to be important in the process of condensation of atmospheric water on snowpacks, and in the moisture relations of frozen soils.

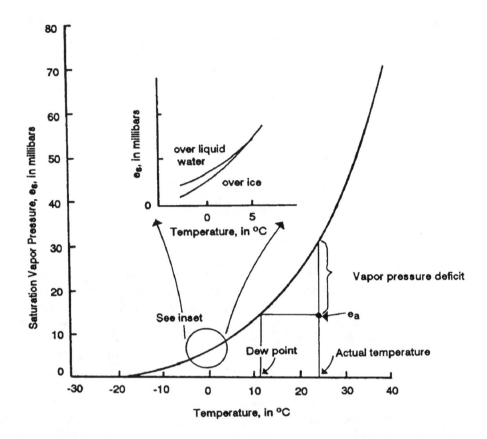

Figure 3-6 Saturation vapor pressure and temperature

Another expression of the degree of saturation is the **relative humidity**. Relative humidity is defined as the amount of water vapor in the atmosphere as a percent of the amount of water vapor the atmosphere could hold at that temperature and pressure. The formula for relative humidity, H_r, is:

$$H_r = (e_a/e_s) \times 100$$

An important measure of the moisture condition of the atmosphere, relative humidity is of special value to climatologists and hydrologists. This general and widely used term aids in the definition of climate and comfort indices. The numerator of the equation is the **vapor pressure deficit**, defined as the extent to which the actual vapor pressure is less than the saturation vapor pressure. This useful concept is, in turn, an expression of the **vapor pressure gradient**, usually used to refer to the potential for moisture to move from a free water surface, immediately above which the air is saturated, to a layer of air some distance above it. The term "saturation" in the definition of the dew point (above) could be replaced by "one hundred percent relative humidity."

If the absolute quantity of water vapor in the atmosphere (useful for detailed atmospheric studies and research) does not change, then as temperature increases, relative humidity decreases, and vice versa. If the temperature remains the same and more vapor is introduced, then the relative humidity increases. And, as air is cooled without removing moisture, the relative humidity also increases. If, in addition, this cooling takes place without loss of heat, the cooling process is said to be **adiabatic**. This is accomplished by lifting a parcel of air through the atmosphere, that is, the temperature is reduced solely because the air is occupying a larger volume (at lower pressure) and not because energy is being removed. This is the primary process by which air is cooled in the atmosphere and by which precipitation is formed. As noted in the earlier discussion of the gas law, the rate of volume change depends upon water vapor content of the atmosphere, specifically, whether the air is saturated or not. If it is not saturated, then the expansion by cooling takes place according to the dry adiabatic **lapse rate** (rate of change in temperature with pressure); if the air is saturated, the expansion by cooling takes place at the moist adiabatic lapse rate. The difference is particularly important to hydrometeorological research and airplane pilots. The change of heat content of a particular air parcel as it is lifted over a mountain barrier produces orographic precipitation and foehn winds, discussed below in the section on precipitation. Normally, the changes of state of water in our natural environment involve exchange of latent heat: this occurs whether the change is sequential (solid, liquid, gas) or bypasses the liquid state as in *sublimation*, going directly to a vapor from the solid state, or vice versa (Dingman 1984).

Other expressions of the amount of moisture in the atmosphere that are of value to meteorologists and atmospheric physicists include **specific humidity** (the mass of water vapor which is present per unit mass of moist air in grams/kilogram), **absolute humidity** (the density of water vapor in moist air in grams per cubic meter), and the **mixing ratio** (the ratio of the mass of water vapor to the mass of *dry* air in grams per kilogram).

Instruments and Limitations

The devices that monitor moisture in the atmosphere are many and quite varied. According to Miller and Thompson (1970), the most accurate way in which to measure atmospheric moisture is to pass a known volume of air through a chemical drying agent and to weigh the amount of water thus collected. The method is also impractical, and consequently, other techniques have been devised.

The standard instrument for monitoring atmospheric humidity is the **Sling Psychrometer**, which consists of two mercury-in-glass thermometers mounted on a calibrated plate. The reservoir of one, the wet bulb, is covered with a close-fitting muslin sock that is wetted before reading. Because evaporation is a cooling process, the wet bulb temperature will be lower than the dry bulb in proportion to how much water evaporates from the sock. If the air is saturated, that is, if the relative humidity is 100 percent, there will be no wet bulb depression. Relative humidity is determined from a table of wet bulb depressions and dry bulb temperatures, and the dew point may be determined from them as well. The psychrometer may be mounted in the instrument shelter along with the other thermometers, where it is ventilated with a fan (about $200, complete). Alternatively, it may attached to a hand-held swivel, and whirled around, thus aspirating the wet and dry bulb thermometers. The standard U.S. Weather Bureau model currently lists for about $43. Upon reading, the dry bulb temperature should be the same as that indicated by the current temperature thermometer and, if they have just been reset, the maximum and minimum thermometers should also read the current temperature. The advantage of locating the instrument in the shelter is that of standardization of readings from one time to the next. An advantage of the hand-held unit is its portability.

The timing of humidity variations is shown on the lower portion of the chart on the hygrothermograph (Fig. 3-5). The pen, in this case, is controlled by expansion and contraction of horse hairs as atmospheric humidity changes. A hygrothermograph is usually installed in the Class A weather station along with the standard instruments.

Several other instruments make use of electrical conductivity of materials that are in moisture equilibrium with the air, or of optical and radiation technology.

A major limitation of humidity readings is that concerned with the relation of the reading to precipitation. For example, a local weather station reading might indicate a relative humidity of 70 percent, but it might be raining. Rain cannot occur unless the relative humidity is 100 percent *in the cloud from which the precipitation originates*. Furthermore, as rainfall occurs, the relative humidity of the air at or near the ground usually increases and reaches 100 percent as a result of splash and evaporation, which can continue during precipitation as long as there is a vapor pressure gradient. Air with heavy moisture content and cooling temperatures may cool below the dew point and develop fog.

Clouds and Fog

Clouds are masses of condensed water droplets or ice crystals. They are formed as moist air is cooled so that the air temperature drops to the dew point or, in the case of super-cooled clouds, either in the presence of condensation nuclei or below -40°C (or F). A large number of different cloud types have been identified. They are associated with different atmospheric conditions, and may or may not yield precipitation. Ice and water clouds may be differentiated with the naked eye: ice clouds appear feathery, whereas water clouds are tufted with sharply delineated boundaries.

Clouds are common because, as Miller and Thompson (1970) point out, 0°C should really be referred to as the "melting point," not the "freezing point": as it cools, water can remain in the liquid state down to -40°C (or F), and "it is common for liquid water drops to exist in clouds at temperatures as low as -20°C." Ice, on the other hand, will always start to melt at 0°C unless it is contaminated in some way. Clouds of water vapor are more common than clouds of ice, in large part, because water may not freeze at 0°C as the parcel of air cools adiabatically.

Clouds are classified (and named) on the basis of their **form** and **altitude** (Miller and Thompson 1970). The nomenclature system involves using the prefixes *cirrus* for high (above 6 km or about 20,000 ft), *alto* for middle (2 km to 6 km; 6500 to 20,000 ft), and *stratus* for low (below 2 km; 6500 ft). For example, the term "nimbo-" used as a prefix means vertical development through two or more of these levels. "Strato-" means layered; and "cumulo-" means tufted, and also may refer to a cloud of vertical structure (for example, the classic *cumulonimbus* thunderstorm). Variations of these terms, and a variety of other terms, are used in combination to describe meteorological conditions and aid in classifying storm systems: in the presence of certain atmospheric conditions, including air mass movement, typical cloud formations occur, thus permitting weather condition determination and forecasting. Clouds, of course, are not static phenomena. They move with the winds or, sometimes, the air moves through them.

Fog is a cloud that touches (or nearly touches) the ground. There are several types, in addition to the fact that fog may be made up of ice or water. Although fog is often thought of as being a "blanket" and, therefore, static, that is true only for **radiation fog**. This type forms near the surface of the ground when there has been strong back radiation that cools the surface of the earth and the immediately adjacent air. The air must be still, as it often is on clear, cool nights when there is little obstruction to outgoing long-wave radiation. Cool, clear conditions often occur under the influence of a full or new moon when atmospheric tides respond to the alignment of the sun, moon, and earth causing high pressure that is in the process of descending; thus it is cool, transparent to long-wave energy, and clear because clouds are not forming as they do in rising air. Radiation fog can be no more than a few feet in height (people can often see one another's heads above it and, upon lying down on the ground, can see one another's feet as well). It typically forms in the early morning hours before sunrise, and it dissipates when the sun penetrates it and warms the ground, causing a rise in the air temperature, and a lowering of the relative humidity.

Another common type of fog is **advective fog** which, as the name implies, results from moisture-laden air being advected (transported horizontally) and cooled. Warm, moist air moving over a cooler surface produces this type of fog, typical of the west coast of the United States, and other locations where a moisture supply and either cool water or land (or snow) surfaces exist.

Steam fog typically forms over the water surface of still lakes, at times when the air is nearly still.[4] Cool air that has drained downslope to the lake depression cannot hold as much moisture as the lake surface is capable of evaporating at the given temperature: that is, the water temperature is higher than that of the air, and potential evaporation is greater than the air temperature can sustain. Both advective and steam fog are dynamic fogs, in contrast to radiation fog. Thus, upon completion of consideration of moisture that is in storage in the atmosphere, it is an appropriate time to consider its movement.

PROCESS

The movement of water in the atmosphere is driven by its global circulation which is, in turn, influenced by the changing seasons, the rotation of the Earth around the Sun and the tilt of its axis, and the forces that come into play during the Earth's diurnal rotation on its axis.

GLOBAL CIRCULATION

The best way in which to approach air movement in the atmosphere is to first consider the completely impossible nonrotating globe with a uniform water or solid earth surface (Fig.

---

[4] Time-lapse photography of stream fog shows clearly that the air is not, in fact, still: small whirlwinds, currents, and recurrent patterns of movement throughout the fog layer are quite apparent.

3-7). The most intense radiation is at the Equator, with no direct solar beam at the poles. Under such conditions, air would be heated at the Equator and, being less dense, would rise and spread poleward at the upper levels of the atmosphere. The air over the poles is cooler and, being more dense, would sink.

When the Earth rotates, this one-cycle depiction gives way to three cycles, as shown in Fig. 3-8. Here, as in Fig. 3-7, heated air rises at the Equator and sinks at the poles, but air also sinks at about 30°N (and S) latitudes, and rises, although less vigorously owing to less heating, at about 60°N (and S) latitude. The sinking air at 30° spreads out, producing the regions around the globe known as the "doldrums," or "horse latitudes".[5] In contrast, at about 60°, cool air from the pole and warm air from the Equator converge at the surface producing what is referred to as the **polar front** near the border of the United States and Canada. This is the region where cyclonic storms are generated, one of the three main storm types discussed herein.

A cyclone is an intense area of low pressure with winds flowing counterclockwise in the northern hemisphere. Hurricanes are one type of cyclone, although they are generated over warm waters in the tropics rather than over land along the polar front. Tornados are also cyclones, often with extremely high winds circulating particularly violently around a very intense low that may be only a hundred yards (or less) in diameter. Cyconic storms produce a major part of the annual continental precipitation. The mechanics of the polar front are helpful in understanding this major source of precipitation, and are shown in Fig. 3-9.

The Coriolis Force deflects winds to the right in the northern hemisphere and to the left in the southern hemisphere[6] (Strahler and Strahler 1973). Thus, as the Earth turns under the lagging winds, equator-bound winds are deflected to the west (designated by the direction of origin, *easterlies*), and pole-bound winds are deflected toward the east (*westerlies*). Thus, the arrows indicate the direction of the masses of air[7] at the surface along the front: easterlies bring cooler, drier air from the arctic; westerlies bring warmer, more moist air from the tropics. The interface between these differing air masses is the front, which goes through a classic and predictable, though not always identical (Gleick 1989), metamorphosis, described below.

The polar front shifts north and south with the seasons. Often, where the air currents aloft divide to turn north and south, a pressure gradient is associated with high-speed air currents at about 18,000 ft to 30,000 ft (5500 m to 9000 m). Discovered in the 1950s when jet aircraft first reached those altitudes and were aided in their west-east flying time, the fast-moving current was called the **jet stream**. Frequently the name is warranted, as the winds are several hundreds of mph, and fairly tightly confined. At other times, however, the jet stream is more of a broad river of air, sometimes with many channels.

---

[5] So called because during early explorations, ships became becalmed, resulting in the death of livestock on board, especially horses, and their being eaten or thrown over the side.

[6] The presentation is confined to the northern hemisphere since the situation is identical but opposite in the southern hemisphere.

[7] Wind direction is designated by the direction the wind is coming from.

Chapter 3 Water in the Atmosphere 49

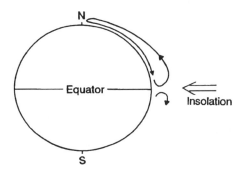

Figure 3-7 The atmosphere on the non-rotating earth

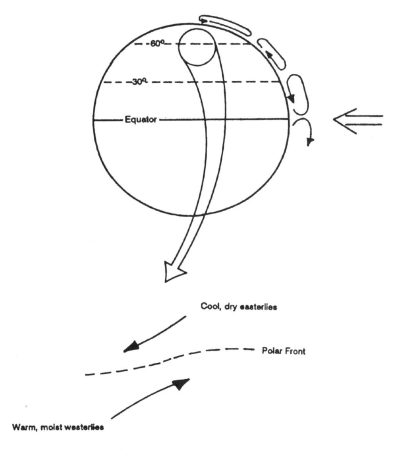

Figure 3-8 Atmospheric circulation of the northern hemisphere
The dense, sinking cool air at the pole generates three to five lobes of high pressure radiating out from the North Pole like a giant hand resting, palm down, on the top of the globe.

These gradually rotate, moving from west to east. Low pressure troughs exist between these "fingers" of high pressure and, when the pressure differences are well defined, the jet stream may take an undulating course along the variably-defined boundary between high and low pressure, duplicating aloft the location of the polar front on the ground. When this happens, conditions are particularly ripe for generation of heavy precipitation and other violent weather.

Satellite imagery in recent years has provided enhanced understanding of the relationships between global circulation and local weather, as well as spectacular pictures of weather systems and even individual storms of different types (Smith, et al. 1986).

PRECIPITATION

Sources of Precipitation

In general, it is the cooling of a parcel of air that produces precipitation or storms. This is normally accomplished by lifting of the parcel of air to regions of sufficiently lower pressure and, therefore, temperature, to reach the dew point. There are several different lifting mechanisms: air mass weather, orographic lifting, and convection. The first to be considered is the air mass interaction which takes place along the polar front, as just introduced above, and which produces cyclonic storms.

Cyclonic Storms

Air movements in the vicinity of pressure systems are actually functions of both forces (the Coriolis Force and the pressure gradient force) and the relative humidity. The greater number of water molecules in moist air *lower* the overall air density (the molecular weight of water is only 18; the molecular weight of dry air is 28), thus making moist air *rise* over dry air. As a consequence, moist air is *less stable* than dry air. Further, as the air rises in the vicinity of low pressure, the temperature is reduced by adiabatic cooling and relative humidity increases, exacerbating the instability. Thus, air moving *inward* to an area of low pressure has to rise and is often unstable, leading to precipitation events that commonly accompany a low-pressure area or system. Air moving *outward* from an area of high pressure tends to subside, yielding the more stable conditions typically associated with low humidity.

Transects through the lower sector of the low-pressure system shown in Fig. 3-9d are depicted in Fig. 3-10. Cooler, dry polar air is being advected from the left, forcing its way under the warmer air to the east and south because the cooler air is more dense[8]. This lifts the warm air, cooling it (both by lifting and by contact with the cooler air along the interface between the two air masses) until it reaches the dew point. If nuclei are present, condensation occurs and, if the drops (or flakes) get sufficiently large, precipitation occurs as well. Warm air already has a tendency to rise, and the cold, dense air slips underneath, helping to lift it. Resultant adiabatic cooling produces cumulonimbus clouds, sometimes to great heights, and often in a squall line, characterized by intense, short-duration precipitation behind the line

---

[8] The implication here is that the described phenomenon occurs only at the polar front: in fact, it happens wherever two air masses that differ in temperature and moisture content meet.

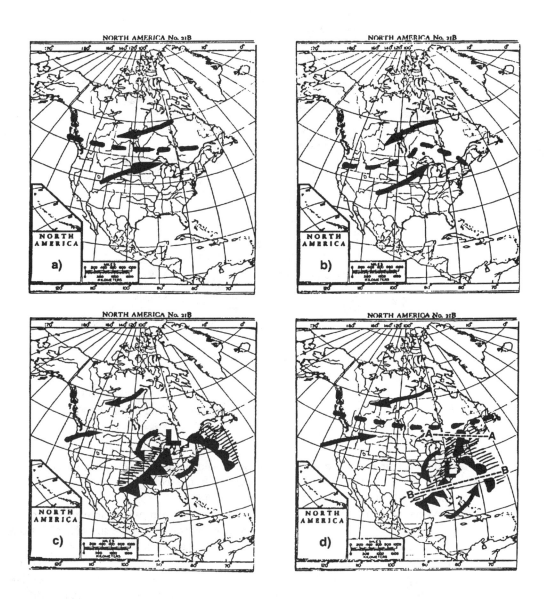

Figure 3-9 Life history of a typical cyclonic storm: winds and cyclonic storm generation at the polar front (AA and BB profiles shown in Fig. 3-10)

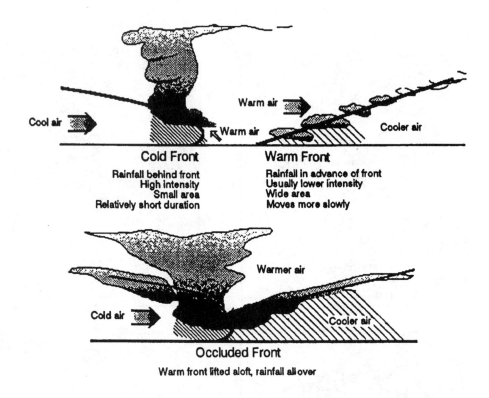

Figure 3-10 Vertical profiles (from Fig. 3-9) through cold and warm fronts (BB), and occluded front (AA)

where the cold air meets the warm air at the ground surface; these storms are also often restricted to a rather small area. This interface is the **cold front**[9] (Fig. 3-10a). The precipitation behind the front is shown shaded in Fig. 3-9.

The line of contact at the ground between the slowly moving warm air in advance of the cold front and the cooler air it is trying to replace is the **warm front**,[10] shown in Fig. 3-10b. Here, cooler air is trapped as warm air rises over it. Again, the adiabatic cooling process produces clouds and, eventually, may produce precipitation. In contrast to the cold front, the gradual trapping of cool air by the advancing, overriding warm air is relatively stable, and stratus clouds usually mark the warm front in contrast to the turbulence associated with the

---

[9] The cold front is indicated on the synoptic weather map by solid (if on the ground) or open (if aloft) triangles representing the teeth of the cold air, and point in the direction of the front's movement.

[10] The warm front is indicated on the synoptic weather map by solid (if on the ground) or open (if aloft) semicircles.

cold front. Warm front precipitation is, therefore, usually less intense. It also tends to cover a wider area, and occurs in advance of the front.

Because the cooler air is trapped, the warm front does not move as rapidly as the cold front, which catches up with it, lifting it aloft to produce the **occluded front** (Fig. 3-10c). Precipitation occurs both behind and in front of the occluded front. The life cycle of the typical continental cyclonic storm looks very much like that shown in Fig. 3-9. These storms may occur at any season of the year, but are most prevalent in winter. Eventually, when the life cycle of the cyclonic storm is completed, the polar front is re-established, and the cycle may begin all over again. The polar front is not, however, always in the position shown: it may have great undulations in it, generating cyclones anywhere.

When the jet stream is fairly straight from west to east across the United States, weather systems move rapidly and do not produce significant precipitation owing to a lack of moisture supply and, typically, an insufficient temperature differential.

When the polar front and the jet stream coincide and run in a southeasterly direction, high pressure will be to the north and east, with generally southeasterly winds "above" the front, and northwesterly winds "below" the front. These conditions, if they develop over the northwestern portion of the United States, will likely produce limited cyclonic precipitation owing to the general blocking of moisture from the south and west and only slight temperature differentials on either side of the front. If the conditions prevail over the central or eastern portions of the United States, there may be no precipitation at all.

On the other hand, if the jet sweeps northeasterly from the Gulf of Mexico and a low-pressure cell (cyclonic storm) forms under it, a major storm system can develop. When cyclones are thus generated on the northeasterly flowing limb of the polar front under the similarly flowing jet stream, vast amounts of maritime moisture are supplied from the south by the southwesterly winds "below" the front. A great dome of continental cool, dry, high-pressure air to the northwest provides the temperature difference, along with northeasterly winds "above" the front, necessary to fuel the continuing dynamics of very strong cyclonic storms. Under particularly great pressure and temperature differentials, violent thunderstorms and tornados may occur, as well as excessive amounts of precipitation.

Similarly, cyclonic storms that sweep in a northeasterly direction into the west coast from the Pacific Ocean will be enhanced by having the jet stream lie along their, and the polar front's, path. If the polar front and the jet stream are lying northeasterly from the Sea of Cortez, strong storms in the Rocky Mountains and in the western and central plains will develop. Soaking rains may inundate the west coast states, and produce large snowfalls inland. As these masses of air swing inland, topography has a profound effect, ramified in the second type of storm considered here.

Orographic Storms

Mountain barriers provide another lifting mechanism that will trigger adiabatic cooling, condensation, and precipitation. Thus, along windward coasts, orographic storms provide a great percentage of the precipitation, especially if there are mountain ("oro") barriers to the smoother horizontal advection of warm, moist air. A classic example in the United States is the mountainous Olympic Peninsula, where annual precipitation may exceed 5000 mm (200 in.).

Similar situations exist on the windward side of many tropical islands where moisture-laden trade winds are lofted and deposit precipitation that supports lush rain forests.

Some examples of orography and the consequences thereof are shown in Fig. 3-11, a transect from Los Angeles to Denver. Here, advected tropical air is raised over the relatively minor topographic barrier of Mount Wilson in the Tehachipi Mountains, increasing annual precipitation to as much as 2000 mm (80 in.). The precipitation occurs primarily on the windward (west) side of the mountain, and the leeward (east) side is in a rain shadow. More important for residents during the fire (dry) season, is the fact that the air has been stripped of some of its moisture, resulting in a release of heat of condensation. Further, as the air descends the east side of the mountain barrier, it is compressed, which raises the temperature even more and, without addition of moisture, lowers the relative humidity to extremely low levels. The result is the hot, dry wind known as the **Santa Ana**. With the high vapor pressure gradient associated with this high-velocity air movement, the region of the rain shadow is extended and creates the desert climate to the east, as well as conditions for a potentially dangerous fire situation in the dry, inland regions of California.

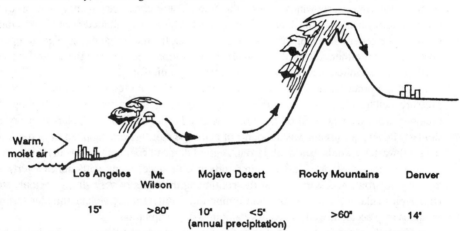

Figure 3-11 Orographic precipitation patterns between Los Angeles and Denver

Picking up whatever moisture is available, the air is lofted once again, although to much greater elevations, over the Rocky Mountains. More water vapor is squeezed from the air, and hot, dry winds again descend on the east slope along the Front Range. Here the wind is known as the **Chinook**. Both the Santa Ana and the Chinook are actually phenomena known as **Foehns**, called by a variety of names around the globe (Kerr 1986).

The hydrologic significance of the Chinook, in particular, is that it is capable of sublimating[11] even thick snowpacks, thus precluding snowmelt runoff and accretion to soil, ground water, and man-made reservoirs. The foehn, as exemplified by a well-monitored Chinook in Colorado (Kerr 1986) is limited in thickness to about 2 km of nonturbulent flow

---

[11] Going from the solid to the gaseous state without turning to liquid.

which can explain the frequently very high velocity that can generate damaging winds, drying out soils, and fostering the conditions favorable to brush and forest fires. These, especially in southern California, leave steep slopes bare and subject to severe erosion during the wet season.

Orographic storms exhibit a wide range of precipitation intensities and durations. They can be brief, afternoon showers in the tropics, and days-long steady rains on the Olympic Peninsula. In complex terrain situations, such as the mountains of western Oregon and Washington, the extent of the relationship between precipitation and elevation is apparently confounded with other factors, such as distance from the ocean and the influence of air mass weather, and is not clearly defined (Schermerhorn 1967). Cooper (1967) reported no such relationship at all for four years of data on a small watershed in the mountains of Idaho, although orographic effects may be clearly seen in many areas by overlaying isohyetal and topographic maps. Farmer and Fletcher (1972) reported a nonuniform increase in intensity with elevation, as well as the expected increase in amount.

Around the Great Lakes and elsewhere, orography contributes to **lake-effect precipitation**. Cold, dry air associated with air mass weather or fronts is advected across the warm lake (especially in winter when the water has not frozen) and, with considerably higher relative humidity after absorbing evaporate, moves over cooler and higher land surfaces. On the lee-side of the lakes, sizable snowfalls occur, but do not contribute a great deal to the local water supply owing to the low density of the snow (Miller 1977). In the heavy snow belt on the lee side of Lake Erie, Webb and Phillips (1973) report that less than 20 percent of the annual snowfall is lake-effect precipitation; over the entire basin, less than 6 percent of the annual snowfall is lake effect. And, since the precipitation is normally derived from the lake, it is not an accretion to the total water supply of the St. Lawrence system. Impacts are not restricted to snowfall: Changnon and Jones (1972) report substantial lake-effect changes in number of cloud days, wind speed and direction, and fog, with major differences between winter and summer as well.

One storm type often behaves like another, and the several categories of storms begin to blend with one another or to actually occur under circumstances that would produce a different type. For example, the extent of the orographic effect on precipitation at the Coweeta Hydrologic Laboratory in the southern Appalachian Mountains varies with original type (cyclonic or convectional) of precipitation and position of the storm with regard to the mountain barriers (March, Wallace, and Swift 1979). Thus, the discussion continues with the third major type of storm.

Convectional Storms

Convectional storms are formed when hot air, warmed by the unequal heating of the earth's surface, rises and cools adiabatically. Thus, darker areas of earth absorb more radiation and are relatively warmer than surrounding lands (or open water): the rising warm air draws air in laterally, often off the water body, and, should the process continue, thunderstorms will develop. The shape of the cloud, cumulonimbus, often with the characteristic anvil top, is the same as that which occurs along an intense cold front and even, on occasion, as orographic storms. The pattern of their occurrence, however, is quite different. Convectional storms occur primarily in summer, are often scattered by 20 mi or more, with downdrafts between them cycling the air that rises in the storm cell (some air descends within the storm cell as well). The rising air cools adiabatically, releasing heat energy when the dew point is reached and condensation occurs. This re-enforcement of the unstable nature of the convection cell produces the extensive vertical buildup.

Convectional storms tend to form in late morning or early afternoon, and start to dissipate when the setting sun terminates the primary driving source of heat. They are quite prevalent in the High Plains, in desert country, and where there is a mix of water and land which absorb energy differentially, such as occurs on the Florida peninsula.

Precipitation from convectional storms is likely to be high intensity, of relatively short duration, over a limited, but scattered area, and may be accompanied by highly variable winds (both in direction and velocity), hail or even sleet, thunder and lightning, and tornados. They tend to move slowly to rapidly, and may grow to be tens of miles in diameter. Their lowest point may be at 300 m (1000 ft), and can, under conditions favorable to their continued growth, extend to 30,000 m (100,000 ft).

Typical convectional storms are quite complex systems. A great deal of research has been conducted on them in the interest of controlling hail (to prevent damage to crops), lightning (to prevent forest fires), and windshear (to aid flying). Building up sufficiently high in altitude to extend past the freezing point, circulating water drops form hail which may be kept circulating by the violent updrafts. Momentum from a downward plunge or their sheer size and weight eventually overcomes the uplifting forces and hail results. Hydrologically, hail is important from the standpoint of removing foliage from vegetation, thus exposing soil to raindrop erosion, and to direct damage by hail. It also takes a while to melt, thereby delaying infiltration; but this may be offset if the infiltration rate is slowed by compaction.

Huff (1975) reports that convective storms are further influenced by urbanization. Urban areas contribute to updrafts both because: (1) a large amount of energy is used for heating, manufacturing, cooling, and transportation, and (2) the dark pavement and buildings absorb and trap heat energy. Huff "concluded that the frequency distribution of heavy [convective] rainfalls may vary significantly between urban, suburban, and rural areas in larger urban-industrial regions." A study by Pani and Haragan (1985) illustrates the feasibility and value of classifying convectional storms into four categories (cell, small cluster, large cluster, and nested cluster) readily determined from radar images and based on storm duration, rainfall volume, and time between events. Ability to predict rainfall amount, and seeding possibility, appears to be related to category, thus allowing more efficient management and forecasting.

Summary

While it is possible to classify precipitation types owing to their origins, the classification system (like most) breaks down when examining individual storm events. The reason for this is that the primary process, adiabatic cooling, always takes place according to certain fundamental physical laws which apply to all condensation-precipitation events, regardless of the circumstances under which they occur. A case in point is that of lake-effect precipitation, where cyclonic and convectional activity combine with orography to produce highly localized precipitation. If temperature differences are sufficient, thunderstorms may develop along warm or stationary fronts and, of course, less intense (or no) precipitation can occur along weak cold fronts.[12]

As might be expected, the fundamental precipitation distribution pattern would be carried through to the runoff process were it not for modifying factors such as snow, soil, depression, and channel storage. Thus, understanding of the imperfect classification of precipitation type is essential to the later comprehension of runoff and stream behavior. For example, precipitation patterns in a typical year are often distributed by seasons, with winter precipitation deriving principally from cyclonic storms, whereas summer precipitation may be convectional.

The patterns of storm distribution in time and space are explained in a useful manner by such classification. Most of the annual precipitation in a given region will be found to be from a particular storm type and, as a consequence, that region may be characterized by this important hydrologic cycle input. Discussed in more detail in subsequent sections, precipitation amount, areal extent, duration, form, frequency, and intensity are the principal parameters that are used to characterize storms and precipitation.

Forms of Precipitation

Precipitation is not evenly distributed throughout the year, nor is it evenly distributed areally. The variation in radiation (Chapter 2) with latitude, season, and time of day combines with proximity to water bodies and orography to cause temporal and spatial patterns that are made even more complex by different forms of precipitation.

Rain

It is dangerous to make sweeping generalities about whether rainfall constitutes the greatest amount of precipitation, or is the greatest percentage of annual precipitation. At any given location, however, it is possible to make such statements, thus starting the process of characterizing the hydrology of a region. The Olympic Peninsula in the state of Washington, for example, clearly gets the greatest percentage of its precipitation as rain. So, too, do the deserts of the Southwest, but the annual total received there is considerably different from that of the Olympic Peninsula. Most areas that experience convectional storms get their moisture in the form of rain. The same is not necessarily true for cyclonic and orographic storms.

Maximum annual rains for the world range from virtually zero in deserts to over 25,400 mm (1000 in.) under combined monsoonal and orographic conditions in Cherrapunji, India (Weather Bureau 1958). In coastal (and some nearby inland) areas, rainfall from hurricanes may provide the largest percentage of the annual precipitation. Single storms along the Gulf Coast in Texas have produced more than 1000 mm (40 in.), easily more than half the annual total, and an even greater percentage of the average annual precipitation.

Raindrops vary in size from 0.4 mm to 7 mm (0.01 in. to 0.25 in.), with a typical size of 1 mm (0.04 in.) (Miller and Thompson 1970). Because drops form when air is moving upward, they tend to remain aloft until either they coalesce into drops that are large enough to

---

[12] The most intense storm I have ever seen was a combination of both convectional and orographic activity, at the Coweeta Hydrologic Laboratory in the mountains of North Carolina. In one 20-min period in the early afternoon (the only time there was any precipitation) the arithmetic average of four reliable rain gages documented 2.50 in. of rainfall.

overcome the friction of the air and the buoyant forces, or until there is a downdraft that propels them to the ground. If subjected to too much friction, the drops will break up. The velocity at which this occurs is the **terminal velocity**, defined as *the rate of fall in still air that will not be exceeded*. This ranges from 170 cm/sec to 900 cm/sec (Miller and Thompson 1970) or over 20 miles per hour. Driven by downdrafts, raindrops can hit the ground with great force, and cause considerable damage to unprotected soils. And, even in the absence of air movement, velocities of droplets reaching the ground can be sufficient to severely alter surficial characteristics of unprotected soil. Since terminal velocity may be achieved in as little as 30 ft of free fall, drops from vegetation can cause soil erosion unless the forest floor litter cover is maintained.

The unit used to measure rainfall is the area-inch, usually reported in simple units of depth such as inches, centimeters, or millimeters. The **nonrecording Standard Rain and Snow Gage** (Fig. 3-12) is the standard instrument used for documenting rainfall, and costs around $185. The instrument consists of four parts (in addition to its support): (1) the receiver, an 8-in. diameter cylinder that has a knife edge to literally split raindrops, and a funnel, recessed to preclude splash of precipitation entirely out of the gage, that leads the collected water to a small hole where it flows into, and prevents water from evaporating out of (2) an inner cylinder, 2.00 in. in diameter and 20 in. high, so that with a cross-sectional area one-tenth that of the receiver, the depth of the precipitation collected is amplified by a factor of ten, (3) an outer cylinder, 8 in. in diameter, which contains the inner cylinder and provides support for the receiver, and (4) a calibrated, cedar measuring stick.

The stick must be inserted into the inner cylinder when a measurement is made because the dimensions of both parts are designed to precisely measure 2.00 in. of precipitation on the receiver. If the inner cylinder is full, insertion of the stick will cause water equal to the volume of the stick to spill into the outer cylinder. The inner cylinder has a soldered bottom plate which will be forced off if the contents freeze between the precipitation event and the time of reading. After the inner cylinder has been emptied, the contents of the outer cylinder (either from spillage during measuring the inner cylinder, from ice formation, or from precipitation in excess of 2.00 in.) can be poured into the inner cylinder for measurement.

The fact that the precipitation in the standard rain and snow gage can be read to the nearest 0.01 in. can be misleading. A pair of gages exposed immediately adjacent to one another will not necessarily (and in all likelihood will not) give the same reading for a given precipitation event. With about one gage for every 350 sq mi, precipitation in the United States is really only sampled at a rate of about 1 in 9.9 $\times 10^{12}$, a very small sample, indeed. Thus, the precision of the gage should be considered in the context of its sampling frequency: as listeners of radio and TV weather reports are well aware, what was recorded at the official station and the site of the listener's picnic are more than likely quite different.

Early research on rain gage form and exposure resulted in the circular form: other shapes have a greater ratio of perimeter to cross-sectional area, and thus are more subject to edge-effect errors. The circular shape has the lowest perimeter-to-area ratio and has been universally adopted. Some vertically-mounted gages were tested that had orifices that were parallel to the slope on which they were situated, leading to an elliptical orifice, the dimensions of which varied with the slope. Experiments into exposure investigated whether the gage should be mounted perpendicular to the slope, or should make the receiver's circular orifice perpendicular to the direction of the precipitation. The former meant that every gage would have a different angle of exposure with regard to the horizontal; and the latter meant that the gage orifice would have to be turned into the precipitation. This could be accomplished (if the rain was borne on winds) with the receiver mounted on a swivel and controlled by a wind vane (Hamilton 1954). The almost limitless and potentially bewildering variety of shape, exposure, and tilting possibilities (Hayes 1944) led to standardization: the cylindrical, standard gage is now exposed vertically (with the orifice horizontal) in a Class A weather station with neither natural nor artificial obstruction to the precipitation which may fall into it. Normally, this means a 45° clearing around the gage site. Where high-velocity winds or turbulence keep the precipitation from falling "cleanly" into the gage, a variety of different types of wind shields have been developed and put into use (Helmers 1954).

Figure 3-12 Nonrecording Rain and Snow Gage (photo from Weather Bureau 1955)

At each Class A weather station, there is also a weighing-type **Recording Rain and Snow Gage** (Fig. 3-13), which retails for about $1500. This instrument also has an 8-in. diameter receiver, but its funnel directs the water collected into a bucket that rests on a platform. The platform is mechanically connected to a pen (and motion damper) that produces a continuous trace on a weekly chart driven by a clock similar to that described for the hygrothermograph. Parameters of rainfall timing are determined from this chart (Fig. 3-14): time of start and end of rainfall, duration, and intensity all may be determined. The *amount* of precipitation, however, is reported from the nonrecording gage. At remote locations, the gage may be equipped with clocks capable of recording for extended periods of time, or for storing the water collected when the gage site is not accessible (DeByle and Haupt 1965). If subject to freezing conditions between readings, a remote gage may be charged with a known quantity of anti-freeze which may be subtracted from the total precipitation noted at the end of the period. Helvey and Fowler (1980) recommended a modified tipping bucket gage for real-time monitoring of both rain and snow at remote locations.

In addition to these two primary instruments, the National Weather Service has approved inexpensive gages used by its cooperators, and several other models, $3 or $4 and up. These inexpensive units enhance the density at which precipitation is sampled around the nation. Research at some of the intensively monitored outdoor laboratories of the Agricultural Research Service and Forest Service has provided insight into how precipitation varies in time and space, and how accurate measurements may be obtained for an entire watershed (Hamilton and Reimann 1958; Leonard and Reinhart 1963; McGuinness 1963).

Figure 3-13 Universal Recording Rain and Snow Gage. Photo courtesy of Belfort Instrument Co., Baltimore, MD

Ice

Precipitation in the form of ice occurs in four principal types (in addition to snow, considered separately, below): hail, rime, sleet, graupel, and freezing rain. **Hail** results from repeated vertical cycling within a cumulonimbus cloud of what initially was a liquid water droplet. The vapor pressure gradient increases as the droplet freezes and, upon falling through lower portions of the cloud, it acquires more liquid. Lifted again, several additional layers may be added, providing concentric shells which record the hail stone's brief life history. Hail stones may grow to the size of grapefruit, but typically are pea-sized. Hail is a prevalent form of precipitation wherever convectional storms build up through the freezing level, such as on the high plains during the summer months. **Rime** consists of small ice granules formed when the temperature is between -10°C and -20°C; it produces an opaque, irregular surface (Miller and Thompson 1970). It has been shown to contribute to the water balance at high elevations

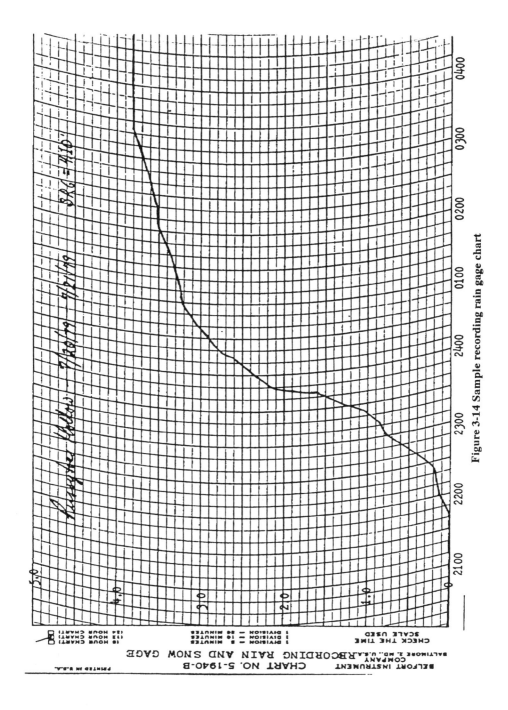

Figure 3-14 Sample recording rain gage chart

(Gary 1972). **Sleet** occurs when a liquid water droplet falls through a subfreezing layer of air before reaching the ground, and **graupel** (or groppel) is the term used for snow pellets that become rimed in a shell of ice condensed from water droplets in a cloud through which the pellet passes. Terminal velocities for ice particles are from 800 cm/sec to 3500 cm/sec or more (Miller and Thompson 1970). **Freezing rain** is formed when relatively large liquid water drops freeze upon contact with a surface that is below freezing. The resultant ice is usually quite clear and smooth. Size and details of formation processes distinguish some other, less frequent forms of precipitation (Miller and Thompson 1970).

None of these types of precipitation are completely and satisfactorily measured by the rain and snow gages described above. The first three types are likely to bounce out of the receiver upon striking the funnel, and freezing rain is likely to clog the hole leading to the inner cylinder. If one of the ice forms of precipitation is anticipated, the funnel may be removed. Although common, these precipitation forms do not, as a rule, make up a significant proportion of the annual precipitation at any given location, and thus special instruments are not normally deployed at regular weather stations to monitor their occurrence; at specialized research stations, or for specific research projects, unique instrumentation may be employed.

Snow

Snow is considered here as a process, perhaps more properly but ambiguously labeled "snowfall." Once it reaches the ground, snow can accumulate into a **snowpack** which may represent a major form of storage on the watershed (considered in Chapter 7).

Snowfall can be as much as 20 ft or more in mountainous areas, especially in the western states (Cox, Rawls, and Zuzel 1975), and contributes significantly to the runoff for the western half of the nation. The depth, however, does not indicate the amount of water in the snow.

Since the ice crystal around which the snowflake forms is already lower in density than water at 4°C (at which temperature water is most dense), and additional growth is in the form of feathery ice crystals, 10 in. of fresh-fallen snow is equivalent to about 1 in. of water, a useful rule of thumb. In other words, its density is about 0.1. Unusually heavy early winter snows may have densities as high as 0.35, but densities of 0.03 to 0.2 are more common. After "settling" owing to freezing and thawing, long- and short-wave irradiation, condensation, and warm rains, the density of the pack increases to about 0.5, under which conditions runoff may occur as described in Chapter 7.

Water actually occurs in all three phases within a snowpack: as a solid, ice; as water vapor, it is present in the interstices between snow crystals; and, as a liquid, it may be present and is analagous to soil moisture, that is, it rests in a porous medium. This latter component is referred to as **snow water content**. **Water equivalent** is the depth of liquid water that can be obtained from *melting* the pack. The latter parameter is of particular interest to the hydrologist: it expresses the amount of water that will be yielded when the pack melts in the spring. It is also relatively easy to measure.

When snow accumulates in a pack in winter, its water equivalent may be monitored by any of several instruments. The measurements are made along a snow course, a transect marked by poles calibrated in feet that can be read directly (and somewhat imprecisely) from the air. The **Mount Rose Snow Sampler** and the **Adirondack Snow Tube** (Fig. 3-15) are used to monitor depth and water equivalent along the transects. By calculation, density of the snowpack layer is sampled. Snow tubes cost upwards of $550. Snowpack water equivalent may also be monitored remotely by **snow pillows** which respond to the weight of the snow above them and transmit the pillow pressure regularly to or on demand from a recording data station (Penton and Robertson 1967; Gray and Male 1981). Snow courses are normally visited on the first and fifteenth of the months when the pack is in existence.

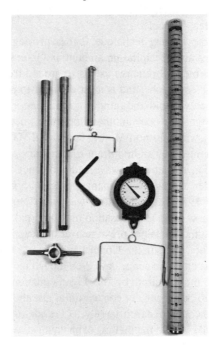

Figure 3-15 Mount Rose snow sampler, disassembled, showing scale, wrenches, driving handle, and cleaning tool on left; Adirondack Snow Tube with weighing scale on right

Snow survey information commenced in 1909 in Maine for predicting runoff for power production (Beaumont 1957), and is carried out by the Bureau of Reclamation, Corps of Engineers, and the Soil Conservation Service, and numerous cooperators. It is most useful in predicting spring floods, and in estimating runoff likely to be available to fill water supply reservoirs. Recent remote sensing research has provided the opportunity to monitor snowpacks more thoroughly than just by the widely scattered snow courses. Satellite imagery (Martinec and Rango 1981) permits determination of snow depth, snow water content, and even soil moisture content under the snow by comparative analysis of aerial images recorded at different wavelengths, and at different times.

Modern, sophisticated statistical analysis techniques have been utilized to evaluate requirements for designing precipitation sampling networks (Bras and Rodriguez-Iturbe 1976). A wide variety of error sources can be accommodated in these processes. In one study, Bastin, Lorent, Duque, and Gevers (1984) reported a practical real-time estimator of areal precipitation but, as was found in other locations, these seem to work best in relatively flat terrain. Statistical analysis also suggests that mountain snowpacks should be widely and proportionately sampled by elevation zone, with sites duplicated (Leaf and Kovner 1972). Since water in snow attenuates natural gamma radiation emitted from the soil, an innovative method for direct snow water equivalent monitoring with a small, portable gamma ray detector was proposed for shallow packs (Bissell and Peck 1973). Environmental radiation may confound results, but an advantage is that the snow is not disturbed and the volume of snow over a large area is sampled.

Another remote sensing technique that is proving useful to the hydrologist is radar monitoring of storms and precipitation amount and intensities (Barge, Humphries, Mah, and Kuhnke 1979). This method requires extensive ground truth for calibration of the radar and existing atmospheric conditions, and is most accurate in relatively flat areas, which often are not in as great a need of rapid and remote storm monitoring as are other, more mountainous areas. Satellite imagery is now another important method of snowpack monitoring and prediction with a Snowmelt Runoff Model (Rango and Martinec 1982; Rango 1988).

Condensation may play an important role in energy relations both above and below the snowpack in addition to net short-wave radiation, net long-wave radiation, convective or sensible heat transfer, latent heat exchange, precipitation, and geothermal heat (Santeford, Alger, and Meier 1972). Condensation was noted to add water to the pack without "precipitation" or snow melt. Condensation may be rapid owing to (1) an abundant supply of water vapor, and (2) the low vapor pressure over the icy snow surface. Warm, unfrozen ground is prevalent under snowpacks in the Upper Peninsula of Michigan, and "condensation onto the snowpack was noted on warm sunny days even though the vapor pressure of the air at 4 ft was less than the 6.11 mb required by the Light equation." The authors propose that "this discrepancy is due to occurrence of condensation just above, rather than at, the snow surface, so that the latent heat is released largely to the air rather than to the snowpack" where, presumably, it would result in the melting of an equal amount of snow.

Snowmelt is a complex process, as there is considerable interaction among influencing factors (Aguado 1985; DeWalle and Meiman 1971; and Male and Granger 1981). For example, the effect of short wave radiation changes with snowpack thickness, surface color (a function of time since snowfall and dry fallout), and angle of incidence. Since pack thickness is affected by areal distribution patterns and factors affecting deposition, it is possible to manage snowpack accumulation and melt (Adams 1976; Berndt 1964; and Swank and Booth 1970). Temperature conditions under which the pack melts influence whether there is infiltration of meltwater and soil moisture retention (Klock 1972). Based on observations of natural deposition, snow fences and cut or planted vegetation can be used to cause snow to be accumulated by design, therefore delaying melt, for example, and improving water yields for our needs. Further discussion of snowpack management is presented in Chapter 7.

Temporal and Spatial Distribution

For all practical purposes, the precipitation measurements made with the rain gages as described are only **point samples**. They do, in fact cover an area of about 50 sq in., but that is so small in relation to the area represented, or for which the data are reported, that little more than a point can be considered as being sampled. Even in the relatively well-sampled United

States, storms "often go unmeasured because the nation's raingage network is too sparse" (Changnon and Vogel 1981). This observation applies more to rainfall than to snowfall, as the latter is fairly accurately measured once it is on the ground and because rainstorms generally cover wider areas.

Further, there is no guarantee that a particular rainfall measurement of a storm will be at the point where the maximum or average rainfall occurs. Even if it were at the point of maximum rainfall, the information on how the rainfall is areally distributed throughout the storm is essential in determining how much water actually falls on the watershed. Research aimed at this information was conducted by the Illinois State Water Survey by placing a large number of rain gages in a systematic network over wide, flat areas during the summer to monitor thunderstorm behavior and rainfall amounts (Huff 1970). In arid areas, such research is more difficult, because "although high rainfall rates are common in the southwestern United States during the summer rainy season, the chance of occurrence at a specific time and place is very low" (Osborn 1983). Diurnal variations in rainfall intensity and duration were similar on two watersheds, but time of day of average occurrence at each was quite different, suggesting a pattern to even the apparently random distribution of precipitation from convectional storms. Osborn and Hickock (1968) reported greater variation for summer than winter rainfall in Arizona. That observation may not hold where winter precipitation is the result of more stable, predictable cyclonic storms than for the irregular convectional activity.

Correlation of rainfall amounts between stations decreases rapidly with increasing distance between the stations (Osborn, Renard, and Simanton 1979), and gives the appearance of randomness in convective storm events. Moss's (1979) conclusion reflects this view in suggesting that both the random nature of events and the uses that will be made of the data should be taken into account in establishing a hydrologic data network: "The efficiency of the data collection and the effectiveness of the resulting information must be integrated to achieve a complete network design." It should be added that data networks may have to be constructed on considerably different bases dependent upon whether they are in arid or humid zones: the methods of management of the water resource in each zone and, consequently, the uses to which the data will be put, can be expected to be different.

Rainfall under a storm cloud will somewhere be at a maximum. Consider a gage placed under this point: at all points away from that gage, rainfall will decrease. If the storm is moving, the pattern of rainfall will be more complex than this simplified model, and obtaining a mean for the storm will be difficult. In fact, the rainfall decreases exponentially as the area covered gets larger and larger (Fig. 3-9). This occurs in part because of the natural tendency for rainfall to be greatest at some central point in a given storm, and in part because larger storms occur less frequently than smaller ones.

This fundamental observation translates into an important characteristic of rainfall that involves the relationships between the **amount** of rainfall (in inches or millimeters), the duration of the rainfall (in minutes, hours, days, etc.), the **intensity**, or rate at which the rain falls (simply the amount divided by the duration, in cm/sec or in/hr), and the **frequency** of occurrence. In general terms, the relationships may be identified as follows:

1. the greater the duration, the greater the amount;
2. the greater the duration, the lower the intensity;
3. the more frequent the storm, the shorter the duration and;

4. the more frequent the storm, the less the intensity.

The reverse of these statements is also true.

Intensity-Duration-Frequency Relations

The collective relationships of the preceding statements are collated and reported for the United States in a series of U.S. county maps for durations of 30 min to 24 hr and for **return periods** (the reciprocal of the frequency) of 1-, 2-, 5-, 10-, 25-, 50-, and 100-yr (Weather Bureau 1961). In addition, 5- to 60-min precipitation frequencies are available for the Eastern and Central United States (National Weather Service 1977).

Selection of a specific site and determination from the maps of the depth at that site, provides data that may be used to construct a table of amounts, durations, intensities, and return periods. With the aid of the area-depth relation shown in Fig. 3-16, useful rainfall information can be worked out for any given area, such as a watershed. An example of the maps from Technical Bulletin No. 40 is shown in Fig. 3-17 for the 1-hr, 100-yr return period. Snowfall data are included in the information that was used to prepare Technical Bulletin No. 40, but since the analyst is usually interested in the more intense or less frequent storm, and low-intensity snowfall and more frequent, gentle storms are at the other end of the data curves, snowfall does not adversely affect such use of the curves.

The maps for Technical Bulletin No. 40 were prepared with data from 200 stations with sufficiently long recording gage records to make reliable estimates of the return periods included, and were enhanced by inclusion of data from 1600 nonrecording gages, and by computer mapping and smoothing. They update and elaborate on the classical study summarized on state (not county) maps published by Yarnell (1935). The 1977 maps extend the durations to periods shorter than 30 min for the eastern and central United States, often necessary to evaluate the flooding and erosional impact of very short, very intense storms.

Given the observation (above) that different regions have different precipitation patterns, one may expect to be able to characterize an area by its intensity-duration-frequency curves. The intensity-duration-frequency relationships can be refined for a particular area that experiences several different types of storms. Thus, the data set for winter cyclonic storms will exhibit some different characteristics (slope and position of the lines, for example) than the data set for summer thunderstorms. And, where enough observations are available, the data set for hurricanes would be still different. One would expect the curves to be more steep (greater intensities over shorter durations) for the thunderstorms, for example, and with longer durations for cyclonic storms and hurricanes. This concept represents the regional, climatic patterns of precipitation that are carried through to the patterns of runoff and are ramified in stream behavior described in Chapter 6.

Areal Distribution

Variability of precipitation over an area of interest, for example, a watershed, has been the subject of much research. Hershfield (1966) pointed out that there are three sources of error in precipitation measurements: instrumental error, "sampling fluctuations," and errors due to network density. The potential for **instrumental errors** is the subject of several comments above in the discussions of the different types of precipitation and the instruments that are used to monitor each.

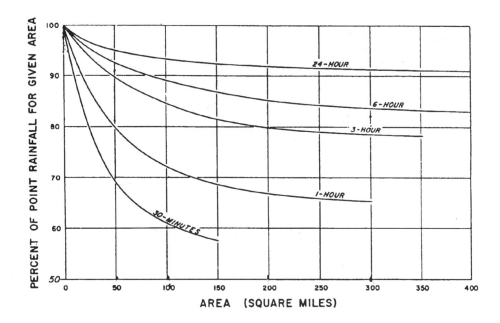

Figure 3-16 Area-depth curves. From Technical Bulletin No. 40 (Weather Bureau 1961)

Sample fluctuations are a function of storm variation, structure, and movement, often interacting with watershed terrain. Network density errors are evaluated by highly complex number theory and based upon observations of gage networks under what are presumed to be relatively uniform precipitation conditions.

Network errors are recognized as distinct from fluctuations in the storm itself. Point samples are subject to error due to wind, which influences catch (Larsen and Peck 1974), and, of course, precipitation varies naturally over a wide area, as evidenced by differing catch records in adjacent gages. Hershfield (1966) showed this to be a function of distance between the gages.

Several methods have been advanced to secure a reasonable estimate of precipitation over a given area. These include the arithmetic average, the isohyetal method, the Horton-Thiessen polygon method, an elevation-weighting method, and the station-angle method. Four of the several available methods are illustrated in Fig. 3-18, and sample calculations of mean precipitation for each are shown in Table 3-2.

The **arithmetic average** is the simplest, and often the least accurate method. If the precipitation were uniform over the watershed, only one gage would be necessary for accurate monitoring. Even a widespread, seemingly uniform winter (cyclonic) storm, however, exhibits great fluctuation from place to place in the amount of precipitation deposited. As more and more area away from the "point" of measurement is included in the measurement, the amount measured decreases, as shown in Fig. 3-16. Rodriguez-Iturbe and Mejia (1974) warn against using this relationship in areas other than that for which it was developed owing to different structures of different storm types: they present a complex alternative, which is not sufficiently

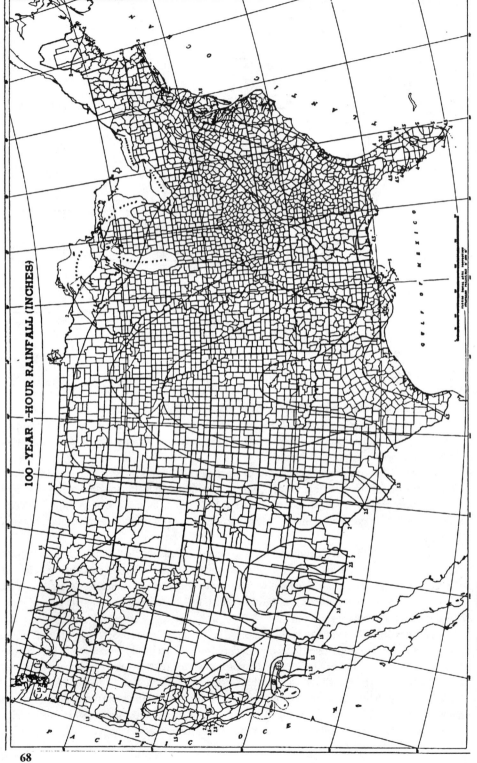

Figure 3-17 Sample of depth-frequency occurrence of precipitation in the United States. From Technical Bulletin No. 40 (Weather Bureau 1961)

different from Fig. 3-16 to warrant recalculation of the complex formulas from which it is derived. An alternative, obviously, is to evaluate areal distribution of each storm directly.

The **Isohyetal Method** is a tedious but accurate method for accounting for storm variation and including the influence of orography in calculating storm totals. Based on large numbers of gage records, the storm, daily, or other period totals are plotted and lines of equal precipitation (isohyets) drawn, often by an experienced analyst who knows how the topography influences precipitation. Other topographic features may also be accounted for, such as distance from a water body or from the edge of a high plateau. The areas between lines of equal precipitation are graphically determined and the resultant weights multiplied times the mean precipitation in each zone; the total of these products is then divided by the sum of the weight factors to obtain the weighted precipitation. The biggest drawback to the isohyetal method is that the process must be repeated for each storm or period to be investigated and, while it is true that different analysts will probably arrive at only slightly differing mean precipitation values, the process is costly in time and money. The influence of elevation may be taken into greater acccount (than simple experience) by a method described by Dawdy and Langbein (1960).

The **Horton-Thiessen Polygon Method** is useful because once the gage network is established and calibrated, it need not be changed and can be used repeatedly for mean precipitation estimates. The method is based on the premise that the precipitation at any given point is best measured by the gage that is located closest to that point. This is accomplished mechanically and objectively by constructing a series of polygons, each with a gage at the center, and delimited by the perpendicular bisectors of lines connecting adjacent gages. The weight factors are determined by the percentage of the total area within each polygon. Multiplied as decimals times the measured precipitation, the total of the products gives the mean precipitation directly. In mountainous terrain, the gages may be located closer together and the resultant polygon network will reflect the differing catch in those gages. If the gages are uniformly spread over the area under study, one might as well use the arithmetic average, since the weight factors will be equal. The network must be recalculated if a gage is added to or deleted.

Determination of precipitation by **weighted elevation zones** is a variation of the isohyetal method, with particular attention given to the terrain. The method, developed for the mountains of West Virginia, can be quite accurate, but requires an extraordinarily large number of gages, which precludes its practical use, as described by Chang (1977). The method is not shown herein, but it should be noted that one of the reasons for the large number of studies of areal distribution of precipitation is that the interaction of weather, climatic, and topographic factors creates patterns that probably cannot be satisfactorily monitored by a single method.

The **Station-Angle Method**[13] (Bethlahmy 1976) is somewhat similar to the Horton-Thiessen Polygon Method in that once the gage network is established, the weight factors are fixed and need only be re-calculated if a gage is added or deleted. That recalculation is much easier, however. The weight factors are the number of degrees subtended by the intersection of two straight lines from the gage to the limits of the quadrant opposite to that of the location

---

13 Also known as the "Two-Axis Method."

of the gage. The quadrants are located by drawing a line through the center of gravity of the watershed along the longest dimension, and a line perpendicular to it at the center of gravity. Using the two perpendicular lines identified for the determination of watershed eccentricity evaluation of watershed shape (see Chapter 6) is suitable and saves nearly duplicate constructions. The sum of the products of the weight factors and the precipitation measurements from each gage is divided by the total number of degrees to calculate the mean precipitation. The gage that would have the highest weight factor in this system would be at the middle of the watershed. Like the Isohyetal and Horton-Thiessen Means Methods, the Station-Angle Method makes use of gages off the watershed, their importance diminishing as distance from the center increases.

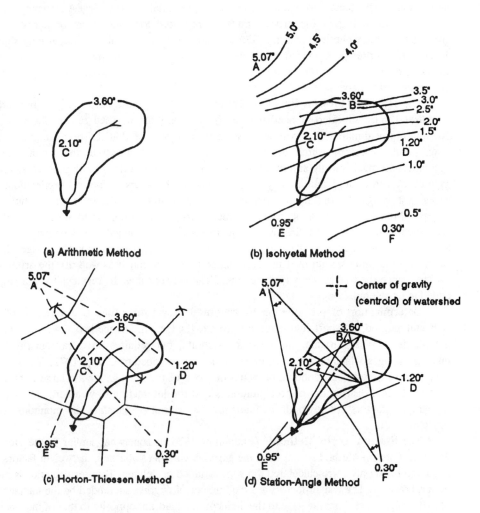

Figure 3-18 Methods for calculating areal precipitation

Chapter 3 Water in the Atmosphere

Table 3-2 Methods of Calculating Mean Precipitation Shown in Figure 3-18

(a) Arithmetic Method

Gage No. (#)	Precipitation at gage (in)
A	-
B	3.60
C	2.10
D	-
E	-
F	-
Total	5.70
Mean	2.85

(b) Isohyetal Method

Range of isohyetal zone (in)	Midpoint of range (in)	Dot grid area (#)	Percent of total area (%)	Weighted precipitation (in)
1.05-1.50	1.28	208	27.1	0.348
1.50-2.00	1.75	224	29.2	.512
2.00-2.50	2.25	180	23.5	.529
2.50-3.00	2.75	96	12.5	.344
3.00-3.50	3.25	40	5.2	.170
3.50-3.65	3.58	18	2.4	.084
Totals		766	99.9	1.986

(c) Horton-Thiessen Method

Gage No. (#)	Recorded precipitation (in)	Dot grid tally (No)	Percent of total area (%)	Weighted precipitation (in)
A	5.07	-	-	-
B	3.60	244	32.1	1.16
C	2.10	408	53.0	1.11
D	1.20	99	12.8	.15
E	.95	16	2.1	.02
F	.30	-	-	-
Totals		760	100.0	2.44

(d) Station-Angle Method

Gage No. (#)	Weight factor[a] (deg)	Precipitation at gage (in)	Weighted precipitation (in)
A	30	5.07	152.1
B	39	3.60	140.4
C	51	2.10	107.1
D	44	1.20	52.8
E	15	.95	14.2
F	22	.3	6.6
Totals	201		473.2
Mean			2.35

[a] Angle subtended by lines to far intersections

The **Reciprocal Distance Squared Method** for estimating areal precipitation is designed for computer use and assumes that the watershed map may be installed on a computer (Wei and McGuinness 1973). (In fact, all of the methods can be accomplished on a computer.) The method (not illustrated herein) makes use of an artificial grid created by the computer program which serves as the basis for location of precipitation estimates based upon a designated gage network.

In a study comparing 13 different methods of calculating and statistically evaluating mean basin precipitation, Singh and Chowdhury (1986) reported minor differences between widely different areas of study, but noted that "there was no particular basis to claim that one method was significantly better than the other, although in a given situation one method might be preferable to another." The arithmetic average was not one of those included.

Elaborate methods for the areal estimation of precipitation over a watershed are useful (and advised) where: (1) the watershed is large; (2) the existing gages are not uniformly distributed; (3) there is a wide range of elevation along with an expected or known orographic effect; or (4) the need is for detailed research information. For most practical precipitation investigation or reporting situations, however, the refinement gained by any of the elaborate methods is likely offset by the inaccuracy of the gage reading by virtue of gage placement, natural storm variability, and the mistaken impression that the standard rain and snow gage is "measuring" precipitation to the nearest 0.01 in.

Limitations

In sum, the utility of the basic rain gage data, the method of mean precipitation calculation, and the intensity-duration-frequency curves derived from them is only slightly restricted by the conditions under which the data were originally collected. In fact, the intensity-frequency-duration curves are probably quite reliable because they tend to smooth out irregularities that are caused by the low sampling intensity and the consequently missed storms. This would tend to make the curves somewhat lower. Or, to put it another way, we can expect to have intensities higher than that shown on the curves for given durations/frequencies. This is true (1) because the data are not based on a 100 percent sample of all storms or precipitation, and (2) because as we observe storms for a longer and longer period of time, we can expect to record a less and less frequent (more intense) event.

The basic data, on the other hand, are limited by questions of randomness and representativeness. The criterion for a random sample of a population, in this case, precipitation, is that each member of the population has an equal chance of being sampled. Since storm tracks are guided by systematic principles of atmospheric physics and modified by immovable topographic features, the location of a gage somewhere on the ground fixes it with relation to that storm track: other tracks simply may not be sampled at all.

Given that a storm does, in fact, move over a rain gage, one must ask to what extent does the record of precipitation in the gage adequately *represent* that storm? That question is, without multiple gage documentation of the variation in the storm, unanswerable. We simply accept the idea that the documentation is representative. It may not be. In all likelihood it is not. It is a good idea to remember that fact when using the data.

Further, precipitation, especially rainfall, is measured to the nearest 0.01 in. by the tenfold expansion of the vertical scale in the standard rain and snow gage. In fact, basic scientific principles dictate that the measurement should be reported to one less significant digit, in this case 0.1 in.. Even this is erroneous, because the original reading on the calibrated stick is to the nearest 0.01 as estimated by eye and not by actual measurement. The 0.01 measurements on the stick are actually 0.1 in. apart, close enough that the meniscus of the water in the inner cylinder that wets the stick could really be read up or down one unit. At best, then, the precipitation measurements are only significant to the nearest 0.1 in. and, considering the lack of randomness and problems of representativeness, ought to probably be reported only to the nearest inch! The consequences of this violation of objective data reporting are probably not serious, unless one considers the false impression such seemingly accurate measurements conveys: Simply put, we tend to put too much faith in the precipitation measurements.

Finally, as will be explained later (Chapter 6), runoff measurements are often reported to the nearest 0.001 in. and analyzed along with precipitation data. This degree of accuracy is often justified because the stream gage (supposedly) monitors 100 percent of the runoff in contrast to the precipitation, which is rather scantily sampled. Statistical relationships evaluated between runoff and precipitation that contain such magnitudes and different sources of error should be subjected to close scrutiny in order to assess the data limitations and, consequently, their utility.

Consideration of these limitations does not preclude making effective use of hydrologic data. The vast majority of the data sets and the relationships between them are valid. The patterns are so logical, well-established, and obvious that they ought not be thrown out just because of some potentially serious data collection limitations. On the other hand, it is necessary to keep the limitations in mind as one proceeds with the consideration of the relationships in the hydrologic cycle.

EVAPORATION

The word "evaporation" has the same root as vaporization and simply means the change from liquid to gas wherever it occurs. Water in the gaseous state is referred to as a vapor because it is so readily condensed (Miller and Thompson 1970). Evaporation takes place from raindrops while they are precipitating, from surfaces wetted by precipitation, from moist soil, and from open water bodies, from miniscule puddles to the oceans.

The Amount

Annual evaporation is highest in the arid southwestern part of the United States, and lowest in the northeastern states and around the Great Lakes. Pan evaporation (see below) ranges from 25 in./yr in Maine to over 110 in./yr in southwest Texas. Direct measurement is difficult without interfering with and thereby influencing the process itself. Measurements of over 155 in./yr were taken in Death Valley and reported by Chow (1964). The amount of observed evaporation from the pan is typically higher than what actually occurs. Pan evaporation is an estimate of the *potential* evaporation, given an unlimited supply of water. The actual evaporation may be considerably less.

The fact that the process takes place while rain is falling and from a wide variety of surfaces means that it is impossible to measure the true amount of evaporation. Since energy is necessary for the process to take place, an upper limit on evaporation due to incoming solar radiation may be calculated. But energy may also be supplied from the water that is left behind (which is why evaporation is considered a cooling process), from substrate, and from the nearby air.

When sufficient energy is applied, the motion of molecules near the water surface exceeds that necessary to break away from the immediate water surface and, if there is an area of lower-than-saturated air above, the water molecules will migrate to that level. The gaseous water molecules follow the **vapor pressure gradient** from areas of high moisture content to that of low moisture content. The greater the gradient, the faster will the water move along it.

Factors Affecting Evaporation

Energy, **water**, and a **vapor pressure gradient** are absolutely necessary in order to evaporate water. In addition to these three primary factors, there are a host of secondary factors that affect each of the three primary ones:

Factors that affect energy available for evaporation
    factors that are properties of the incoming energy:
    latitude
    elevation
    time of day
    time of year
    degree of cloud cover
    temperature of the water
    temperature of the substrate
    temperature of the adjacent air
factors that are properties of the water but affect the energy available for evaporation:
    depth of the water
    color/turbidity of the water
    color of the bottom of the water body (if shallow)
factors that are properties of a wetted surface:
    color
    roughness
    temperature
    exposure to advection of warm air
    orientation with respect to solar radiation
Factors that affect the amount of water present
    water body characteristics (e.g., stagnant or flowing), including basin depth and volume
    frequency of supply of water to a wetted surface (e.g., by storm precipitation)
    continuous supply of soil water, by capillary action
Factors that affect the vapor pressure gradient
    elevation
    atmospheric pressure
    extent and continuity of the gradient
    moisture content at levels of interest
    temperature of the air at levels of interest
    wind (including exposure of evaporating surface, above)

Dalton's Law

The most important factor, the vapor pressure gradient (one assumes that the energy and water are available), is the focus of Dalton's Law:

$$E_a = b\,(e_w - e_a)$$

where E_a is the evaporation in in. per hour, e_w is the vapor pressure at the evaporating surface, e_a is the vapor pressure at some finite level above the surface, and b is a coefficient that represents all other factors. There are several adaptations of Dalton's Law, each seeking to relate the amount of evaporation to the vapor pressure gradient and to other pertinent factors that have been identified as significantly affecting the process. In the Thornthwaite-Holzman equation, for example, the value of b is refined by including wind velocity and temperature at, and the distance between, the two levels. Manning (1987) points out that Dalton's Law is not really verifiable and, therefore, cannot lead to reliable estimates of evaporation.

Instruments, Limitations, and Measurement

The standard instrument used to measure evaporation at the National Weather Service Class A weather station is the **evaporation pan** (Fig. 3-19). Made from galvanized iron, and retaining that grayish color, the cylindrical unit is 10 in. high and 47½ in. inside (4 ft outside) diameter, and is mounted on a slatted (ventilated) platform made of 2x4s that rests on a 6 in. mound of earth. The water is maintained 2-3 in. below the top of the pan. A brass stilling well, also cylindrical, 8 in. tall and 3½ in. outside diameter, rests on a triangular platform in the pan, thus permitting water to enter through the bottom of the cylinder. The platform support must be leveled and the cylinder plumb.

Figure 3-19 Standard Class A Evaporation Pan with cup anemometer. Photo courtesy Weather Bureau (1955)

A very precise, micrometer-controlled **hook gage** (Fig. 3-20) is used to monitor the water level inside the stilling well. The gage consists of a tripod support that rests on top of the stilling well, and a calibrated, threaded vertical shaft, the position of which is controlled by the micrometer screw assembly. The gage is operated by lowering the hook below the water surface and raising it until the tip of the hook just "dimples" the water surface. This dimpling is easily seen as one looks down into the stilling well against the reflection of the sky above the observer. The micrometer can be read to the nearest 0.001 in. The difference in successive water level readings, corrected for precipitation since the last reading, will give the evaporation from the pan. The entire assembly (pan, support, hook gage, and stilling well) costs around $1200.

The standard evaporation pan provides a precise and accurate measurement of the evaporation from a 4-ft diameter, galvanized iron pan, that is 10 in. deep, located on the ground. The evaporation from lakes, ponds, wetlands, puddles, and the soil and other wetted surfaces, however, may be substantially different. Nevertheless, a standard (again) is needed in order to provide a temporal and spatial benchmark. There are variations on the Class A evaporation pan. The pan may be sunk in the ground so that the water surface is the same as the land surface, or the pan may float on the surface of a lake or reservoir. A "pan coefficient" of 0.7 is recommended for conversion of the evaporation pan reading to lakes and reservoirs since the rate of evaporation is greater from small areas than from large ones (Chow 1964).

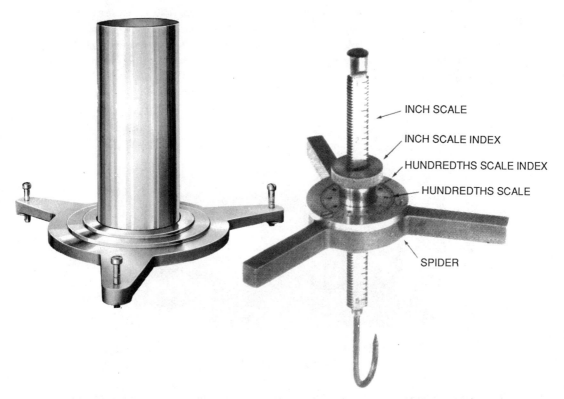

Figure 3-20 Hook gage and Stilling Well. Photo courtesy Belfort Instrument Company, Baltimore, MD

Another device for measuring evaporation is the **Livingston atmometer**. Black and white porous spheres on the ends of glass capillary tubes are exposed to ambient and radiative energy levels. The lower ends of the tubes are immersed in distilled water, and the change in water level will provide an index to evaporation between readings. The white bulb rate is very close to that of the Class A evaporation pan, but the instruments are subject to clogging and the white bulb must be kept absolutely clean. This and several other types of atmometers are particularly useful to plant physiologists and are mostly used in laboratories (Chow 1964).

Other methods[14] of determining evaporation are indirect, such as by Dalton's Law (above). These numerous equations make use of the vapor pressure gradient and refine the estimate through restricting the conditions under which the evaporation takes place (e.g., from a reservoir) or through accounting for more and more of the factors that affect the process. As with the overall process, monitoring temperature and humidity in small micro-layers over an evaporating surface in order to establish the magnitude of the vapor pressure gradient is a tedious, frustrating, and expensive proposition that is also not very productive of better evaporation estimates.

In addition to these, there are some other indirect approaches to the problem of evaporation estimation.

Water-Balance Method

An annual water balance may be derived from the observation that the hydrologic cycle is a closed system and the constraint that, during the period in question, there is no net change in storage, that is, the storage at the end of the period is the same as that at the beginning. The water budget may be expressed generally as follows:

$$P = RO + E + I + T \pm S$$

where P is precipitation, RO is runoff, E is evaporation, I is interception, T is transpiration, and S is storage. Since evaluated for a year, during which time S is supposed to be zero, this term drops out of the equation. If the other terms can be determined, evaporation may be calculated by solving the formula for E. Unfortunately, the other evaporative-type processes of I and T are also nearly impossible to measure. They may, however, be combined with evaporation into **evapotranspiration**, thus:

$$E_t = E + I + T$$

and the original equation may be written:

$$P = RO + E_t$$

Solving the resulting equation for E_t:

$$E_t = P - RO$$

While this process does not provide a means of isolating and evaluating E, it has provided a means of evaluating the combined evaporative-type processes. It is used extensively in watershed research, as described in Chapters 4 and 7.

Energy-Balance Method

A large number of researchers have investigated the energy budget as a means of predicting evaporation. As is the case with direct measurements, none are completely satisfactory:

---

[14] The general outline and, as cited, some details are abstracted from the excellent presentation by Chow (1964), which contains complete references to many of the methods summarized herein.

Like the water-balance method, the energy-balance method is complicated by the difficulties of evaluating the needed items for the solution of the energy-balance equation, including such items as atmospheric radiation, long-wave radiation from the water body, and energy storage. The conduction of sensible heat to or from the body of water is also a difficult item to evaluate. With the energy-balance equation, it is possible to obtain the sum of energy conducted as sensible heat and energy utilized by evaporation. (Chow 1964)

The Bowen Ratio, R, which expresses (see Chapter 2) the ratio between the sensible heat and latent heat may be constructed to use air temperatures and vapor pressures (Satterlund 1972). Specifically, R may be determined from atmospheric pressure, P, and temperature, T, and vapor pressure gradients, e:

$$R = (0.61P/1000) \times (T_w - T_a)/(e_w - e_a)$$

The value of R (theoretically) varies from -1 to +1, but is usually between 0.20 and 0.30 for a 24-hr day. The value of 0.61 is reported to adequately represent normal atmospheric conditions. This method is useful for short periods of time, and produces errors of ±5 percent for periods in excess of about a week (Chow 1964). Storr, et al., (1970) report that the Bowen Ratio "was found to be very dependent on wind direction, sky conditions, and basin aspect." In a short-duration (19-day) study, estimates of daily E_t were from 0.06 in. to 0.31 in., which were judged "reasonable" when compared with other estimates of E_t. A negative Bowen Ratio means that moisture is being transferred from the atmosphere to the Earth, as occurs when dew or frost forms (Satterlund 1972), and when there is a temperature inversion (air temperature increases with altitude, rather than decreases as is normally the case).

A detailed energy-balance study of evaporation at Lake Tahoe showed that a pan coefficient of 0.85 was close to modeled values in summer, but too low in winter and fall, when the energy-balance model actually showed evaporation to be highest (Myrup, et al. 1979). The authors attribute this apparent anomaly to a characteristic large upward flux of sensible energy from the lake in fall, associated with excellent mixing conditions in the atmosphere.

Fleming (1970) used an energy balance approach to derive a dimensionless evaporation index, EVI, the ratio of actual evaporation rate to average daylight evaporation. For clear days, a plot of EVI against time from sunrise divided by day length, DHOUR, was "virtually independent of time of year and of total evaporation." There were insufficient data to evaluate cloudy days on other than a "preliminary" basis.

Mass-Transfer Method

The term mass transfer refers to the mixing of layers (masses) of air with different properties and parameters. The Thornthwaite-Holzman equation is of this type. The emphasis is on *all* the factors that influence the phenomenon under study, in this case, evaporation, not just on the molecular level. Thus, long- and short-wave energy sources, energy conducted upward through the ground (or water), advective energy, and various forms of stored energy must be accounted for in addition to latent heat transfer.

A combination approach, involving a surface energy balance, water vapor and sensible heat transfer was developed and tested by Van Bavel (1966). The complex equation accurately reproduces daily evaporation from sites where water is not a limiting factor.

Hage (1975) developed another model from Dalton's Law that "appears to be accurate to within ±5 percent in warm season months but fails in winter when diurnal variability is negligible in comparison with air mass changes . . . . Covariance and nonlinearity errors do not compensate each other in winter with the result that averaging errors can reach 20-25 percent." "A simplified form of the equilibrium mode of evaporation predicts evaporation from six subarctic and tundra surfaces with an accuracy of 8 percent," report Rouse, Mills, and Stewart (1977):

> The shallow lake and wet sedge meadow have similar energy budgets. They both represent efficient evaporating surfaces, though the lake is a somewhat more efficient evaporator, because of its higher net radiation. The mature lichen woodland has the highest evaporation of any of the upland surfaces. The lichen heath, new burns, and old burn all have similar evaporation rates and display a strong surface resistance to evaporation.

Like precipitation, evaporation is the ramification of many factors interacting to produce a phenomenon that is characteristic of a particular location and is used to characterize the climate of an area.

CLIMATE

Climate traditionally has been classified on the bases of annual or seasonal heat and moisture. Thus, humid cold areas may be differentiated from humid hot ones, or from arid areas. The great climatic regions of the Earth are caused by the movement of heated air from the tropics toward the poles, as modified by the topography of the area over which the air is passing and ocean and atmospheric currents.

Koppen's climatic classification (published in 1900) was based principally upon thermal zones: these were modified by the number of rainy days, converted into a complex moisture index (Mather 1978). The moisture index was the forerunner of the Thornthwaite climatic classification, based upon the water budget, which is described more fully in Chapters 4 and 5, as appropriate to vegetation and soils.

Because vegetation is dependent upon adequate and sufficient moisture supplies, these areas are associated with broad vegetation associations (Blumenstock and Thornthwaite 1942). For example, in general terms, Table 3-3 shows the relationship between vegetation and climate.

Soil development is also dependent upon several climatic factors, as is vegetation, so all three are closely interrelated. As a consequence, *watershed management must be based upon comprehension of the complex interrelationships among vegetation, soil, and climate, as modified even further by geologic forces, and tempered by the influence of civilizations.*

Considerations of heat (and radiant) energy and moisture relations are thus important to understanding the natural hydrologic cycle. And that understanding is essential for successful management of the water resource. The use of climatological data for evaluation of the water balance, and for prediction of runoff, is discussed in Chapter 6. When such models are presented it is important, too, to recall the limitations of the instruments used to collect the data upon which they (and we) depend.

Table 3-3 Climatic Region and Vegetation Type

Climatic region	Vegetation
A -- Superhumid	Rain Forest
B -- Humid	Forest
C -- Subhumid	Grassland
D -- Semiarid	Steppe
E -- Arid	Desert

Source: Blumenstock and Thornthwaite (1942).

SUMMARY

The two generalized processes treated here, precipitation and evaporation, balance over the course of several years, providing the basis for examination of the water budget. Deviations from this balance are ramified in and/or are due to changes in storage.

Precipitation occurs in a variety of forms from several different categories of storms. The combination of form, type, and season of occurrence provides the opportunity to begin to classify hydrologic regions. Temporal and spatial variation of precipitation exhibits characteristics that are summarized in intensity-duration-frequency relations. Because most of the factors influencing precipitation are randomly distributed, intensity-duration-frequency relationships are nearly normally distributed. The patterns of water movement and storage thus engendered are the basis for patterns of runoff and stream behavior.

Evaporation is also regional, dependent primarily upon the energy available and vapor pressure gradients above the evaporating surface. It is measured by a rather artificial device, and may be estimated by a variety of other methods. It is often combined with the other two evaporative processes, interception and transpiration, into evapotranspiration, which can be evaluated by the water budget method (discussed in Chapters 4 and 7).

Precipitation is sampled, and at a rather low frequency. The potential error thus introduced is neither normally acknowledged nor accounted for. It must be kept in mind when using precipitation data for any of the many purposes for which it must be used in hydrologic research and water resources management. Evaporation is sampled on an even less-intensive basis.

Assessment of water moving through and temporarily stored in the atmosphere is an inexact science based on extensive sampling and often inaccurate, though precise, measurement.

PROBLEMS

1. What is the simplest way to use Fig. 3-1 to illustrate the Coriolis Force in the Southern Hemisphere?

2. On a single sheet of graph paper (to facilitate visual comparison) with the horizontal axis in hours of the day and night, and the vertical axis temperature, show your estimate of two curves representing (1) the summer temperature inside a standard Weather Bureau shelter and (2) the true temperature. Repeat for winter curves.

3. Using Fig. 3-17, construct a graph of intensity-duration curves for the Olympic Peninsula, Minneapolis, Miami, and Albany, NY. How do the curves differ for the different regions the cities represent? Why?

4. Using a series of 100-yr frequency curves in Technical Bulletin No. 40 from which Fig. 3-17 is taken, construct a set of intensity-duration curves for a selected area, such as your county.

5. Convert vertical axis on Fig. 3-16 to in. depth in a 100-yr, 1-hr storm for a selected area (county) shown on Fig. 3-17. How might this information be used?

4 Water in the Vegetated Zone

*In the vegetated zone,
movement of water between the atmosphere and the soil
plays a diversified role
in the storage capacity of
and the complex relationships between
all three regimes*

The ratio of water stored in the biosphere to that in circulation (Table 1-1) is over 17:1. Even though more water is stored in it than moves through it each year, as is the case with the atmosphere, there has been considerably more research effort dedicated to the processes of water movement in the biosphere than to storage phenomena. Indeed, even the water stored in vegetation is in constant movement, involved in a variety of complex life-support functions Consequently, there is even more confounding of storage and process here than in the atmosphere. For that reason, process and storage are separated in only part of this chapter.

PROCESSES

INTERCEPTION

Interception is both a process and an amount. A diagrammatic representation of the process is shown in Fig. 4-1. As a process, interception may be defined as the interruption of the downward movement of precipitation and its redistribution by vegetation (and other objects). The amount is the depth of water lost through this process. Because some of this amount is (temporary) storage, it is impossible to separate the discussion of interception process and amount, which is often referred to in the literature as "storage."

Interception of precipitation by vegetative surfaces has three distinctly different effects: mechanical, quantitative, and conservational. The mechanical effect is protection of the soil; the quantitative effect is a reduction in precipitation that reaches the soil; and the conservational effect refers to the fact that several aspects of the interception process favor conditions under which soil moisture may be maintained at high levels.

Mechanical Effects of Interception

Vegetative surfaces protect the soil by reducing the force with which raindrops hit the exposed surface. This reduces erosion and soil loss. As rain falls on the forest canopy, the leaves absorb the shock of high velocities and temporarily restrain the water from reaching the soil surface. When a new drop forms on the leaf to begin its fall to the ground, it may not have as far to go before reaching the soil as it did when it fell from a cloud. It is, therefore, less likely to reach its terminal velocity [a limiting rate dependant upon raindrop size, water temperature (which affects viscosity), height of fall, and vertical air movement]. However, where height to the base of the canopy is in excess of 25 ft to 30 ft, and/or there is significant downward movement of the air and an unprotected or thinly protected forest floor, raindrop impact may be severe even under forest cover. In addition, multiple canopies of understory vegetation, and low ferns and other herbaceous plants, along with the forest **litter** (dead vegetative material on the ground) each again interrupt the downward movement of and redistribute the precipitation.

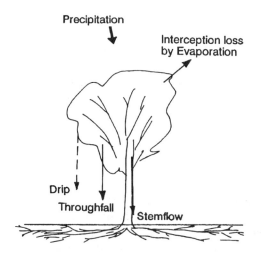

Figure 4-1 The interception process in the forest

Litter layers can intercept a quarter of an inch of rainfall or more. Since water so intercepted is not as exposed to the drying effects of wind as is moisture intercepted by canopy vegetation, it may not repeatedly contribute to interception "loss."

Unprotected soil surfaces are subject to three erosional effects. First, fine materials may be washed into the larger pores in the soil surface, slowing down the rate of infiltration and, as a consequence, causing surface runoff where none existed previously owing to the usually high infiltration rate of undisturbed soils. Second, soil particles may be actually moved by the impacts of raindrops, starting the erosion process. And third, exposure of the soil to greater temperature ranges and desiccation causes the soil **aggregates**, combinations of mineral soil and organic material, to break down. Soil aggregates play an important part in the water relations of natural forest soils, and destroying them makes the soil more vulnerable to erosional forces (Chapter 5).

Quantitative Effects of Interception

Interception is often termed a "loss." Thus, the storage portion of the definition expresses the fact that there must be a certain amount of precipitation before water reaches the ground under a forest. The leaves must be sufficiently wetted first for, except for the relatively few and variable number of drops that fall directly through the canopy without touching any vegetative parts, all the precipitation water may be intercepted. In the absence of violent shaking of the leaves by wind or by the impact of the raindrops themselves, this evaporated water is equal to the amount left on the leaves after the rain stops and, owing to evaporation, is "lost" to the soil reservoir.

In addition, as long as the atmospheric relative humidity is below saturation, that is, as long as there is a vapor pressure gradient immediately around the wetted leaf, evaporation can continue during the storm. That water, then, is also "lost" to the soil and to runoff, and is to be added to the amount "stored" on the leaves and other vegetative parts.

The amount of water that can be stored on individual trees varies from about 0.01 in. to 0.36 in. (0.25 mm to 9.1 mm). The data for 39 studies of interception by different species at different ages and in different seasons in the United States were reported by Zinke (1967) and are abstracted in Table 4-1. Note that considerably more water may be "lost" during a storm as long as intercepted water continues to evaporate.

Table 4-1 Summary of Interception Storage on Vegetation

Vegetative type	Special conditions	Range (in)
Conifers,	rainfall	0.01 — 0.30
	snowfall	0.02 — 0.36
Hardwoods,	summer	0.02 — 0.05
	winter	0.01 — 0.03
Shrubs		0.01 — 0.07
Grass		0.04 — 0.06
Forest Floor Litter		0.02 — 0.44

Source: Zinke (1967).

If a forest consists of several overlapping canopies and a thick litter layer, there may have to be up to an inch (24 mm) of rain before water will reach the soil. Typically, about 0.05 in. (1.3 mm) of rain is required before the leaf surfaces are wetted and collected water starts to drip from them. This is higher than the apparent "average" in Table 4-1 because some of the intercepted water may evaporate, some may be absorbed in plant tissues, or some may start to run down the stem (see below). Interception appears to be lowest for rainfall on hardwood forest stands in the dormant season, and for very thin forest canopies, that is, those that consist of tree species that are relatively intolerant of shade and have sparse crowns.

Litter on the forest floor can also retain considerable moisture. Clary and Ffolliott (1969) reported that water distribution in the three layers of organic material above the soil (the litter, fermentation, and humus layers) under a Ponderosa Pine stand was proportionate to their mass; 85 percent was retained in the well-developed humus layer. It is likely that with different distributions of litter and causative fermentation processes, water distribution would also vary. With a thick litter layer under eastern hardwoods, as much as 0.5 in. (12.7 mm) or more of precipitation must occur before any water reaches the soil. Results of an interception study at the San Dimas Experimental Forest near Los Angeles showed that the products of average height and percent of ground cover were determinants for the water storage capacity of the grass cover. Helvey (1964) estimated that interception by litter was about 3 percent of the annual rainfall under hardwoods in the southern Appalachians: measurement is difficult owing to the possibility of and the diificulty of measuring continuing evaporation from the litter during and between storms.

Some of the water that falls on the leaves collects to drip directly to the ground. This is **leaf drip**. The drops that form on and fall from the leaves may be of considerably different size than those of the storm's raindrops: in a fine, misting type of rain, drops from the leaves may be a great deal larger and capable of doing more damage at the soil surface if that surface is unprotected. On the other hand, the large drops of a driving thunderstorm are often broken up and reconstituted into smaller droplets by the vegetative surfaces. The overall effect is that vegetation seems to buffer the extreme effects of rainfall impact.

The water that falls directly through the canopy is called **throughfall**. The amount of throughfall will depend upon canopy density and several other factors identified and discussed below. Throughfall consists, in fact, of both raindrops falling directly through the canopy and leaf drip. It may be possible to determine by eye whether the drops coming down are from the foliage or directly from the storm by drop size, but the precipitation gage cannot distinguish between them. Thus, the gage used to measure precipitation reaching the ground under a forest canopy is called a **throughfall gage**, although it measures leaf drip as well. An early type of throughfall gage was rectangular,[1] but was found wanting owing to loss of catch due to splash and edge effects (Reigner 1964).

The areal distribution of throughfall is highly variable, as is to be expected from the high degree of variability introduced by the forest canopy. This variation must be accounted for in order to relate it to **gross precipitation** (the precipitation measured in the open, or above the forest canopy): Helvey and Patric (1965), using data by Black (1957), suggest that in order to have comparable observations, it takes about twice the number of gages in the growing season as it does in the dormant season under deciduous hardwoods, owing to greater variability of the forest canopy when leaves are present. Roth and Chang (1981) used 15 gages to "sample throughfall with 90% accuracy." The number is dependent upon the amount of throughfall (decreasing with increasing throughfall) in order to obtain a given percent standard error of estimate so that the throughfall measurement has the same degree of error (or confidence) as the gross precipitation (Table 4-2). The distribution of throughfall is important to the distribution of soil moisture (Eschner 1966) and, consequently, to the availability of water to the roots of the tree, the foliage of which interrupted its delivery.

Interception of precipitation in young eastern pine stands was reported by Helvey (1967); he concluded that:

> the older the stand, the greater the loss. Comparison of estimated interception losses under climatic conditions prevailing in the mountains of western North Carolina indicate that interception loss from old growth white pine may be as much as 100 percent greater than loss from hardwoods. Results also indicate that during winter months, interception loss from pine and hardwood greatly

---

[1] The round-bottomed, metal trough varied from 4 in. to 9 in. (10.16 cm to 22.86 cm) in width, and from 2 ft to 80 ft (0.61 m to 24.4 m) in length. The longer dimension was to ensure integration of the high degree of variation in water that reached the ground.

exceeds potential evapotranspiration calculated by the Thornthwaite method [see below].

Table 4-2 Number of Gages Needed for Throughfall Measurement to Keep Standard Error under 5 Percent of Mean Throughfall

Throughfall Range (in)	Number of Gages	
	Dormant	Growing
< 0.19	24	46
0.20 — 0.39	6	18
0.40 — 0.59	6	14
> 0.60	6	13

Source: Helvey and Patric (1965).

Although as much as a third of the snow falling on Douglasfir and Western White Pine saplings was intercepted, "80 percent of the snow initially caught in the crowns ultimately reached the ground..." (Satterlund and Haupt 1970). Intercepted snow is noteworthy immediately following the storm that deposits it on the branches, but wind during and after a storm, especially if accompanied by a shift in wind direction and strength, dumps most of the snow on the ground before it has a chance to evaporate directly to the atmosphere. Hoover and Leaf (1966) report that wind action after (and during) a storm affects interception amount, and that cutting history and irregularities in snow deposition and snowpack characteristics affect interception as well.

Increased snowpack in openings in thinned Wyoming Lodgepole Pine stands was due to decreased interception; however, snowpack reductions downwind of the cuttings offset the increases in the clearings (Gary 1974). Wind/stand dynamics play a complex role in interception loss and process in coniferous stands.

In southeastern stands of Longleaf, Slash, Loblolly, and Shortleaf pine, throughfall was from 80 percent to 86 percent in 28 winter storms, and from 84 percent to 91 percent in 16 summer storms: the winter average was 83 percent, and the summer average was 86 percent (Roth and Chang 1981).

Interception loss increased as hardwoods were converted to white pine (Swank and Miner 1968), and was highest (22 percent) in Loblolly Pine at 10 years of age, increasing from 7 percent at 5 years, and decreasing to 18 percent at 20 and 30 years. Simultaneous research showed that thinning Loblolly Pine stands permitted an extra 4 in. (102 mm) or 7.7 percent of precipitation to reach the ground under a 25-yr plantation (Rogerson 1968).

Some of the water collects and runs down stems, reaching the ground as **stemflow**. This usually does not amount to a great percentage of the annual precipitation, or even of a storm, and it often does not start until 0.20 in. (5 mm) or more of gross precipitation has occurred (Helvey and Patric 1965). This is water that is delivered directly to the soil and in the vicinity of the plant roots, however, so it is particularly effective for plant growth.

The amount of stemflow will depend upon the configuration, especially the branching angle, of the tree. This is usually a characteristic of the species. American Elms, for example, are urn-shaped and water that collects on upper vegetative parts collects and runs down the trunk quite readily in large quantities. Some conifers, on the other hand, have drooping branches, so that water may not collect to run down the stem at all. Rather, the intercepted water is dispersed all around the tree, often where the feeding roots of the tree are located. Hoover (1953) reported that young Loblolly Pine showed "sharply upthrust" branches that funnelled collected precipitation down the stem: he speculated that older Loblolly Pine might have different stemflow regimes owing to different branching habits. Many older conifers exhibit differential branching angle between the newer growth near the top of the tree and lower, older branches. The latter tend to shed intercepted precipitation out further from the stem where the feeding roots of the older tree are located.

Thus, Lawson (1967) reported that a 6-in. (15.2 cm) diameter hardwood produced 12.0 liters (3.2 gal) of stemflow in a 3-in. (7.5 cm) storm as contrasted with a pine of the same size that produced only 6.8 liters (1.8 gal). Both position of tree in the crown (which affects branching habit as well as number, shape, and orientation of leaves) and basal area of mixed oak stands were reported to be correlated with stemflow (Brown and Barker 1970); black oaks yielded more stemflow than white oaks, although throughfall was similar on all plots; storm size was the only storm characteristic correlated with the amount of stemflow, and stemflow was much greater during the dormant season. Rogerson and Byrnes (1968) similarly found that hardwood stemflow was about twice that of conifers.

Stemflow is measured by securing a metal or plastic trough (Fig. 4-2) around the tree so that no water can pass between it and the bark, and only a minimum amount can fall directly into it (a potential source of error). The water thus collected is funnelled into some receptacle in which it can be readily measured and mathematically "spread" back on a horizontal projection of the crown to find the depth of water appearing as stemflow. Typically, this amounts to little more than 0.1 percent of the annual precipitation, although for some species it may reach 1 percent or 2 percent. For an individual storm, stemflow can reach 7 percent of the gross precipitation, dependent upon the species and a number of factors that affect stemflow. The actual quantity of water can be staggering: a medium-sized [10 in. (or 254 cm) dbh] Tulip Poplar at the Coweeta Hydrologic Laboratory in the mountains of North Carolina produced more than 60 gal (227 l) of stemflow water in a singlestorm of 3 in. (76.2 mm) (Black 1957). Note that from a given tree, 100 percent of the stemflow is measured, whereas the precipitation and throughfall are sampled.

It is apparent that the list of factors affecting evaporation from free water surfaces presented in Chapter 3 needs to be expanded to include characteristics of the biosphere that will help account for the interception and stemflow processes:

Vegetative characteristics that affect the amount of **interception**:

type of vegetation (tree/shrub/grass)	bark roughness
deciduous/evergreen foliage	branching habit of the vegetation
density of vegetative canopy	number of vegetative layers (in the forest)
position of tree crown in canopy	cutting history and land use of the area

Figure 4-2 Stemflow gage

Some of the separately listed factors could be grouped under "type of vegetation" since they are genetic characteristics of a particular species. This includes a long list of characteristics that are not itemized, such as whether leaves are positioned horizontally or vertically in the canopy; leaf roughness; stability of the leaf when wet; strength of the petiole which holds the leaf to the twig; and whether the leaf is a "sun" or "shade" leaf, because certain leaf properties vary within a tree dependent upon leaf position in the crown. If the leaf is held horizontally by a petiole that does not flex readily, for example, it will hold more water than a vertically positioned leaf. Change in the angle that the leaf presents to precipitation can also occur during the day as evapotranspiration stresses the tissues and turgor pressure affects petiole support.

Several of the atmospheric characteristics listed in Chapter 3 that were not identifiable as affecting evaporation from a free water surface must be included here because they interact with certain features of the biosphere that influence interception process and the amount. For example, if a storm were to produce only 0.05 in. of rain and all of it went into wetting the leaf surfaces, there would be no throughfall.[2] Furthermore, if there were sufficient time between ten successive storms each of 0.05 in. (1 mm) to permit the foliage to completely dry out, there would be no throughfall at all, even though there was a total rainfall of 0.5 in^3 (12 mm). From such consideration, the concept of **antecedent moisture conditions** (the amount and location of water in storage at the start of the hydrologic event or time period under study) is derived. Proper identification of all the factors that affect interception as an evaporative process, then, demands expansion of the list of atmospheric features, too:

Atmospheric factors that may interact with vegetative characteristics to affect the amount of **interception**:

season
    form of precipitation
    leaf condition

storm characteristics
    amount
    intensity
    duration
    frequency
    time of day

The last factor is necessary because, were the precipitation to occur in late afternoon, the vegetation might not dry out before the next morning. At that time, another storm might produce throughfall nearly coincident with the start of gross precipitation since the interception storage capacity of the canopy would be already filled. On the other hand, with sufficient time between storms, and sufficient energy, the storage capacity will be restored by evaporation. The entire list of factors affecting evapotranspiration (from Chapter 3 and the foregoing paragraphs) is shown in Table 4-3.

The now-long list of factors influencing evapotranspiration was classified by Tanner (1957) into three broad categories:

1. the climate (heat available from solar radiation and the air);

2. the soil moisture available (capillary conductivity of the soil, soil moisture stress, and soil water content);

3. the physiological reaction of the plant to the difference between the moisture availability and the "evaporation demand."

Considerable overlap among these categories persists, however, owing to the complex interactions between climate, soil, and vegetation.

---

[2] This assumes that all the throughfall is from raindrops falling directly through the canopy, and that there is no leaf drip. This is a questionable assumption for all but the most dense of forest canopies. For the purposes of analyzing the factors affecting the process, however, it is an entirely appropriate assumption.

[3] If this were to persist, there would not be sufficient moisture in the ground to grow trees in the first place, so it is a rather spurious argument: again, it is a useful one in evaluating the factors that affect interception loss. The argument may be expanded to include interception by lower vegetative canopies within a forest stand, and by the litter layer.

Table 4-3 Evapotranspiration Factors

Factors affecting **ENERGY** available for evaporation
properties of the incoming energy:
    latitude
    elevation
    time of day
    time of year
    degree of cloud cover
    temperature of water
    temperature of substrate
    temperature of adjacent air
properties of the water:
    depth
    color/turbidity
    color of water body bottom (if shallow)
properties of a wetted surface:
    color
    roughness
    temperature
    exposure to advection of warm air
    orientation with respect to solar radiation

Factors affecting amount of **WATER** present
characteristics of the hydrology:
    frequency of supply of water to a wetted surface
    supply of soil water by capillary action
characteristics of the water body:
    basin depth
    basin volume
    stagnant/flowing water

Factors affecting the **VAPOR PRESSURE GRADIENT**
    wind
    elevation
    atmospheric pressure
    extent of gradient
    continuity of gradient
    air temperature at surface
    air temperature above surface
    relative humidity at surface
    relative humidity above surface
    exposure of evaporating surface

ATMOSPHERIC factors that interact with vegetation
seasonal characteristics:
    form of precipitation
    leaf condition
storm characteristics:
    amount
    intensity
    duration
    frequency
    time of day

VEGETATIVE characteristics affecting interception
    type of vegetation
        tree
        shrub
        grass
    deciduous/evergreen foliage
    density of vegetative canopy
    position of tree crown in canopy
    bark roughness
    branching habit of vegetation
    number of vegetative layers (in the forest)
    cutting history and land use of the area

The relationship between gross precipitation, P_G, throughfall, T, and stemflow, S, can be formulated to introduce and define the term **net precipitation**, P_N:

$$P_N = S + T$$

and interception, I, may be formulated as

$$I = P_G - P_N$$

or as

$$I = P_G - (S + T)$$

Throughfall was reported by Kittredge (1948) to be a linear function of gross precipitation. Thus, generally,

$$T = a + bP_G$$

The value of the Y-intercept, a, is negative, representing the necessity for some gross precipitation to occur before throughfall commences. The b coefficient varies from about 0.74 to 0.94. Regression of throughfall on gross precipitation generally yields very high values for the regression coefficient. This indicates that most of the variation in throughfall can be accounted for by measuring precipitation and, in order to minimize error due to natural variation, throughfall may be estimated from the regression equation. Seasonal differences owing to form of precipitation and foliage condition (in deciduous species) should be taken into account by having separate regressions for each season rather than installing numerous gages.

Finally, it is important to specify whether precipitation is actually a gross or net figure. *Effective precipitation* is what the vegetation, the soil, and runoff depend upon, not the total.

Conservational Effects of Interception

Interception provides conditions whereby water may be added to the soil, and conserved. Advective, moist air has been shown to add materially to the soil in a process called **fog drip**. On the windward sides of vegetation exposed to advective fog, several inches per year may be added to the soil moisture reservoir (D. H. Miller 1957; Oberlander 1956). Precipitation at high elevations is also enhanced by this phenomenon, when advected, cooled, moisture-laden air (and clouds) deposit substantial quantities of water on vegetation near the crests of mountains (Kittredge 1948). The author observed such "misting" on the Continental Divide in the summer of 1984 (Black 1985) when El Niño was causing streams of moist, tropical air to flow northeastward over Arizona and the Colorado Rockies (Fig. 4-3). Miller (1957) refers to the process as "fog precipitation."

If the temperature is between -10° and -20°C, accumulation of moisture on vegetation (or other objects) is known as **rime** (see Chapter 3). Ice buildup of this sort can be quite heavy and, like fog drip, can add a significant amount of water to the soil, especially at high elevations. Up to 10 percent of snowpack water content was found to be contributed by rime ice in a study in Colorado, and "up to 60 percent of the water content of the annual snowpack in the mountains near Steamboat Springs, Colorado could be due to rime ice deposits" (Hindman, Borys, and DeMott 1983). The energy relations of condensation of water vapor on ice are quite complex. The heat released by the condensation process may melt some already-frozen water at or near the surface, for example. The alternate freezing and thawing that results aids in creating a crust on the snowpack, in addition to that created by diurnal, ambient air temperature fluctuations through 0°C.

The water which reaches the soil through fog drip is precipitation that, in the absence of the vegetation, would not occur. Therefore, in those regions where fog drip regularly occurs, removal of vegetation may reduce precipitation. Significant quantities of water reaching the ground under vegetation exposed to advected, moisture-laden air have been reported for many years (Vogelmann 1976). Harr (1982) reported that fog drip from old growth Douglas-

(1) Tent site during misting event

(2) Same location, ten minutes later

Figure 4-3 a "Misting" at timber line in the Colorado Rockies

Figure 4-3 b Collection of water on grass and trees from misting at tent site

fir could add 35 in. to the 85 in. of annual precipitation measured in a nearby clearing, and that "standard rain gages installed in open areas where fog is common may be collecting up to 30 percent less precipitation than could be collected in the forest .... Long term forest management (i.e., timber harvest) in the [Bull Run, Oregon] watershed could reduce annual water yield and, more importantly, summer stream flow by reducing fog drip." Fritschen and Doraiswamy (1973) recorded dew accumulations under Douglas Fir in a lysimeter and noted that the amount per tree was 15 to 20 percent of the evaporation from the tree, adding "thus dew formation could represent a large part of the hydrologic balance of fir forests."

The measurement of fog drip is difficult. Gages located under the canopy of an intercepting tree may be positioned under a drip point and indicate considerably higher moisture than actually occurs (Oberlander 1956). The uncertainty is due, in large part, to the fact that identification of the area upon which the water that flows to the drip point originally fell is generally not possible. Thus, although gages have recorded tens of inches of fog drip, translating the volume to a catchment area in order to report an inches-depth-of-precipitation-figure is likely to be misleading. D. H. Miller (1957) reported on research results where upwards of 25 in. (677 mm) of fog drip have been registered along the California coast under vegetation. At the other extreme, a study at the Fernow Experimental Forest reported only very small amounts of dew, 0.055 in, the significance of which is not known Hornbeck (1964). Because of the difficulty of translation of the volume of water caught by a gage to an inches-depth equivalent, and because whether the gage itself induces the fog drip or is under vegetation which collects it, the significance of such accretion to the hydrologic cycle cannot be precisely assessed.

Research has been conducted by the Aeronautical Icing Research Laboratory to quantify collection efficiency of various diameter objects and vegetative parts (Ettenheim 1962). Exposure of devices that duplicate vegetative collection efficiencies and exposures is intricate and almost any measurements obtained are subject to criticism. A highly detailed "general equation" to predict water accumulation from either rain or fog on forest foliage was presented by Massman (1980). The complex equation reportedly unified the findings of several other researchers. Complete identification of the storage on the foliage is necessary to its successful use, however.

Vertical-wire harps and cylinders (Goodman 1985) and artificial "leaves" have been used to "catch" and produce fog drip. Narrow, cylindrical coniferous needles have been shown to be the most effective "collectors" of airborne fog particles and, perhaps, water vapor as well (Ettenheim 1962). This needle form is typical of the Coast Redwoods, which survive in an area of only about 35 in. of annual precipitation, a surprisingly low figure considering the apparent water demands of these giant trees. Eucalyptus, cedars, and other conifers also have a leaf shape that is effective in collecting water from the air (Oberlander 1956), and they thrive in the same type of climate. Other types of vegetation are generally less effective in collecting droplets or vapor.

Merriam (1973) suggested that the high cost of installing and maintaining artificial devices in order to extract water from fog would likely outstrip the less efficient but managed natural vegetation. Ingwersen (1985) reported that "analysis of streamflow data from the Fox Creek Experimental Watersheds in the Bull Run Municipal Watershed in Oregon indicates a significant recovery from the impacts on summer water yield due to the loss of fog drip upon timber harvesting .... Recovery begins about five or six years following harvest, possibly due to renewed fog drip from prolific revegetation."

The amount of water collected through fog drip is important in another way, too. This water, along with that left on the vegetative surfaces following a storm, will evaporate, given sufficient energy and time before the next storm. And it takes energy for that change of state. If the vegetative surfaces were not wet, that energy would be used for transpiration, which involves removal of water from the soil. Thus, "interception, by substituting for transpiration in conifer stands, can appreciably increase soil moisture and plant water" (Nicolson et al. 1968). In coastal areas where frequent storms and fog keep the vegetation wet a great deal of the time, soil moisture is thus conserved. Again, in forest types such as the Coast Redwoods, this is an especially important aspect of the species' ecology (Black 1967). The magnitude of the effect of wetted vegetation surface has been tested under controlled conditions by Thorud (1967), who reported that "for thirty-six 2-hour periods the average transpiration reduction was 14 percent, or 9 percent of the applied water .... Higher savings were observed with more moderate weather."

As noted above, fog drip provides water to the soil which would not occur if there were no vegetation on which the fog impinges and deposits moisture: the same is true for the gages, and it is thus impossible to precisely monitor fog drip.

Fog in many coastal areas, of course, also keeps the radiant energy from even reaching vegetative surfaces during the day, thereby further reducing the vegetative transpiration demand on soil water. And, if fog keeps the foliage wet, then there is little or no storage capacity available in the canopy and, at the start of a rain, throughfall will occur immediately.

TRANSPIRATION

As a process, transpiration is of great importance to the physiologist; its physiological role is, however, of less concern to the hydrologist. The transpiration process is important to the hydrologist because it represents the movement of a significant amount of water from the soil to the atmosphere. Incidentally along the way, its physiological role is undeniable.

Transpiration is governed by the same physical relationships that govern evaporation. It is also influenced by the same factors — often modified by the nature of the evaporating surface — plus some others that are characteristic of the vegetation through which it moves and in which it is stored.

The Movement of Water and Nutrients

Water is absorbed by the roots of vegetation from the soil. It carries with it nutrients that are distributed throughout the plant. Water also moves down *in* the plant, carrying starch to the roots. Under conditions where hydrostatic forces are strong, dissolved substances move by mass flow, that is, are simply carried along by the moving water in a rapidly moving transpiration stream. Under conditions of slower movement and lower tension, the dissolved substances are also carried from areas of high concentration to areas of low concentration along a diffusion gradient, and the water moves by a combination of osmotic pressure and hydrostatic forces.

D. E. Miller (1977) points out that opposite the forces exerted by a vapor pressure gradient, molecular forces *within the plant tissues* exert resistance to movement from one cell to another and from the plant to the atmosphere:

> The resistance of vapor moving from leaf to air is as much as 10^7 times the resistance to movement in 1-m length of stem, and 10^6 times the resistance to movement of water into a root. This internal resistance is added to the resistance located between leaf outer surface and air, which depends on wind speed and other factors.

An inefficient process for conveying such materials, a large amount of water is needed to accomplish this distribution, and the process is speeded by having some of the water leave the top of the plant through the microscopic openings in the leaf called **stomata**. The stomates (which also facilitate the movement of other gases such as carbon dioxide and oxygen) are usually located on the underside of the leaf, and are opened and closed by the expansion and contraction of **guard cells** which respond to moisture content, regulated by relative humidity levels and complex physiological characteristics (R. Lee 1967). Guard cells respond in a complex manner to temperature, CO_2 content, light intensity, past history, and, perhaps, certain chemical contents as well (Kramer and Kozlowski 1979). If the relative humidity near or at the leaf surface is low, the guard cells tend to shut down and conserve plant and soil water. If the relative humidity is high, they open and permit water to move out of the cells in the process called **transpiration**. At very high relative humidity, the vapor pressure gradient is controlling, and transpiration will not occur. Guard cells also regulate the flow of gases in and out of the leaf, and respond to gaseous concentrations as well.

Obviously, then, a major factor affecting transpiration is relative humidity. Others include those factors that affect relative humidity such as temperature and atmospheric pressure, and wind, which affects the gradient. Certain characteristics of the vegetation may be added to the already-long list of factors that affect evaporative-type processes, including genetically controlled ones that determine stomate number, size, and distribution, guard cell opening characteristics, and habitat of the plant itself; those plants that thrive in water-rich environments have more water available for the transpiration process than do those in more arid areas.

Two observations may be generalized from the foregoing: (1) that relative humidity immediately above the leaf surface is "sensed" by the soil moisture by a continuous but variable tension gradient that responds to changes anywhere along it, but especially to changes at the leaf surface, and (2) that there is a confounding of temporal, vegetation, and atmospheric factors affecting relative humidity.

As a consequence, it is appropriate to think in terms of "environmental water," rather than of water in its separate realms. A parallel ecological mistake in observation that mankind has made, as pointed out by Bates (1960), is that by virtue of our walking on the surface of the Earth, we naturally look up at the forest and down at the sea and we thus fail to see the continuous nature of our environment.

The concept of environmental water, along with the concepts of infiltration and antecedent moisture conditions, are three of the most basic, important, abstract, and difficult-to-define concepts in wildland hydrology.

Measuring Transpiration

Measuring transpiration is difficult, if not impossible. A variety of approaches are used to obtain estimates, which is all that can be ultimately expected. In general, and often in overlapping terms, the methods may be divided into direct and indirect groups. But the words are used with some reservation, because most attempts to evaluate transpiration have been either intentionally or inadvertently combined with estimates of all or part of the collective evaporative-type losses. The intimate linkages between the several evaporative-type processes in the real world make it necessary to consider the measurement problems and results in that light and to simultaneously review the research on the topic. Most of the research, in fact, has estimated *evapotranspiration*, not transpiration alone. Thus, the discussion of measurement of transpiration is continued below in the section on measuring evapotranspiration.

EVAPOTRANSPIRATION

Evapotranspiration is the term used to describe the combined losses to the atmosphere through interception, evaporation, and transpiration. In the annual water balance approach, as presented in Chapter 3 (p. 77), evapotranspiration (E_t) may be equated to precipitation-minus-runoff (*P-RO*), as shown, too, at the top of Figure 1-2. This oversimplification masks the complexity of how water moves from lithosphere to atmosphere. There is considerable opportunity for variety in the manner in which water returns to the atmospheric reservoir because (1) the evaporative potential of the atmosphere establishes the upper limit of upward water movement in any given time period, (2) sufficient water must be available for this potential to be realized, and (3) any of the three E_t components may be available for upward water movement.

Dunne and Leopold (1978) point out that "evapotranspiration dominates the water balance .... More than two-thirds of the precipitation falling on the conterminous United States is returned to the atmosphere by evaporation from plants and water surfaces .... Although evapotranspiration is necessary for plant growth, it is usually viewed as a 'loss' from the water budget in that it reduces the amount of streamflow, lake storage, and groundwater available for direct human use." All the complex details of the processes are not fully understood; conflict over measurement and interpretation of several commonly observed evapotranspiration phenomenon persist. For example, the limit on the process whereby plants withdraw moisture from the soil (the **wilting coefficient** as discussed in chapter 5) to meet transpiration demand is referred to by the plant physiologists as a property of the soil and by the soil physicists as a property of the vegetation: Kramer and Kozlowski (1979) observe that "neither the field capacity nor the wilting percentage are physical constants, but merely convenient regions on the water potential-water content curve." Goodell (1963) suggests:

> that the vaporization of rain or snow intercepted by vegetation may result in a water loss from that site that is less real than apparent because of a compensating reduction in transpiration. The magnitude of such compensation under field conditions is questioned in view of (a) transpiratory regulation by plants under the stress of limited water availability, (b) probability that the energy available for evapotranspiration from wet leaves is greater than from dry leaves, and (c) the

possibility that under winter conditions in cold climates transpiration is limited more by the availability of water than of energy and that interception increases the quantity of water favorably exposed to the energy supply.

Eagleson (1982) and Eagleson and Tellers' (1982) attempt to numerically evaluate soil-vegetation systems in order to derive and test a relationship between average evapotranspiration and optimum vegetation canopy density are another case in point: while the research is directed at situations under water stress, the tests were successful only under humid and limited semiarid conditions. The complex ecological processes and interrelationships may work for one end of the moisture-available spectrum, but tempting extrapolation to areas where such relationships are not true is a dangerous consequence of such oversimplification. The reason the relations work for the unstressed end of the spectrum is that available moisture is not a limiting factor: when it is, the E_t processes are not quite so simple.

The importance of E_t in the overall water balance is obvious: without it there would either be no recycling of moisture or some other mechanism would have to exist. But E_t also performs other functions. Evaporation is a cooling process: the energy used to do the work of vaporization often is heat energy taken from the material from which the water is being evaporated or from the adjacent air. The physical laws governing vaporization may thus be used to predict evaporation from a water body (Dake 1972).

In addition, dissolved materials are conveyed in water translocated within plant tissues; excess salt may be left behind in the process of evaporation. Thus, E_t plays a role in water quality, nutrient movement, and vegetative growth and development. Local climate is also affected by E_t as ramified in varying absolute and relative humidity measurement in or near vegetated and nonvegetated surfaces.

With such a large quantity of water involved, the potential exists, too, for control of vegetative cover so as to alter E_t and, as a consequence, manipulate streamflows to meet our needs. Furthermore, vegetation usually may be manipulated concurrent with its beneficial economic management, making vegetation management as a means of control of the water resource on wildlands particularly attractive.

In order to evaluate this potential and to practically implement management, research was conducted regarding the nature of E_t simultaneously with fundamental investigations. Thus, much of this research and, of necessity, a great deal of the basic E_t theory and measurement methodology stems from the results of attempts to evaluate E_t for practical management. Major research on the topic has been carried out by the Agricultural Research Service, Forest Service, Soil Conservation Service, Tennessee Valley Authority, and many colleges and universities. Investigations have focused on individual tree stems, small plots, homogeneous soil/vegetation systems, and experimental watersheds.

An example of the complex interrelationships among the several processes and factors is reported by Singh and Szeicz (1979): they measured evaporation rates from wet canopies "several times" that of the transpiration rate from a dry canopy. Clearly, availability of moisture for evaporative-type processes interacts with the evaporative potential of the atmosphere. And, responding to the frequently presented observation that "some investigators suggested that energy used to evaporate intercepted water must necessarily come from that which would have been used for evapotranspiration," Murphy and Knoerr (1975) present a detailed analysis, and assert:

> From this analysis it is clear that precipitation intercepted by vegetation evaporates at a greater rate than transpiration from the same type of vegetation in the same environment.... Our analysis also allows us to identify the energy source for the more rapid evaporation of the intercepted water as being an increase in net radiation through a decreased long-wave reradiation and a decreased or even negative sensible heat flux. Furthermore, the model demonstrates that the enhanced evaporation of intercepted water can occur for forests of large areal extent, where horizontal advection may be negligible. Thus it reaffirms the conclusions from empirical experiments that interception of precipitation does represent a loss of water to the soil and to the streamflow under field conditions.

This observation is consistent with that of Tajchman (1971), who calculated E_t and convective exchange by turbulent diffusion and energy budget methods. In spite of "satisfactory agreement" of predicted E_t with that observed for field crops but not for forest, and under conditions where soil moisture was not limiting, Tajchman concluded:

> Since the net radiation and ventilation rates in a forest canopy were greater than those of the field crops, it would be natural to assume that there was greater evapotranspiration from the forest. In a similar way Baumgartner [1966], analyzing the net radiation of various cover types, concluded that the evapotranspiration of forests, especially coniferous forests, must be greater than that of agricultural lands. But the differences between the actual evapotranspiration of forest and field calculated in this study were not large. In fact, whereas the potato field vaporized about 14% less water than the spruce forest, the alfalfa used 4% more water than the forest.

Measuring Evapotranspiration

"Direct" Approaches

It is possible, of course, to install a meter in the trunk of a tree, but doing so would break the water column, thereby interfering with the transpiration process. Placing an isolated plant in a pot and controlling the amount of water provided to the soil is a method that was originally used to gain some idea of the amount of weight a plant gains. Isolating the plant, however, removes competition (which the average plant experiences in the real world) and therefore more water is available for transpiration than is needed, resulting in higher-than-actual transpiration. In addition, this type of instrument, known as a **phytometer**, may hold excess water in the interface between the walls of the pot and the soil, which again increases the amount of water available for transpiration. Neither of these methods is completely satisfactory for evaluating the amount of transpiration.

Putting dissolved materials such as salts into the soil water and evaluating the rate of translocation produces results that are suspect because the emplacement of dissolved materials increases the nutrient diffusion gradient which, in turn, undoubtedly influences the transpiration rate. Similarly, applying a **heat pulse** to the tree stem and twigs has been used along with subsequent temperature measurements along the stem to determine how rapidly the heated translocation water moves. Here, insertion of the temperature probes introduces air that interferes with the transpiration process. There is a limitation on the size of stem that can be monitored with any reasonable degree of confidence by this method. Measurements made on twigs appear to be well correlated with reliable results from other methods. Transfer of these transpiration rates in twigs to whole trees is difficult and risky (Swanson 1972).

One method that has been used to successfully measure transpiration has been covering the plant with a plastic **evapotranspiration tent**, in which the temperature, pressure, and relative humidity are all carefully controlled. In an experiment in Arizona (Mace and Thompson 1969), it was determined that mature Tamarisk and Mesquite transpired from 0.016 in. to 0.67 in. per hr (4 mm to 17 mm). The method had been subject to considerable criticism owing to potential influence of the tent itself on the transpiration rate, but it was reported by these authors that careful control of and testing of the incoming and outgoing air, as well as maintaining the tent fully occupied by vegetation, kept the error term (which varies with certain atmospheric conditions within the tent) quite low. This type of installation is quite expensive, difficult to control, and has not been reproduced widely.

"Indirect" Approaches

Adaptations of the energy balance method (Chapter 3) have been derived to accommodate transpiration as well as evaporation. The "method is complicated by the difficulties of evaluating the needed items for the solution of the energy-balance equation, including such items as atmospheric radiation, long-wave radiation from the body of water, and energy storage" (Chow 1964); it is complicated still further by the difficulty of evaluating these characteristics within and surrounding forest stands (and other vegetation). A nomogram for estimating evapotranspiration and, consequently, determining how much irrigation of field crops is necessary, was presented by Follett, Reichman, Doering, and Benz (1973). A simple, bookkeeping format that displays accumulated soil moisture depletion is suggested.

Penman (1948) derived an equation for the estimate of daily transpiration, E_t in mm, based on the slope of the saturated vapor pressure curve, A, the accumulated degree-days (sum of the number of days above 32°F times the number of degrees between 32 and the measured high temperature), H, and the daily evaporation, E in mm:

$$E_t = (AH + 0.27E) / (A + 0.27)$$

This is one of several such equations that have been developed for the purpose, summarized by Chow (1964).

The Blaney-Criddle method (Blaney and Criddle 1950) is developed for vegetated sites, and therefore makes use of vegetation type in the determination of the estimated evapotranspiration. It is normally applied to agricultural crops. Estimated evapotranspiration, E_t in inches, is determined from the summation of m monthly values, of the percent of daylight hours (tabulated by month and latitude) during the period, P; the mean monthly temperature, T in °F; and an annual or seasonal consumptive use coefficient (tabulated for the type of crop and length of frost-free season), K:

$$E_t = K \times \sum_{1}^{m} PT$$

A simplified approach was proposed by Decker (1963). The availability of computers has permitted variation of the original Blaney-Criddle method, and other less confusing methods have also been developed. A major shortcoming of the Blaney-Criddle method is that it is restricted to agricultural crops and, especially, irrigated situations in the western United States; it is particularly useful for irrigation planning (Dunne and Leopold 1978).

Penman's approach (Penman 1948) makes use of evaporation from a sunken evaporation pan, ignoring the exchange of heat through the pan walls (either to the air or ground). While this can be in part corrected through the use of pan coefficient, the method does not make use of the fact that evaporation from soil slows down as the soil gets drier. As a result, estimates of E_t by the Penman method tend to overestimate actual rates (Dunne and Leopold 1978).

The 1957 Thornthwaite and Mather model (see below) incorporates reduced evaporation rates as the soil storage declines. All these methods make use of the fact that the "vapor pressure deficit is the dominant meteorological factor directly controlling forest evapotranspiration" (McNaughton and Black 1973).

Combining Direct and Indirect Approaches

Another approach is to combine direct and indirect methods, utilizing one of two forms of the water balance: runoff and other measurements on a real watershed, and a water budget. The water balance equation (Chapter 3) has several terms which may be evaluated: precipitation and throughfall may be sampled by gage installations; evaporation may be determined from pan evaporation and pan coefficients or energy-balance formula; and runoff is measured in its entirety on an **experimental watershed** (one on which there is intensive instrumentation and control, often including a second, paired watershed for comparison of treatment effects on one). The only unknown, then, if the budgeting is done over a period of time when there is no net change in soil storage, is the transpiration term, for which the equation or budget may be "solved." The several uncertainties involved in sampling the precipitation and the throughfall, and in the estimation of evaporation, cast some doubt on the validity of results from this approach, too. However, verification for a particular watershed under study is possible with proper instrumentation. Once a model is fabricated and tested, it's use may be extended to other watersheds, but only with caution, as there may be no chance for verification at the new site.

The "Thornthwaite Water Budget," as it is generally known,[4] makes use of both direct and indirect approaches by combining detailed monthly accounting and a complex equation for predicting potential evapotranspiration on the basis of mean monthly temperature and latitude. The equation, developed by Thornthwaite, et al. (1944), estimates potential evapotranspiration, E_t in cm, from mean monthly temperature, T in °F, a temperature efficiency index, TE, and the sum of the 12 monthly values of i, thus:

$$i = (T/5)^{1.514}$$

and

$$E_t = 1.6\,[(10T)/TE]^a$$

---

[4] Originally, the "Thornthwaite and Mather Water Balance," as developed (at the Drexel Institute Laboratory of Climatology at Centerton, NJ), published, and referred to previously.

where *a* is a logarithmic, third-order polynomial expansion of *TE* with correction factors based on day length, as determined by a series of 12 monthly equations[5] based on latitude (Thornthwaite and Mather 1957). The Thornthwaite Water Budget is quite complex and has been computerized by Black (1989b).

The Thornthwaite water budget method is based upon this complex formulation and is useful because: (1) annual budgeting eliminates or minimizes the influence of soil moisture storage change, (2) it allows for ready availability of and ease of handling monthly data, and (3) it eliminates the influence of vegetation by consideration of the single estimated term as *potential* evapotranspiration: if the method can effectively predict what the monthly runoff would be *in the absence* of vegetation, then introduction of effects of vegetative manipulation on a watershed can be evaluated (below).

The experimental watershed may be used in two other ways as well. First, the evaporative-type losses may be combined in the term "E_t" and the expression "*P-RO*" (Chapter 3) may be used to evaluate them all together over a period of time during which change in storage (*S*) is presumed to be 0 (chapter 3). This approach has a severe drawback because it assumes that all precipitation that does not go into runoff will go to evapotranspiration; some of it may at least temporarily go into storage on the watershed in any of a variety of forms (snow, depression, soil, or ground water). If monitored, however, the effect of change in storage can be accounted for in the *P-RO* expression.

The second way involves the use of deliberate modification to the watershed in such a way as to affect only the transpiration rate (Goodell 1966). This is what was done at the Coweeta Hydrologic Laboratory. On a 38-acre (15.4 ha) watershed, all the forest vegetation was cut, but not removed at the start of the growing season (so, presumably, the intercepting vegetation remained the same for at least the first summer or until the leaves dried and fell off). The watershed runoff had been monitored for 6 years to calibrate the relationship between precipitation and runoff under natural conditions. The change in runoff could then be measured, compared with the unaffected runoff on the control watershed, and evaluated statistically. Results from this study indicated that about 15 additional in. (381 mm) of runoff were added to the annual streamflow. Annual precipitation at the Coweeta Hydrologic Laboratory is about 80 in. (2032 mm), almost all in the form of rain, and there obviously is an abundant supply of water. Varied and conflicting interpretations of these (and other) experiments were the result of imperfection in data collection, analysis, and reporting; Penman (1963) set forth several views that disagreed with the published results, but failed to draw his own conclusions, perhaps because the situation as he was able to tell from the reports was not completely clear.

In another study, converting Chaparral to grass in Arizona resulted in a 12-in. (305 mm) increase in streamflow (Hibbert 1971). These results are reliable for the particular watershed that was studied, and not necessarily valid for nearby watersheds (as was shown by experiment), much less on other areas. It is, of course, rather impractical to conduct this type of study on every watershed to determine transpiration. Usually, a combination of methods is used to arrive at a satisfactory estimate of the amount of transpiration.

---

[5] Not shown. The coefficients for the equations were determined for computer application of the Thornthwaite Water Budget, and may be found in Black (1987).

As pointed out earlier, a large amount of the information on transpiration comes from interpretation and extrapolation of the data obtained from research of this type. Vegetative cutting and land use studies have been conducted at almost all the major experiment stations around the Nation and in other countries. Additional studies have been conducted at university experiment stations as well. Practical use of this information has been increasing stream yields for years (Goodell 1966; Rich 1972). For example, commercial forest operations are practiced on many municipal watersheds throughout the country. Conversely, increased forest transpiration "greatly speeds streamflow recession" (McGuinness and Harrold 1971; Federer 1973): "rapid recession occurs when transpiration over the whole watershed removes soil water that would otherwise drain and become streamflow." The seasonal timing of soil moisture is an influencing factor (Hornbeck 1973) and is elaborated upon in Chapter 5.

USING THE WATER BUDGET

Another generalized approach, as suggested above, is to monitor or model the water budget (based in part on the water balance approach) discussed above and to evaluate how closely the runoff predicted by the model emulates the observed streamflow. Three particularly successful methods have been developed to carry out this type of analysis.

Plots

Monitoring the soil moisture on an intensive basis on an homogeneous plot, and correcting for percolation (downward movement of soil water after it has infiltrated the soil), deep seepage (movement to ground water, if any), and precipitation inputs, can provide information on how much and how rapidly water is removed by the vegetation. Until the development of the "high tech" neutron soil moisture meter with its high degree of accuracy (Chapter 5), soil moisture measurements involved removing volumes of soil for moisture content determinations. These not only destroyed the site and interfered with subsequent soil moisture movement and storage, they were time-consuming and subject to considerable error (Hewlett and Douglass 1961). If the method is used on a controlled watershed, however, the entire water balance equation may be utilized to aid in the evaluation of the transpiration component.

One such study at 8400 ft (2560 m) in Utah concluded from soil moisture depletion data that removal of aspen might reduce annual evapotranspiration by about 6 in. (152 mm) in a 9-ft (2.7 m) soil profile (Johnston 1970): annual evapotranspiration losses from bare, herbaceous, and aspen/herbaceous plots were 11, 15, and 21 in. (279, 381, and 533 mm), respectively.[6]

Lysimeters

A combination of plots and experimental watersheds is the concept behind the lysimeter, a large, self-contained, artificial "pot" or a glorified phytometer (which usually has only one plant whereas a lysimeter may have many). A total of 26 lysimeters were built flush with the

---

[6] Note that 6 in. equals ½ af/A/yr, no small amount!

ground surface during the Depression at the Forest Service's San Dimas Experimental Forest in Southern California (Patric 1961). Surface dimensions were 10.5 ft x 20.8 ft (3.2 m x 6.4 m), and each concrete unit was 6.0 ft (1.8 m) deep. Carefully replaced soils contain a variety of soil moisture sensors, and surface- and subsurface runoff flows into containers that are serviced via an access tunnel.

The extensive research program at San Dimas has yielded important results:

1. Regardless of amount of rainfall, the bare soil never absorbed more than 9 in. (229 mm) of rainfall.
2. Each plant cover tested has increased infiltration. During the dry years scrub oak was most effective, but in the wet years infiltration was greatest under pine.
3. Seepage yields invariably were much greater under grass than under any other plant cover tested.
4. Annual evapotranspiration losses from the grasses tested varied but little from the 16-in. (406 mm) average shown in the tables, though Colman-unit[7] readings indicated that additional water was available in the lower 2 ft (3.6 m) of lysimeter soil.
5. The experiment has not revealed how much water the woody plants might use if soil moisture were available throughout the year (Patric 1961).

But water held in even these large lysimeters would not be held in their absence. Thus, even here, with elaborate and expensive installation and monitoring facilities, the method of research affects the result by influencing the amount of water available. As a consequence, the type of vegetation that will grow is affected: even without trying to alter the vegetative types, the vegetation that established itself on the lysimeters was different from the surrounding native species. For example, Mustonen and McGuiness (1967) reported that lysimeter evapotranspiration was higher than that measured from natural watersheds owing to the artificial nature of the boundaries of the lysimeters themselves.

Lysimeter studies with woody species agree closely with estimates and measurements of other methods, but again artificiality and assumptions of translation of data to an areal basis provide a continuing problem with precise determination and utility (Fritschen, Hsia, and Doraiswamy 1977). During the three years reported (1972-1974), evapotranspiration was 55 percent of total water applied (precipitation and irrigation) to Douglasfir and western white pine, evaporation was 21 percent of precipitation, and the evaporated portion of amount intercepted was 32 percent.

Experimental Watersheds

Modeling the water budget on an experimental watershed may be accomplished with the aid of the Thornthwaite equations (above) developed originally for agricultural vegetation on a daily basis in southern New Jersey. Using Thornthwaite's budgeting process (Thornthwaite and Mather 1957) for estimating resultant runoff from the estimated evapotranspiration and other water balance components, one can make specific use of the fact that the equations do *not* include vegetation effects. Thus, by calculating estimated evapotranspiration on bases

---

[7] A device used to monitor soil moisture by electrical resistance, see Chapter 5.

other than vegetation influences, and by observing or manipulating vegetation intentionally, the difference between estimated and actual runoff can be used to evaluate the effects of the vegetation change and, as a consequence, used to estimate transpiration. Theoretical evaluation of the ability of vegetation to extract moisture from soils in the root zone by modeling of the root/soil interface have supported observed transpiration rates (Molz and Remson 1970).

The experimental watershed approach may be extended to field watersheds that have been gaged and calibrated. R. Lee (1970) calculated that "observed water balances probably are accurate only to ±13-30 percent of yield" and that, in fact, the uncertainty may be greater than those limits, and argues for combining approaches to hydrologic research. Hewlett (1971) points out that "many basic questions about hydrologic processes on watersheds will never be answered by catchment experiments," but that with a better balance of research input from other types of facilities, "perhaps the catchment experiment remains the surest way to furnish each region with practical knowledge of local forest-water relations." The method can also be used to evaluate the influence of changes in land use on runoff behavior (Black 1968).

The Thornthwaite Water Budget is useful in research, management, and in the classroom. A flexible computer adaptation (Black 1989b) allows computation of either the mean annual water budget or continuous annual water budgets, with a component summary for the period of record (up to 40 years). Both versions allow (arbitrary) characterization of the watershed based upon various land use and hydrologic characteristics such as primary location of the land use, slope, shape, and included water bodies. The mean annual version provides optional plots of the climatogram and comparison of predicted and actual streamflow in addition to recalculations with new effective soil depths and latitude, which reflects average watershed aspect. The method's major deficiency is its temporal imprecision, which can be up to two months: heavy precipitation at the end of one month, for example, may not show up as runoff until the following month, a fact not necessarily reflected in the tabulated values.

The formulas for the computations of the several water budget components are shown in Table 4-4, and a sample output for Poughkeepsie, NY is shown in Table 4-5. The climatogram for these data are shown in Fig. 4-4. An important observation by Thornthwaite from the climatogram is that the annual monthly march of precipitation does not match the annual monthly march of evapotranspiration. Any deficiency in potential evapotranspiration must either be made up from water held in storage or is not met at all. Usually, there is sufficient moisture in the soil to meet the potential evapotranspiration demand, but the water budget shows that in the event that sufficient moisture is not present, and that annual evapotranspiration exceeds annual precipitation, the soil moisture normally does not get recharged: there is, during the normal year, no excess of moisture available.

Identification of the **hydrologic seasons** is an important consequence of analysis of the climatogram. They are based on the **Water Year** or **Hydrologic Year**. The accounting period commences on October 1 throughout the temperature zone in the Northern Hemisphere, and is the time at which, following the summer period or *Season of Maximum of Evapotranspiration* (also known as the *Season of Maximum Soil Moisture Utilization*), the soil moisture is drawn down to its lowest value of the year. The value of the moisture remaining in the soil on this date is relatively constant from one year to the next; thus, it provides a useful date upon which to base accounting for the annual water budget. The Geological Survey reports its runoff data in Water Year format.

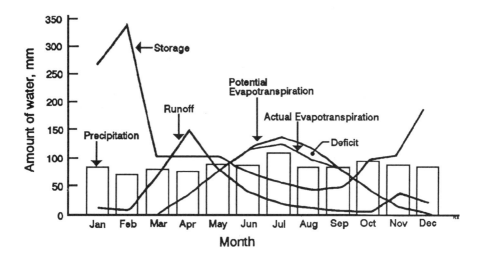

Figure 4-4 Climatogram for Poughkeepsie and Wappingers Falls, NY (data from Table 4-5)

The hydrologic year actually commences with the *Season of Soil Moisture Recharge* that lasts until (a) the soil moisture removed during the summer to make up any deficit is replenished, or (b) until a new *Season of Maximum Evapotranspiration* begins in the following year. After the soil moisture is replenished, there may be sufficient excess moisture to provide either runoff or ground water recharge, hence the third hydrologic season is known as the *Season of Maximum Runoff* or the *Season of Ground Water Recharge*.

There are always seasons of soil moisture utilization and recharge, but if the latter is not completed, there may be no season of maximum runoff. The Thornthwaite model shows, however, that it is possible to have insufficient annual precipitation to meet the evapotranspiration demand (annual PMPET is negative) and still have the soil completely recharged, depending upon temporal distribution: such is the case for the 1966-1970 period for Bountiful, Utah, where heavy snowmelt provides enough water to recharge the soil in spring, yet leave a deficit in summer (Black 1989b). Other configurations are possible as well.

The degree to which annual precipitation exceeds evapotranspiration and vice versa is related to vegetative type along a transect at 41°N in the United States, as shown in Fig. 4-5. The X-axis represents the degree to which the combined effects of annual precipitation and evapotranspiration, constitute a **Moisture Index** (I_m), identified more specifically as an **Aridity** (I_a) or **Humidity Index** (I_h). Timing, sequence, and amount of the relative deficits and surpluses play a role in finally characterizing the relationship between vegetation and available moisture.

Table 4-4 Sources, Formulas, and Explanations for the Computer Calculation of the Thornthwaite Water Budget

Term	Program Label	Source, formula, or explanation
Temperature	TDEGF	Mean monthly, in degrees F
Precipitation	PPTIN	Mean monthly, in inches
Precipitation	PPTMM	Mean monthly, in millimeters
Runoff	MROIN	Measured runoff, in inches (optional)
Temperature	TDEGC	Mean monthly, in degrees C
Heat index	HEATI	$i = (T/5)^{1.514}$
Unadjusted potential evapotranspiration	UNPET	= antilog $[0.012 - 0.0245\, I + (0.46745 + 0.01702\, I) \log T]$ (if $T < 0°C$, UNPET = 0; if $T > 26.5°C$, UNPET = 4.5; if $0 < T < 26.5°C$, use formula)
Correction factors	CORFA	$= a + b\, LAT + c\, LAT^2$

[Month	$a =$	$b =$	$c =$	Month	$a =$	$b =$	$c =$]
[JAN	31.4	-0.16542	-0.00170	JUL	31.1	-0.12850	-0.00118]
[FEB	28.2	-0.04519	-0.00087	AUG	31.1	-0.09894	-0.00029]
[MAR	31.2	-0.01318	0.0	SEP	30.7	-0.01346	-0.00066]
[APR	30.6	-0.02313	-0.00107	OCT	31.2	-0.05682	-0.00170]
[MAY	31.2	-0.10770	-0.00117	NOV	30.3	-0.08208	-0.00126]
[JUN	30.2	-0.11555	-0.00165	DEC	30.5	-0.04299	-0.00298]

Term	Program Label	Source, formula, or explanation
Potential evapotranspiration	POTET	= CORFA x UNPET
Precipitation-minus-E_t	PMPET	= PPTMM - POTET
Accumulated potential water loss	ACPWL	Dependent upon Soil Depth (S), air temperature, and PPTMM<>POTET
Soil storage	STRGE	= antilog $[\log S - (0.525 / S^{1.03710}) \times \text{ACPWL}]$
Change in storage	DELTA	Sequential difference in values of STRGE
Actual evapotranspiration	ACTET	= PPTMM + DELTA
Deficit	DEFIC	= POTET - ACTET
Surplus	SURPL	= PPTMM - ACTET (or, if excess goes to replenish STRGE, = PPTMM - (ACTET + DELTA)
Water runoff	WATRO	Half of current + half of previous months', etc.
Snow runoff	SNORO	Ten percent of current + half of previous months', etc.
Total runoff, mm	TROMM	= WATRO + SNORO
Total runoff, in.	TROIN	Total runoff, in inches (to compare with MROIN)

Source: Black (1989b)

Table 4-5 Thornthwaite Water Budget for Poughkeepsie and Wappingers Falls, NY

MEAN ANNUAL WATER BUDGET

Poughkeepsie and Wappingers Falls, NY
1928 to 1967

```
APL Version, 1985, based on          Weather Bureau Station No.: 3725
Thornthwaite and Mather, 1957        Streamflow gage No.        : 6817
(Data in mm. or in. indicated)       Latitude, degrees North    : 42
(Zero-values suppressed)             Soil storage capacity, mm. : 100
```

Component	Jan	Feb	Mar	Apr	May	Jun	Jul	Aug	Sep	Oct	Nov	Dec	Year
TDEGF	22	24	30	45	55	65	70	68	60	50	39	27	
PPTIN	3	3	3	3	4	3	4	3	3	4	3	3	40.4
MROIN	3	2	4	5	2	1	1		1	1	2	2	23.6
TDEGC	-6	-5	-1	7	13	18	21	20	16	10	4	-3	
HEATI				2	4	7	9	8	6	3	1		38.9
UNPET				1	2	3	4	3	3	2			
CORFA				33	38	38	39	36	31	29	25		
POTET				36	76	115	136	119	79	44	12		617.0
PPTMM	84	71	79	76	89	86	109	84	84	94	86	84	1026.2
PMPET	84	71	79	40	13	-29	-26	-35	5	50	74	84	409.2
ACPWL						-29	-55	-90					
STRGE	268	339	100	100	100	75	57	40	45	95	100	184	
DELTA						-25	-18	-17	5	50	5		
ACTET				36	76	112	127	101	79	44	12		586.7
DEFIC						3	9	18					30.3
SURPL			79	40	13						69		200.7
WATRO	9	4	42	41	27	13	7	3	2	1	35	17	200.6
SNORO			24	107	54	27	13	7	3	2	1		238.3
TROMM	9	4	65	148	81	40	20	10	5	3	36	18	439.0
TROIN			3	6	3	2	1				1	1	17.3

EVAPOTRANSPIRATION RESEARCH

Much of the research on evapotranspiration and vegetative management reviewed here has actually been conducted with a variety of approaches. Therefore, the review that follows is not classified according to the physical type of research *facility*, but according to General Studies, Vegetation Manipulation, Field Investigations of Hydrologic Impacts, and Regional Management Potential for Water Yield Manipulation.[8]

General Studies

A two-year replication of July, August, and September pan evaporation and bog evapotranspiration as estimated by Penman showed that while the two agreed closely, Penman's estimation of potential evapotranspiration did not fit (Sturges 1968); the complex interrelationships between insolation, vegetation development, and size of lake may result in

---

[8] Since many of the research reports presume a wider knowledge of hydrology than the first reading of this book has so far provided, the reader may wish to skip (or skim) this section and return to it after having read Chapters 5, 6, and 7.

reduced evaporation from large lakes (Idso 1981). Errors of all kinds have been noted to lead to erroneous results in lake water balances (Winter 1981). E_t from the bog was "never less than pan evaporation, and averaged 27 percent more than pan evaporation on the 6 measurement days." Reasons given for the discrepancy, and for nonconformity with other studies, included the fact that the peat surface is a good energy sink, there is greater peat-atmosphere surface contact owing to micro-relief, and E_t was limited by atmospheric conditions, not water availability (Sturges 1968).

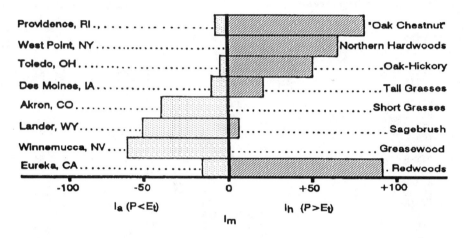

Figure 4-5 Moisture Index and vegetation along 41°N latitude in the United States

Conclusions of interception and evaporation rate studies in a mixed evergreen forest canopy in New Zealand (Pearce, Rowe, and Stewart 1980) were:

1. The fact that nighttime wet canopy rates were not significantly different from daytime wet canopy rates suggests that "wet canopy evaporation is driven by advected energy and not by the radiant energy balance."
2. Annual nighttime interception was 40-50 percent of total gross precipitation, and about 30 percent of the total evaporation: it was about 9 in. (229 mm).
3. Net interception loss was about 81-84 percent of gross precipitation, and about 52 percent of total evaporation.
4. In areas where substantial amounts of annual rainfall occurs at night, reduced convective activity may enhance the amount of the loss.

Evapotranspiration was estimated by a mass-transfer equation for prairie potholes in North Dakota (Eisenlohr 1967). These features, common to a large part of the state, are often filled with emergent vegetation that exhibited E_t rates higher than by evaporation alone from clear ponds when the vegetation stood above the water surface, but could be less during May, July, and August owing to shading and "by reducing air movement at the water surface." Water loss monitoring was confounded, however, by the possible increase in infiltration through the bottom of the ponds caused by the presence of vegetation.

Evapotranspiration rates approximate theoretical rates if the amount of water root zone is not limiting. This has been the conclusion of several studies, including those on Douglasfir stands (T. A. Black 1979), and for vegetated surfaces, generally, as compared with bare soil (Eagleson 1978b).

Proximity of vegetation to streams and location of precipitation has been reported to be of importance in protecting stream water quality and in buffering increased peak flows that are produced by disturbed areas (Curtis 1966). This concept is built into many states' forest practice act legislation and/or regulations, and into Federal guidelines for Best Management Practices (see Chapter 7).

One of the pitfalls of hydrologic research in general is that many measurements are only taken and observations made during normal working hours: not a great deal is known about what happens during the other two thirds of the day. Further, variance of soil moisture is likely to be greatest during the day owing to varying demand by evaporative processes on the water in storage in the soil: thus, daytime data for E_t estimates are, in all likelihood, subject to greater error. Interest in how soil moisture variation varied between day and night prompted the 24-hr study in the Coast Redwoods reported below in the section on Water Stored in Vegetation.

Sampling all possible times and environmental conditions is clearly impossible. One of the research challenges, then, is to identify those conditions which will result in extremes of hydrologic behavior and variation of the parameter under study, and determination of relevancy of varying influencing factors under those particular conditions. This may be a serious limitation on all hydrologic research, and should be continuously kept in mind.

Vegetation Manipulation

The relationships between vegetation and runoff have been written about for more than 100 years, and intensively studied for considerably less. Marsh (1874), for example, made extensive observations around the world on how removal of vegetation led to increased runoff, erosion, and sedimentation.

Deforestation was a principal reason for the increased runoff that resulted in the 1881 Johnstown Flood in Pennsylvania. Protection of water supply was a prime force behind the creation of the Adirondack and Catskill Forest Preserves in New York in the nineteenth century. And the importance of forest vegetation and management was written into the 1911 Weeks Forest Purchase Act (36 Stat 961) that launched the Federal government into the forest management business. In addition, much of the justification behind early forest fire control and prevention, and "good" forest management practices, were aimed at improving (or maintaining) the wildland water resource. The research that lent support to the Weeks Act was conducted at Wagon Wheel Gap in Colorado long before its official publication (Bates and Henry 1928): cutting the forest vegetation on one watershed caused an increase in runoff as compared with a control on which no vegetation was cut.

As a consequence of early promising research results, articles appeared as early as the late 1940s and early 1950s on how small forested watersheds might be managed for improved water yield (e.g., Wilm and Dunford 1948; Weitzman and Reinhart 1957). Hewlett (1958) pointed out that the Coweeta studies were of importance to the water supplies of over fifty million people for the possibility of changing vegetative cover might improve water yields without degradation of water quality. The many locations of such research in the United States were identified and discussed by Storey (1959), who also pointed out that in some cases the water management goals might be in conflict, notably flood control which requires maximum storage potential while E_t reduction increases on-watershed storage for later runoff. The opportunity for simultaneously managing water and timber resources, long officially sanctioned by virtue of the Weeks Act, was given added impetus in the 1960 Multiple Use and Sustained Yield Act (74 Stat 215) for the National Forests (Woods 1966).

The role of partial cutting to increase water yields from eastern hardwood forests was discussed by Tryon (1972). Under such conditions, he pointed out, there are two problems: one is that full stocking will be re-attained by the residual stand, and the second is that reproduction may re-occupy the site. Both of these will, of course, reduce the newly available light, nutrients, space, and water which were released by the partial cutting. The first problem might be alleviated by more frequent cuts, or "deep" cuts into the stand's basal area, or by keeping the residual stand in as large a size class as possible. The second problem might be partially resolved by "establishment of a dense grass sod."

The effect of vegetative manipulation, especially forest cutting, on stream behavior can be classified into effects on the storm hydrograph (see Fig. 1-4), effects on the annual hydrograph and water budget, and effects on water quality. Some of the studies reviewed here involved investigations of both storm flow and annual hydrograph or water budgets. In much of the literature, "storm flow" is also referred to as "quickflow," often as one word. It may be a combination of surface and subsurface runoff, channel and bank storage, and perhaps include some rapid response from the ground water reservoir as well. The effects on storm flow are less likely to be verified by statistical methods than on either the water budget or water quality.

Effects on Storm Flow

Typical of the many studies on the impact of forest cutting on stream behavior is that reported by Verry, Lewis, and Brooks (1983). Clear-cutting aspen on upland portions of watersheds in Northern Minnesota caused snowmelt peaks to increase from 11 percent to 143 percent. Rainfall, which will reach the watershed outlet much more rapidly than snowmelt, produced peaks up to 250 percent higher. The volume of storm flow increased as much as 170 percent. Increased volumes decreased to pre-harvest levels by the third year, while increased peak flows persisted for nine years. Similarly, clear-cutting and clear-cutting with wildfire produced increases in peak snowmelt rates of 42 percent and 30 percent, respectively (Megahan 1983). On an adjacent watershed which burned only, no increases were detectable, even though the unsaturated soil moisture content was increased substantially.

In a study on a 108-acre (43.7 ha) hardwood-forested watershed with adjacent control at the Coweeta Hydrologic Laboratory, no forest material was removed. Comparison of runoff behavior before and after treatment and with the control watershed showed only a slight increase, about 7 percent, in peak discharge rate, while storm flow volume increased a statistically significant 11 percent (Hewlett and Helvey 1970). Hornbeck (1973) reported somewhat similar results from the Hubbard Brook studies in New Hampshire, where "the absence of the hardwood canopy ... caused earlier and more rapid snowmelt and affected most spring storm flow events involving snow water. In contrast, storm events occurring after soil moisture recharge (see Chapter 5) in the fall and before the start of spring snowmelt were unaffected by forest clearing."

These results support the findings of Rothacher (1965), who reported that the increases in runoff following forest cutting occurred when it rained, not necessarily during the period of high E_t between the storm events and when, presumably, the increased moisture available for runoff became available through reduced transpiration. Troendle and King (1985) and Rothacher (1970) also point out that annual increases due to forest cutting occur during the runoff season, not necessarily during the season of maximum E_t (Chapter 5) when the actual change in transpiration occurs.

Changes also occur in the timing of peak flows. Results from the Mountain Farm and other land use studies at Coweeta show dramatic decreases (from 45 minutes to 10 minutes or less) in the lag time between start of precipitation and peak flow occurrence for cleared mountain watersheds (Dils 1953). Troendle and King (1985) indicate that the most persistent change in stream behavior 30 years after the cutting studies on Fool Creek in the Colorado Rockies is the advance of the annual peak flow, an average of 7.5 days. In an hydrologic environment where a great percentage of the annual flow occurs within a very short duration, typically two to three weeks, that much of a change can mean considerable opportunity for control over streamflow (see Chapter 6). Patric and Reinhart (1971) also reported increases in instantaneous (peak) flows following deforestation in West Virginia.

To illustrate the complexity and uncertainty of this type of research, note that: (1) E. L. Miller (1984) reported that timber harvesting in Hawaii resulted in *lower* storm volume runoff, probably owing to post-timber cutting treatment and regrowth that transpired more soil water; (2) Ziemer (1981) reported that selection cutting and tractor yarding of 85-year-old second-growth Douglasfir and redwood had *no* effect on large storm peaks, although "the first streamflow peaks in the fall ... were increased about 300 percent ..."; and (3) Harr and McCorisin (1979) reported that annual peak flow from a small watershed in western Oregon exhibited a *reduction* of 32 percent in peak storm flow following clearcut logging of a 450-year-old Douglasfir stand, and a delay of all peak flows of 9 hr. It may be that the porous volcanic soils played an important role in this study. All three studies highlight the difficulty of generalizing the results of research on forest cutting and the impact on storm flow characteristics. The same, by and large, may be said for the effects on annual water budget components. Throughout, basic principles appear to be well founded, but it is difficult to fully identify and evaluate the factors that affect the particular situation. It is apparent from the many studies that effects of forest cutting is greatest for relatively short time periods, on small watersheds, and in the local area.

Effects on Annual Water Budget

The study of clear-cutting and continued slash cutting with no removal of forest products on Watershed 17 at the Coweeta Hydrologic Laboratory was the first watershed experiment of its kind: it provided an extensive period of calibration of both control and treatment watersheds, and thorough instrumentation of meteorological and hydrologic phenomena in the vicinity. Instrumentation and calibration commenced in 1934, and the first treatment was done in 1940. Increases in annual E_t amounted to approximately 40 percent over pre-treatment yields the first year, and were sustained at about 16 percent over the period of the study. Most of the increases which occurred in the months following soil moisture recharge in the fall. This was a north-facing watershed, however, and rarely received direct solar beam radiation.

In contrast, a 108-acre (43.7 ha) south-facing watershed, clear-cut in 1963 and with slash reduced rapidly to simulate levels as that on Watershed 17, showed an 11 percent increase in storm flow. Added up over the year, however, this was a barely detectable increase in annual yield (Hewlett and Helvey 1970). On this south-facing watershed, solar energy was received either on the tree crowns prior to cutting or on the forest floor or herbaceous vegetation after cutting. Since E_t is dependent upon energy available, the E_t losses did not change much after cutting. In contrast, on north-facing Watershed 17, the solar energy was effective only when it was received on the forest canopy prior to cutting when taller trees near the bottom of the slope transferred the energy received to the soil water reservoir.

Other studies at Coweeta showed consistent results, with forested watersheds showing 50 percent of the increased yields obtained on Watershed 17 if 50 percent of the basal area[9] were cut, etc. On watershed 13, which was cut simultaneously with Watershed 17, the forest was allowed to grow back, which caused the increased annual streamflow to decrease gradually to its pre-treatment level by the time the forest canopy had closed, about 20 years (Kovner 1957). The decline in increased streamflow following treatment was found to be a logarithmic function of years since cutting. The decline in streamflow increases following the second cutting was found to yield about half the increases of the first cutting, and the logarithmic pattern of yield increase decline was confirmed by Swift and Swank (1981).

The same principles were observed in the cutting studies on the Fool Creek Watersheds on the Fraser Experimental Forest near the Continental Divide in Colorado. Here, however, the precipitation is largely in the form of snow, in contrast with the rainfall of the southern Appalachians of Coweeta. The original studies at Fraser were to evaluate the effect on snowpack water equivalent (water available for runoff), not for the effect on runoff itself (Wilm and Dunford 1948). In the 30-year evaluation, Troendle and King (1985) reported a persistent 9 percent increase in average peak water equivalent over the entire watershed. The increase is attributable to both changes in the depositional patterns of snow during precipitation events, and to the change in E_t following cutting. The former does not appear to have changed greatly over the period of (slow) forest regrowth, and the lion's share of the persistent change is, therefore, attributable to the deposition pattern. As with other studies, this is testimony to the difficulty of evaluating E_t!

The major cutting studies at Hubbard Brook in the White Mountains of New Hampshire have been reported by Hornbeck (1973). Forest cutting was followed by applications of herbicides, and the Hubbard Brook studies are more known for their controversy over nutrient relations following such treatment than for their contribution to forest hydrology. The reduction in transpiration and interception loss as a result of the cutting resulted in wetter sites that are less capable of storing rainfall and snowmelt. With no annual soil moisture deficit, no increase in annual yield was detected (Hornbeck 1973).

On drier sites, on an Arizona watershed, overstory kill of Pinyon-Juniper with herbicides caused an increase in streamflow of 157 percent, and streamflow returned to pre-treatment

---

[9] The sum total of cross-sectional area in sq ft of all woody stems over a certain limit (e.g., 0.5-in. diameter, breast height) per acre.

levels after 8 years (Baker 1984). Conversion of 55 percent of an Arizona chaparral watershed to grass in a mosaic pattern produced an increase of 68 mm of runoff per year for 4 years; consideration inherent in the mosaic pattern was given to wildlife habitat, water quality, and landscape esthetics (Hibbert and Davis 1986).

Cutting studies at the Fernow Experimental Forest in West Virginia included clear-cutting, a diameter limit cutting, and extensive- and intensive-selection cuts. The results are summarized in Table 4-6. In addition to changes in annual yield, there were reductions in the quarter, half, 1 percent-, and 5 percent-flow intervals, all of which indicate a **flashier** streamflow, that is, one in which highs are higher and lows are lower than normal. Similar results have been reported in a variety of other studies.

Table 4-6 Summary of Changes in Streamflow Under Different Conditions of Timber Cutting at the Fernow Experimental Forest, West Virginia

Water-shed No.	Treatment	Flow Interval (days)		Flow Interval (%)	
		quarter-	half-	1%	5%
1	Clear-cut	3.6	9.8	31.9	37.0
2	Diameter Limit	3.2	1.5	23.6	21.8
3	Extensive Selection	4.0	2.7	20.5	5.9
4	Intensive Selection	1.0	3.6	3.2	6.3

Sources: Reinhart, Eschner, and Trimble (1963); Troendle (1970).

Increases in water yield following commercial timber harvesting on a 237-acre (95.9 ha) watershed at the H. J. Andrews Experimental Forest in the Cascade Range of Western Oregon were about 18 in. (457 mm) per year, with 80 percent of the increase occurring during the soil moisture recharge period between October and March (Rothacher 1970). A watershed that was partially cut showed a proportional increase in annual yield. Streamflows were affected, too, by extensive erosion following the logging and slash burning that is typical of the practices in the area.

Vegetation conversion studies have been conducted at several locations. Swank and Miner (1968) reported on the conversion of mature mixed southern Appalachian hardwoods to Eastern White Pine. Following closure of the crowns, "streamflow steadily declined at a rate of 1 to 2 in. (25 to 51 mm) per year. . . . Most of the reduction in water yield occurred during the dormant season and was attributed mainly to greater interception loss from white pine than from hardwoods." The differences were expected to increase as the white pine matured, which was confirmed and reported (Swank and Douglass 1974): streamflow reductions occurred in every month, from 0.59 in. to 1.37 in. (15 mm to 35 mm). The reduction in runoff on reforested watersheds differs from that on abandoned lands in both degree and timing of crown closure and, consequently, on interception loss (Schneider and Ayer 1961).

Conversion of oak woodland to annual grass resulted in a 4.5-in. (114 mm) increase in water yield in the Sierra Nevada foothills in California (Lewis 1968), and the increases from half-cutover watersheds of eastern hardwoods on the Fernow Experimental Forest were 6 in. and from the completely cutover watershed, 10 in. (254 mm) (Patric and Reinhart 1971). Moisture available for runoff was increased by logging in Engelmann Spruce (Hart and Lomas 1979) and 130-year-old Douglasfir (Harr, Levno, and Mersereau 1982) by 10 in. to 16 in. (254 mm to 406 mm), respectively. Peak flows were not affected in the Douglasfir cuttings, and the detected changes persisted for only about 4 years. In the Engelmann Spruce cutting, changes are expected to persist for up to 50 years. Seven- and ten-year changes in water yield were reported for varying levels of thinning and clearing of Ponderosa Pine in Arizona (Baker 1986).

In a study of three watersheds in "poorly drained flatwoods" of Florida, a 250 percent increase in water yields was observed from lands harvested with "maximum disturbance," while increases of only 117 percent were observed from the lands harvested with "minimum disturbance" (Riekirk 1983). The third watershed served as a control, and the reported water yield changes persisted for only one year.

Ground water levels were shown to be affected by strip cutting of Jack pine in Michigan (Urie 1971). Increases in water available for and measured in wells amounted to an average of 3 in. (76 mm) per year over four years. Runoff was increased after native meadow in Nebraska on a 4-acre (1.6 ha) watershed was converted to cultivation; and, conversely, surface runoff was "significantly reduced" following conversion of "marginal cultivated fields" to perennial grass (Dragoun 1969). The return to pre-cultivation level was complete after only three seasons. On the other hand, impacts of vegetation manipulation on annual water budgets in the lake states lasted for at least 12 to 15 years (Verry 1986).

In a study of the effect of weed removal from a 7-year old red pine plantation in Wisconsin, Lambert, Gardner, and Boyle (1971) report that E_t was reduced from 0.62 in. to 0.35 in. (16 mm to 9 mm) during a period when only 1.06 in. (27 mm) of precipitation occurred: "But nearly all of this excess was lost from the root zone by a 61 percent increase in drainage associated with weed removal, so the increase of water available to plants was not great." This natural root grafting (so that what appear to be many separate individuals actually share a common root system) and the results of several other studies confirm that selection cutting, thinning, or other means of partial cutting to increase water available for streamflow is not likely to succeed if the goal is increased water yield: the residual vegetation quickly takes over the site, making effective and efficient use of the excess light, space, nutrients, and water.

Effects on Water Quality

Comprehension of the magnitude of the water quality problems of wildland management was quite incomplete when the major studies of the effects of vegetative manipulation on hydrologic characteristics commenced in the 1930s. Most of the original concern with wildland vegetation management and water quality focused on sediment production, especially from intensive land uses in highly erodible areas such as the southern Appalachian mountains.

Research at the Forest Service's Coweeta Hydrologic Laboratory was initially aimed at the magnitude of the effect of uncontrolled (and sloppy) land use practices on both hydrology and sediment production (Dils 1957). Thus, the Mountain Farm, Mountain Grazing, and

Common Logging watersheds served as important demonstration projects for the thousands of annual visitors to the 5000-acre outdoor laboratory. By the time everyone knew that such practices were bad, the research thrust was shifted to provide information on how to rehabilitate such abused lands. At the same time in the 1930s, research into evapotranspiration losses was begun on Watershed 17 and its neighbors, as noted above.

It has only been rather recently that additional research has been directed at the impact of logging and other vegetative manipulation on water quality. With the exception of Hubbard Brook, most of the major research stations established around the same time emphasized sediment production and, on occasion, stream temperatures.

A detailed study of the effects of different types of forest harvesting techniques in Texas showed that increased sediment persisted for up to 4 years following treatment, although the level decreased to below acceptable standards before that (Blackburn et al. 1986). Most of the movement of mineral soil came about from overland flow or surface runoff, which did not occur in the control (uncut) watershed. Differences between the several harvesting methods decreased following the first year. Similar research on the Alsea Watershed Study in the Coast Range in Oregon showed dramatic increases in sediment upon installation of roads, but no significant increase in sediment after yarding and felling harvestable timber (Brown and Krygier 1971). The irregular effects of large storms and high variation in sediment yields characterized the results. A field test of planned and careful logging operations on the Chatahoochie National Forest in North Georgia showed that commercial logging operations could be carried out without producing sediment in streams draining the logged area (Black and Clark 1959). And Heede (1987) reported that guantities of sediment and overland flow during five post-harvest years were insignificant. The overall conclusion of these and other studies is that increased sediment production from carefully controlled commercial forest harvest operations need be neither damaging nor long lasting.

Vegetation conversion has shown both changes in annual water yield and nutrient budgets. For example, conversion of chaparral to grass in Arizona increased both runoff and nitrates (Davis 1984). Nitrate concentrations increased up to 100-fold, 29-fold over the 11-year study. Peak runoff occurred earlier and was higher than on the untreated watershed, and base flow recession was quicker. The changes in nutrient concentrations lagged behind the changes in streamflow, and persisted for a longer period of time as well.

How nutrients move out of a watershed (or change within the watershed ecosystems) on which vegetation has been treated has been the subject of much research. A typical relationship between a stream hydrograph and pollutograph is shown in Fig. 4-6. Actually, there seem to be few predictable patterns, although the flushing action of a stream tends to push the pollutograph to its peak prior to the peak of the hdyrograph. Typically, temperature naturally decreases as a flush of cool bank storage water influences overall stream temperature on the rising limb of the storm hydrograph. Following removal of riparian vegetation that shades the stream channel, stream temperatures may be considerably higher (Brown and Krygier 1970). Flushing action has also been described by Kunkle and Meiman (1968), Zebuhr (1968), Likens (1970), Troendle and Nilles (1987), and others. The fact that increases in streamflow following varying degrees of vegetation removal occur at the start of the runoff-causing event support this concept.

The sequence of nutrient removal from a managed stand varies with type of treatment (method of harvesting, slash disposal methods, and use of fire) as well as timing and sequence of operations and vegetative type. For example, nitrate nitrogen concentration in the soil solution increases from 3 to 4 orders of magnitude to as much as 50-fold (Pierce, Hornbeck, Likens, and Bormann 1970) following harvesting in general, accompanied by a decrease in calcium and phosphorous, which tend to increase later as the forest is restored (Haines 1984), as shown in Fig. 4-7. Nutrient losses tend to be low and persistent on heavy-textured (clayey) soils and, if only woody parts are removed, "nutrient shock" may occur owing to shortages of sodium, copper, and zinc (Stark 1980).

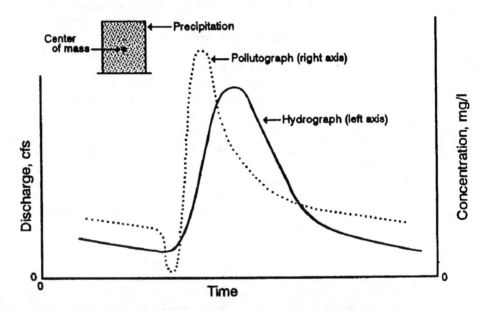

Figure 4-6 Relation between typical hydrograph and pollutograph

The results of nutrient budget studies following harvesting at Hubbard Brook were reported by Martin and Pierce (1979): they included the observation, following removal of all vegetation 5 cm dbh and larger, of up to 16 times reference levels, with both nitrate and calcium returning to reference levels within 5 years. In a detailed study in the lower Coastal Plain in Florida, the most significant nutrient in runoff waters following logging was potassium: it and other nutrients rapidly returned to pre-harvesting levels following treatment (Riekirk 1983). Cutting studies at various sites illuminated "mechanisms that underly forest ecosystem response to disturbance" (Vitousek et al. 1979):

> A systematic examination of nitrogen cycling in disturbed forest ecosystems demonstrates that eight processes, operating at three stages in the nitrogen cycle, could delay or prevent solution losses of nitrate from disturbed forests. An experimental and comparative study of nitrate losses from trenched plots in 19 forest sites throughout the United States suggests that four of these processes

(nitrogen uptake by regrowing vegetation, nitrogen immobilization, lags in nitrification, and a lack of water for nitrate transport) are the most important in practice. The net effect of all of these processes except uptake by regrowing vegetation is insufficient to prevent or delay losses from relatively fertile sites, and hence such sites have the potential for very high nitrate losses following disturbance.

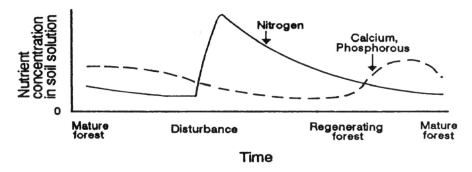

Figure 4-7 Nutrient loss after forest disturbance

Recent research at Coweeta has been directed at determining nutrient budgets of natural as well as disturbed watersheds (Day and Monk 1977; and Monk et al. 1977). Nitrate nitrogen depletion following logging was reported by Swank and Caskey (1982) and illustrates the complexity of the nutrient chemical environment:

> The amount of NO_3-N exported in a second-order mountain stream draining a clearcut and logged mixed-hardwood forest was studied over a 4-year period. Calculations based on measurements of stream chemistry and discharge rates indicated a within-stream depletion of NO_3 from the upper reaches of the stream to the watershed outlet. Within-stream depletion the first year of treatment was 127 percent of total NO_3-N discharged from the watershed outlet and declined in succeeding years to 99, 42, and 5 percent.

The 3.9 Kg/yr of NO_3-N lost by within-stream depletion was found to be more than twice the loss from instream sediment losses.

At Coweeta, Swank and Caskey (1982) also reported that the concentration in runoff waters of nitrate nitrogen and sulphur is a direct function of the basal area removed, with 89 percent of the variation explained by a combination of the basal area removed and the radiation. Swank also reported that small losses of NO_3, PO_4, and NH_4-N occurred in cut white pine stands, and moderate increases in Cl, K, Na, Ca, and Mg.

Predicting nutrient losses is far more difficult than predicting impacts on storm flow or annual water budgets owing to the complexity of factors affecting nutrient concentrations in plant, soil, and stream (Swank and Caskey 1982). Again, nutrients were observed to exit the watershed during runoff events in a natural flushing action (Webster 1984). Swank and Caskey further cite the influence of (a) dynamics of below-ground processes, where a large portion of a site's biomass may be located and not removed in harvesting operations, (b) the role of canopy/atmospheric interactions, (c) incidentals such as debris in the stream and on the forest floor, and (d) nitrification rates, which may be further influenced by (e) rates of regrowth and species uptake rates and (f) the amount of nutrient remaining in the soil, which is a function of natural and disturbed leaching rates. Being called upon to predict impacts of water quality prompted this comment by Fralish (1977):

> I am trying to emphasize that even if forest ecosystems were left undisturbed for the next century, water quality would not improve substantially. In many parts of the eastern United States, the forest is a minor part of the landscape and we may be stepping on ants while being overrun by elephants.

Observations on some Montana clear-cutting studies provide a summary that seems to fit the results of most commercial clearcut operations:

> Clear-cutting and removing the logs using a jammer or similar skidding device from carefully constructed, widely spaced roads, and broadcast burning of the remaining residue result in temporary impairment of watershed protection and attendant increases in overland flow and erosion of soils derived from Belt series rocks on gentle-to-steep slopes in the larch-Douglas-fir type of western Montana. Rapid vegetal recovery ensues, and conditions return to near prelogging status within 4 or 5 years. During the denuded period, a small increase in losses of plant nutrients occurs in the overland flow and sediment. This increase is not considered a hazard to water quality and represents a small fraction of the nutrients available for plant growth on these sites.

Aubertin and Patric (1974) support these findings:

> Clear-cutting had a negligible effect on the stream's temperature, pH, nonstorm turbidity, and concentrations of dissolved solids, Ca, Mg, Na, K, Fe, Cu, Zn, Mn, and NH_4^+-N. Storm-period turbidity, nitrate-nitrogen, and phosphate concentrations showed slight increases, while the sulfate concentration decreased. .... Success in avoiding damage to water quality was attributed to careful road management, retention of a forest strip along the stream, and rapid, lush vegetative regrowth after clear-cutting.

Other studies also conclude that effects are short-lived and often of little or no significance to the receiving aquatic system, including Coweeta in the southern Appalachian mountains (Johnson and Swank 1973), the Cascade Range in Oregon (Beschta and Taylor 1988), the Bull Run watershed in Oregon (Harr and Fredriksen 1988), and central Colorado (Leaf 1974). In summary, the effects of clear-cutting are mimimal, especially with protective and mitigative practices (see Chapter 7) built in. In spite of this conclusion, the high degree of variation of results led Brown (1985) to conclude "it is difficult to predict the exact change that timber harvesting will have on the concentration of nutrients in a stream."

In summary, impacts of timber harvesting on storm flow may be 10 years or less, dependent upon recovery measures. And impacts of such drastic practices as clear-cutting on water quality tend to be rather short-lived in forested watersheds, with effects also likely to be less than 10 years. Overall, forest cutting has more dramatic, more complex, and less persistent effects on water quality than on storm flow or annual water budgets.

Field Investigations of Hydrologic Impacts

Given the large number of gaged watersheds and meteorological stations in the United States, it is possible to study the impacts of land use changes and vegetative manipulation on a large, if somewhat uncontrolled, scale. This is in marked contrast to the detailed control and replicability usually available under conditions of planned research. But the trade-offs include more information on a wider range of sites, greater practicality, and real-world situations. Completely uncontrolled natural events can occur on well-instrumented installations, but these are rare. More promising are the numerous opportunities for the careful researcher to find massive land use and vegetative changes at fortuitous locations.

Natural Events

Documenting natural events requires that the event occur on an instrumented watershed. Hopefully, the event takes place where there has been no or minimum other disturbance, the effects of which could be confounded with the event under study, and on a watershed where there has been a sufficient calibration pre-treatment period. In addition, the natural event should completely cover the gaged watershed. While such conditions do not occur frequently, they have been observed in scattered locations in the United States. Some of them are identified here, as any natural removal of vegetation or foliage without simultaneous disturbance of the site makes it possible to estimate transpiration or net E_t on a broad scale.

Water yield increased dramatically following an outbreak of bark beetle on Engelmann spruce in western Colorado in 1939 (Bethlahmy 1974): extensive blowdown over two of three watersheds that had long-term streamflow records permitted comparing runoff from the uninfested watershed with that from the infested watersheds. Streamflow increases were highest in the fifteenth year following the outbreak, and have shown statistical significance for more than 25 years. Similarly, a mountain pine beetle outbreak in lodgepole pine/alpine fir on the Jack Creek watershed in Montana resulted in 35 percent timber mortality, and a 15 percent increase in water yield (Potts 1984). In addition, comparison of four years of records prior to and five years of records following the outbreak and kill showed the annual hydrograph was advanced an average of three weeks, with only a little increase in peak flows, and a 10 percent increase in minimum flows.

In contrast, forest decline on the island of Hawaii is reported to have resulted in "no significant change in either annual streamflow or peak streamflow" (Doty 1983). It is speculated that the "reduced evapotranspiration resulting from the loss of the overstory may have been offset by increased growth in the understory and a reduction in fog drip."

The well-instrumented San Dimas Experimental Forest in the mountains east of Los Angeles suffered a major fire in 1960 (Crouse and Hill 1962). The effects of removal of vegetation by wildfire on stream behavior in that location are, as a consequence, well-documented and summarized by DeBano, Dunn, and Conrad (1977): it was found that "although some plant nutrients, such as Ca, Mg, and Na, are only released and deposited on the soil surface, others, such as nitrogen and potassium, are volatilized and lost." There were also changes in soil pH and other factors.

Effects of fire on the hydrologic environment by both prescribed and wildfires is reported in the proceedings of a symposium in 1978 (Tiedemann et al. 1979). The results reported

exhibit a high degree of variability in timing, magnitude, and detection of effects, summarized as follows:
1. *Rainfall interception*: "No studies were found that provided direct measures of change in interception loss by fire."
2. *Infiltration*: The most extreme effect here is noted on chaparral soils where, without the protective vegetation, soils become so dry they are unwettable, and all precipitation runs off the surface causing excessive loss of topsoil and sedimentation of streams.
3. *Soil moisture storage*: Destruction of foliage tends to reduce transpiration, increasing soil moisture in the vicinity of the roots, thereby decreasing infiltration and increasing surface and subsurface runoff.
4. *Snow accumulation and snowmelt*: removal of vegetation would accelerate increases in snow accumulation and melt, although no direct studies of the impacts on runoff have been made.
5. *Overland flow*: As noted above, surface runoff is increased following destruction of vegetation, but the results vary with soil texture, slope, and severity of the fire.
6. *Surface erosion*: Sediment production by removal of soil and soil material varies from minimal to extensive, dependent upon form, intensity, and season of precipitation as well as interrelated factors.
7. *Mass erosion*: The gravity-induced movement of large volumes of soil is influenced by fire, especially under conditions of nonwettability of soils. Results may be catastrophic.
8. *Water quality*: Sediment and turbidity loads "are the most dramatic and important water quality responses associated with fire," caused by overland flow and removal of protective vegetation. Stream temperatures are likely to increase, and almost all nutrients and trace metals found in forest ecosystems will increase in both the soil solution and the draining stream.[10]

Man-Made Events

Reforestation of several hundred thousand acres of abandoned farmland in New York (and other parts of the Northeast) has resulted in considerable reduction of streamflows, according to Ayer (1968), who summarized as follows:
1. Reforestation reduced flood peaks on streams draining the study areas for the period November through April an average of more than 40 percent.
2. Total streamflow (runoff) from the three forested areas for the 6-month period November through April was reduced an average of 26 percent of the 24-year average, or about 1½ percent per year.
3. No significant changes with time were found in:

    a. peak discharges for the summer months.

    b. base-flow recession rates.

---

[10] Detailed summary and discussion of these effects are presented in Chapter 6.

c. volumes of direct or overland runoff.

d. annual minimum daily flows.

Results consistent with the foregoing were reported by Black (1970): the long-term reduction in farmland in Central New York resulted in a detectable reduction in flood peak magnitude and frequency, and a shift in time of occurrence that was consistent with soil moisture recharge differences between active farmland and brush lands.

In apparent contrast, Black (1968) reported that there was a long-term *increase* in streamflow from a 180-square-mile watershed in southeastern New York following farmland abandonment. This was attributed to the fact that the abandoned lands were not systematically reforested, but were subject to "natural" reforestation that resulted in widely scattered brush and scrub second-growth that enhanced snowpacks and delayed melt. The results are consistent, however, with changes noted in patterns of snow accumulation and melt in the region as reported by Eschner and Satterlund (1963) under different cover types: scattered brushy hardwoods tend to accumulate the greatest pack and, presumably, have the greatest and most delayed runoff.

Protection of forests in the Adirondack Park in New York State over a documented 39-year period has caused a decrease in annual stream yield of 7.72 in. (196 mm), "largely as a result of a 5.18-in. decrease in dormant season runoff" (Eschner and Satterlund 1966). At the opposite corner of the country, field studies and research studies for means of increasing water yields in Arizona were reported by Ffolliott and Thorud (1977): simple vegetative clearing and complex silvicultural treatments were summarized for increasing water yields in mixed conifer forest, and in Ponderosa pine forests, but limited results were envisaged for Pinyon-Juniper woodlands, and there had been no positive results in aspen and alpine forests. The opportunities for increasing water yields through cutting riparian (streamside) vegetation where greater abundance of water holds promise of greater increases are discussed below (see Phreatophytes).

The results of limited research on impacts of man's activities on stream behavior and hydrology are consistent with the detailed research on small areas at the many experiment stations throughout the United States. The results are encouraging, too, for expanding wildland management to beneficially affect the water resource.

Regional Management Potential for Water Yield Manipulation

In an article introducing a series of reports in the *Water Resources Bulletin* on the state of the art of this topic, Ponce and Meiman (1983) identify four major issues relating to successful implementation of the research on water yield control through vegetative manipulation. First, it has been shown that water yield can be changed through vegetative management. Second, economic values in the joint management of timber and water can be determined so as to permit analysis and planning of management. Third, a variety of models have been developed to aid in the complex task of identifying sequence, magnitude, and location of treatments. And, fourth, legal constraints will likely limit practical management schemes to public lands.

The Sierra Nevada Mountain range of California is highly productive of water and thus has maximum potential for controlling runoff. Small (less than one-half hectare), narrow north/south openings in red fir-Lodgepole pine and in mixed conifer forests have produced up to 40 percent increases in snow water content. Since the openings have to be made in a wider-ranging forest setting, consideration of the entire managed area means increases of from ½ percent to 2 percent. The authors estimate that on the average, a 1 percent increase in streamflow can be realized from National Forest lands by intensive watershed management. Perhaps more importantly, one of the characteristics of snow management in this region is that the peak runoff rate is delayed, a result of spreading out the increased runoff over a longer-than-natural time period. The authors point out that "delayed snowmelt will probably reduce total yield, but more water will be available for storage and use" (Kattelmann, Berg, and Rector 1983).

In the Rocky Mountain Region, as is the case in the Sierras, most of the runoff is produced on a relatively small proportion of the entire region. The runoff occurs over a short time of the year, that is, during the spring/summer snowmelt period, often less than a month, and water yields per producing acre are extremely high. Overall, Troendle (1983) estimates that annual yield could be increased by slightly more than 1 in. "The optimal harvest design appears to consist of small openings, irregularly shaped, and about 3 to 8 tree heights in width parallel to the wind." In his summary, Troendle states that "water yield can be increased significantly through forest management while minimizing the detrimental effects in terms of peak discharges, erosion and sediment production, and channel integrity" (Troendle 1983). In a report focused on the subalpine zone, Troendle and Kaufman (1987) summarize the basic principles of vegetation management to influence water yield as follows:

> Forest vegetation is important in the hydrology of subalpine ecosystems. First, leaf surface area of the canopy presents a massive intercepting surface to both rain and snowfall, much of which subsequently is evaporated back to the atmosphere. The same canopy biomass also transpires significant amounts of water, thereby depleting soil water reserves and increasing storage capacity for subsequent rain or snowmelt that reaches the soil. Reducing the canopy biomass by either partial or clear-cutting decreases the interception loss, decreases overstory transpiration, increases understory water use, decreases soil water depletion, and may increase total streamflow, peak flow, and base flow. Increased stand density has the reverse effect.

A third major water-producing area of the nation is the mountain ranges in western Washington and Oregon. Here, research indicates that "estimated sustained increases in water yield from most large watersheds subject to sustained yield forest management are at best only 3-6 percent of unaugmented flows" (Harr 1983). This is due to the already high runoff as a result of orographic precipitation: the climate here already dominates the hydrologic patterns, and little can be done by man to augment water supplies.

The eastern hardwood forests, especially those in the Appalachian Mountains, are also major producers of runoff, although not of the same magnitude as the western mountain ranges. Here the change in water yields from forest management have been much more variable than for the western studies due, in part, to the greater forest ecosystem complexity and variability, and in part to the fact that precipitation occurs in the form of both rain and snow (in the inland western mountains, it is mostly snowpack that is managed, and that form is much better buffered from the extremes of weather that might affect eastern runoff).

A complicating problem, states Douglass (1983) is that most of the forest land that might be managed is in private, not public ownership, and "there is no incentive to expend money to increase production of a product (water) that the landowner cannot sell." However, the lands upon which management for increased water yields might take place are often municipal or industrial watersheds, managed by the owner/operator or by a private water company. Thus, concludes Douglass (1983), "generally high rainfall and extensive forests in the East combine to produce excellent potential for managing forests for increased water yield. ... However, because of the diverse land ownership patterns and the economic objectives of owners, realizing the potential will be difficult at best."

While the western rangelands would not appear to be major water producers, the potential for beneficial water yield management is present for certain areas. Hibbert (1983) points out that surface runoff can be increased by slowing or preventing infiltration, although this can result in considerable degradation of water quality. "An attractive alternative, where applicable, is to replace vegetation that uses much water with plants that use less .... Probably less than 1 percent of the western rangelands can be managed for this purpose." Some of the management potential in areas of phreatophyte control are discussed below.

In their summary article, Ponce and Meiman (1983) observe that:

> The greatest potential for water yield augmentation appears to be on carefully selected watersheds that have the biophysical potential to produce water that is used for high value purposes, and can be managed under sound multi-resource management. The first such watersheds to be managed in all regions are those that supply municipal-industrial water directly or contribute directly or significantly to hydropower generation. However, management of these types of watersheds on any large scale will require changes in approaches to economic analysis and in water laws.

If one accepts the high degree of variability and an associated high degree of unpredictability, water yield control through vegetative management is certainly a viable management practice. Even if increased water yields are not the goal of management, forest management can be conducted to provide revenues that aid in the maintenance of the water delivery system, and can be effected without degradation of water quality. Considering the large number of intensively managed watersheds, and the growing need for high-quality water supplies, the primary problems seem to be social, economic, and political, not technological (Black 1982).

Water yield changes as a result of vegetative manipulation tend to be relatively short duration, perhaps only a season, years, or, on occasion, tens of years. The primary effects are observed during runoff events or runoff seasons, not at the time of the reduction of evapotranspiration, and are local in scope, being masked by other portions of the larger watershed as one moves downstream into larger drainage areas. The observed change in yield may actually be a net combination of offsetting effects.

Owing to complexity and confounding of interacting factors, differences in harvesting methods are not likely to produce significant or even major or persistent, detectable differences in water yield quantities, patterns, or in water quality characteristics. Large increases in water yield from small areas need to be "spread" out in time and areally on the larger managed unit, which mathematically reduces the effect, although the quantity of water may be high. Management efforts will be most successful on high-value water yielding lands, such as municipal and industrial watersheds, and on lands in public ownership.

Summary

The foregoing does not mean that vegetative manipulation will not have an impact (especially local impact) on local receiving waters. It is here that the effects will be most noticeable, although likely short-lived. The high degree of variability in plant-soil-water relations in certain regions suggests a high degree of unpredictability of results.

MODELS

Simulation and optimization models for the purposes of evaluating evapotranspiration and other water balance components have been developed and tested widely since the widespread introduction of computers in the 1970s. However, any conceptualization of a natural process in mathematical or visual form is also a model, and these have been around for a long time. For example, Dalton's Law, expressing the relationship between vapor pressure gradient and evaporated moisture, is a model. Applications and adaptations of it are also models. The Bowen Ratio is a more complex model expressing the ratio between sensible heat and energy utilized by evaporation (Chow 1964).

In the past, and often when they were created, computations involving some of these descriptive models were generally accomplished by hand, or calculator. The advent of the modern computer has provided the opportunity to create elaborate, sophisticated models that make use of the basic laws (simpler models) in natural process simulation. High speed computing makes it possible to run the model over and over under a variety of constraints and environmental conditions, thereby providing insights into model capabilities and providing important opportunities for sensitivity analyses and error determination on the models themselves: "Sensitivity analysis combined with instrument error variance offers the researcher a tool to help him in evaluating the performance of the model" observe Coleman and DeCoursey (1976). Both instrument error variance and prediction (model) error variance must be added to determine overall system variance: "In model development, the best model will be one whose sum of these two variances is a minimum if there is no prediction bias."

Despite the refinement and extension of models for predicting hydrologic (and water quality) characteristics of complex natural systems such as single observed watersheds calibrated by climatological variable analyses, Nik, Lee, and Helvey (1983) concluded that "the paired watershed approach is expected to remain the preferred method for determining the effects of forest management on the water resource." Their observation is substantiated by the finding of Austin (1986) that "eighty-six percent of those responding [to a water resources model-use survey] indicated they have used mathematical models in the last year. Lack of appropriate data, inadequate time and funding to do the modeling and lack of models that represent the 'real world' situation were the most frequently mentioned constraints to model use." Some of the research in model use on water balance and evapotranspiration predicting and simulation are summarized, as follows.

Using inputs of mean daily temperature and precipitation, Federer and Lash (1978) evaluated possible differences in transpiration on streamflow with BROOK, a deterministic, lumped-parameter model. Timing of fall color change, rooting depth distribution, and leaf diffusion resistance showed changes in simulated streamflow from as little as 0.6 in. to as much as 4.7 in. (15 mm to 119 mm) per year. "On a deep residual soil the differences in streamflow were spread through the year, but on a shallow till soil the differences were restricted to the months in which there were changes in transpiration." This observation conflicts with the observation that increased streamflows from changes in transpiration tend to occur during the period of maximum runoff (storms or season), not when the reduction in E_t occurred (above): it shows the importance of storage potential since, in the shallow till soil, there would be better storage available.

Evaluation of amount and timing of differences in E_t for mature and clear-cut deciduous forests and a young pine plantation at the Coweeta Hydrologic Laboratory were reported by Swift et al. (1975). The model used was PROSPER, a "phenomenological model of water exchange between soil, plant, and atmosphere." Testing was accomplished by comparing simulated drainage to measured streamflow: "in a year of unusually high precipitation the simulated annual drainage was within 1.5 percent of measured streamflow .... Simulated evapotranspiration during the summer was nearly identical for hardwood and pine forests, while winter and early spring water loss was greater for pine. Simulation suggests that the greater evapotranspiration by pine was due to increased interception in all seasons and increased transpiration during the dormant season. For the clear-cut area, simulated evapotranspiration was considerably less than it was for the pine or hardwood forest and thus caused simulated soil moisture contents to be greater during the summer season." These results are consistent with the major findings of Murphy and Knoerr (1975) discussed earlier.

In the process of developing a soil and atmospheric boundary layer model using both energy and moisture balances, Camillo et al. (1983) identified limits on daily evaporation as being restricted to 3.2 mm to 3.8 mm (0.13 in. to 0.15 in.). They also noted that the model, calibrated with soil surface temperature data, was quite stable. In another study, Ben-Asher (1981) cited potential error in estimating annual E_t of as much as 52 percent because "a major source of error lies in the fact that the model neglects the effect of surface moisture content on net radiation." The importance of this model input has been identified in other studies, cited above as well. Barry et al. 1990 report limits to the application and sensitivity of an energy and mass balance model of snow cover in a fir forest. The model was sensitive to water retention.

Spittlehouse and Black (1981) developed a model in which interception is calculated from daily rainfall. The model was tested on two Douglasfir stands which exhibited differing stand density and leaf area indices. "The coefficients used in the evapotranspiration submodel were found to be the same for both stands. It was also found that over 20 percent of the growing season rainfall was lost through interception."

Some of the difficulty in evaluating E_t for different types of vegetation lies in the net radiation relations in forests as opposed to lower crops (discussed above in the sections on evapotranspiration and also in this section) and some in application of the Penman equation, which "provides a good estimate of unstressed transpiration for short vegetation but not for forests" (Federer 1982): "Even when atmospheric variables and the Penman estimate are held constant among forest canopies, unstressed transpiration can vary by a factor of two because of variation both in the maximum value of leaf conductance and in the ratio of canopy conductance to leaf conductance." Federer suggests that the best model would involve "depth variation of soil water potential and of root and soil properties ... the ratio of available water in the root zone to maximum available water." The simulation model in which this is used "may soon allow accurate transpiration estimates for any green cover ...."

In testing three computer models for E_t and streamflow predictions from a mixed conifer watershed, Baker and Rogers (1983) observed that "outputs of all precipitation-runoff models are subject to errors which may be random or systematic." Allowing, too, for input data errors, they noted for the area studied that the model that accounted for the most variation in monthly streamflows, ARZMLT, also produced the most bias in predicted monthly flow. That model, developed from WATBAL (an alpine snowmelt model) was least sensitive in distinguishing between zero and low monthly flows, overestimated larger monthly flows, and assumed that all excess water reached the stream by the end of the day it precipitated, an unrealistic result for snowmelt regimes. Another model, BURP, provided a good balance of agreement between simulated and observed monthly flows on the one hand and the least amount of runoff simulation bias. High flow years were predicted best by BURP, but the model cannot handle zero flows. The results of a third model, ECOWAT, "fell in between those" of the other two.

In clearly identifiable snowmelt zones the job of prediction of streamflows based on precipitation and E_t estimates is much more reliable. Rasmussen (1970) reported a correlation coefficient "between winter accumulation of water and the April through March runoff" of 0.84 over the upper Colorado River Basin for the winter seasons 1957 through 1963. While still crude, the author noted that to obtain such a high r-value over such a large watershed was encouraging. Varied results were found for making 1-month-ahead streamflow forecasts in New Jersey: Alley (1985) found good prediction fits for spring runoff months, but noted that "for the rest of the year the improvements in forecasts over those obtained using the simpler autoregressive models were either very small or the simpler models provided better forecasts."

The next step in application of models is to combine hydrologic or water quality behavior models with those of various land management practices, especially timber harvesting. With the incomplete knowledge concerning the former, any attempt to combine the two, especially over the long period of time necessary for timber crop growth, is risky, at best. The foundation for such models has been set forth by Black (1963), Betters (1975), and Leaf and Alexander (1975) for timber and increased water yields from Rocky Mountain watersheds. Even more detailed and complex models have been developed: Ryan, Morison, and Bethel (1974) present an ecosystem model of a forested river basin in which different watersheds may be identified and managed for forest production and other resource management goals, with attendant long-term hydrologic and water quality impacts. Water quality characteristics include suspended sediment, temperature, dissolved oxygen, nitrate, and other constituents.

Numerous applications of ecosystem models for forest growth have been developed and are in current use throughout the United States. In addition, considerations of wildland water (and other resource) values are included in FORPLAN, a widely used National Forest linear programming resource allocation model (Dress and Field 1987).

Two recent studies with hydrologic system models suggest caution. Alley (1984), using a monthly water balance model with attendant large time errors, urges "extreme caution ... in attaching physical significance to model parameters and in using the state variable of the models in indices of drought and basin productivity." And Mawdsley and Ali (1985), using an equilibrium approach, note that "the model can be applied using only regularly recorded data" and, even then, the model does not provide a full description of nonpotential evapotranspiration."

As models become more complex and more sophisticated, their usage may have to be restricted to the region of their testing in order to meet constraints on localized input parameters. Thus, models may not be currently capable of being used as widely as their creators desire. Ultimately, as we become more knowledgeable about the real world and the models we create to simulate it, differences between models may be resolved into super models that can handle a wide variety of environmental conditions. And, if we were to know enough about the natural environment to create a fully accurate, precise, and reliable model for a particular watershed, we wouldn't need it.

PHREATOPHYTES

A phreatophyte is literally a water-loving plant. Chow (1964) defines the term as being plants "that grow mainly along stream courses, where their roots reach into the capillary fringe overlying the water table." The implication is that the phreatophyte must have its roots in the capillary fringe. However, D. E. Miller (1977) points out that phreatophytes actually can survive with their roots in the saturated zone for greater or lesser periods of time, having "accommodated to a low oxygen concentration." Thus, phreatophytes can adapt to any area where they can put roots down to water, often into the ground water aquifer, as well as along streams where alluvial soils provide an ideal environment for mechanical support and access to water without interference or competition from other vegetation. Hughes (1968) noted that 75 species were considered as phreatophytes; that they may be found along man-made water facilities such as irrigation canals, and that they may modify the flow regime of rivers and, consequently, sediment deposition and river morphology.

Many of man's water management practices, such as flood control and river regulation, particularly in the southwest, have helped spread phreatophytes by creating habitat favorable to their development (Miller 1977). Native vegetation that is identified as phreatophytic includes the willows, cottonwood, red alder, and many species of grasses. Virtually all plants that grow in and around wetlands are phreatophytes. Alfalfa, which can send its roots down 40 ft (12.2 m) and more is also a phreatophyte. The introduction of the number one nuisance species, the ornamental tamarisk (also known as salt cedar) in the early twentieth century in Arizona, has also played a major role. The species has now spread widely throughout the western states (it has been found in southern Oregon and as far east as the High Plains) and is the cause of great concern in those arid states where waste of water by (economically) worthless plant life is not acceptable.

The magnitude of the problem is tremendous: estimates of tamarisk E_t range up to 9 or 10 acre-feet per acre (30,480 m^3 per ha) of water lost each year. Tamarisk alone is estimated (Davenport, Martin, and Hagan 1976) to occupy 1,235,200 acres (500,000 ha), evapotranspiring 4,863,600 acre-feet of water each year (about 4 AF/A/Yr). Losses from another widespread species, the water hyacinth, are estimated to cause a loss of over 2 million acre-feet (2467 m^3) from Texas reservoirs alone (Benton, James, and Rouse 1978). Since the basic meteorologic factors that control the evaporative-type processes also act on phreatophytes (Eisenlohr 1966), one might conclude that that amount of water will be transferred to the atmosphere whether vegetation is present or not. However, in the absence of vegetation, movement of water will stop if the capillary fringe cannot intersect the soil surface via simple evaporation or if the water table is too deep to be reached by plant roots. Thus, plants with access to deeper water supplies will, in fact, increase the transfer of water from terrasphere to atmosphere, even at night (Davenport et al. 1982a). The exact quantity of water thus removed is difficult to assess, and there is considerable controversy over whether phreatophyte control can be effective in reducing water loss in the western states. Survey methods were published by the Forest Service (Horton, Robinson, and McDonald 1964). Estimates of savings must be tempered by noting that the phreatophytes occupy only a small percentage of the total floodplain, not to mention the watershed or entire arid region of the nation, and thus this loss must be "spread" out over a larger area to be put in proper perspective.

Costs of control are high, and savings need to be in the 50 percent range of annual losses of 5 to 6 AF/A/Yr (about 18,000 m^3/ha) in order to be economically feasible (Horton 1972). Campbell (1970) reported an estimate that "3 million acre-feet can be salvaged by eradicating all riparian vegetation along major streams." For perspective, this is one-fourth to one-fifth of the annual flow of the Colorado River. On the other hand, Van Hylckama (1970) reports that the estimates of water savings are grossly overestimated, and that Tamarisk "still may thrive but use comparatively little water."

Control of phreatophytes can be effected by mechanical or chemical means. Physical removal of the vegetation, especially Tamarisk, is accomplished either by a large bulldozer raking the vegetation directly out of the soil, or by a chain dragged between two bulldozers. Chemical control can be either by herbicidal treatment of the vegetation or by use of an antitranspirant (Davenport et al. 1976) that for varying periods restricts the movement of water vapor through the stomata. The water savings claimed for both methods may be severely overestimated simply because results are compared with areas that are bare of vegetation, or with the area after it has been cleared and before some new transpiring crop develops.

As recently as 1979, Davenport et al. (1979) noted that the cost of wax anti-transpirants was too high for economical application and treatment, and that possible adverse side effects, notably on wildlife, were uncertain. However, if the price for water increases, the use of wax-based antitranspirants may become economical (Davenport, Martin, and Hagan 1982b). Given sufficient time, the same or other species will reoccupy the site (D. E Miller 1977), greatly reducing the savings and, perhaps, eliminating them altogether. Even subterranean water movement over short time periods confounds the ability to accurately assess the effectiveness of phreatophyte removal (Hibbert 1969). Finally, incomplete phreatophyte removal may result in residual vegetation taking over the evaporative potential of the site and "pumping" the same amount of excess water into the atmosphere (Chow 1964).

There are other losses due to phreatophyte removal: modification of local temperature and humidity, disturbance to terrestrial wildlife habitat, change in water temperature and, as a consequence, aquatic habitat modification with attendant aquatic species changes, and increased erosion and sedimentation all may result from such practices (Campbell 1970).

Chapter 4 Water in the Vegetated Zone 129

Summary

The processes examined within the vegetation zone involve: (1) the redistribution of precipitation when it strikes vegetative (or other) surface; (2) evaporative activity enhanced by transpiration; and (3) limitation of evapotranspiration by meteorological factors and water availability. The presence of vegetative interception and transpiration, on balance, increases the likelihood that the evaporation potential of the atmosphere will be attained. Man has the capability to manipulate the movement of water at the air-soil interface, but the effects are temporal, lasting a period measured in years or tens of years, and perhaps less for water quality changes. The research into this aspect has also been used to estimate both transpiration and evapotranspiration. Side effects of vegetative manipulation on terrestrial and aquatic wildlife habitat, and erosional processes may be adverse, and are limited areally.

STORAGE

WATER STORED IN VEGETATION

The quantity of water retained in vegetation has not been the subject of much research. It is unlikely that a great deal more water may be stored in subterranean vegetative tissue than would be normally stored in the soil alone. Above ground, however, the amount of water stored in vegetation may be substantial.

The water content of various parts of a flood-felled Coast Redwood at Bull Creek Flat in California was analyzed by Zinke in 1956. Water content of the woody stem was found to average 25 percent, and the bark, 50 percent. Assuming the average diameter of a 250-ft-high tree to be 10 ft, including 1 ft of bark, volumes of 12,566 cu ft of wood and 7069 cu ft of bark, respectively, contained approximately 6676 cu ft of water. Adding a reasonable 25 percent that is held in the crown yields a total of about 8345 cu ft of water per tree. Using an average of 12 trees to the acre, typical on the alluvial flats in the heart of the Coast Redwood range, it is calculated that approximately 100,140 cu ft of water per acre (about 7000 m^3 per ha) is stored above ground. This is about 2.3 acre-feet per acre, considerably more than the 1.5 acre-feet per acre that are normally found in retention storage[11] in the upper 5 ft (1.5 m) of soil. While the Redwoods no doubt represent an extreme (perhaps the extreme) case, it is clear that significant amounts of water can be held above ground *in the vegetation* in forest stands.

Further, on-site evidence in support of the large volume of water stored above ground in Coast Redwoods forest vegetation is the large number of sites that have been cut over and not re-vegetated and which are now water-logged: since the vegetation is not present and holding the water *above* the ground, it is *on* the ground. Springs can often be observed in other previously forested areas, such as subdivisions, where none were observed prior to development.

---

[11] Water held against the force of gravity, typically 25 percent to 33 percent by volume in the alluvial soils of the Redwood Region.

Because the atmospheric and lithospheric realms of storage are joined by vegetation, it is appropriate to think of soil and vegetative moisture together, as **environmental water.**

There is a continuum of water pressure and water quality parameters which, by diffusion pressure gradients and osmotic pressure help regulate water movement and storage from the water surfaces in the stomata at the tops of the trees to the water surfaces within the soil pores. Under such conditions, it would be desirable to avoid sampling soil moisture, especially for the purpose of evaluating transpiration draft, during daylight hours when E_t is at a maximum and, as a consequence, variation of soil moisture would be greatest; soil moisture sampling might be most accurately accomplished in the hours just before sunrise.

Indeed, the entire water column does respond to changes in water vapor pressure at the transpiring surface, as evidenced by the results of the study on an old growth redwood at Dyerville Flat (the Founders Grove) as shown in Fig. 4-8 with instrumentation outside the stand taken at South Fork, then a log deck/train stop east of the grove. The 270 soil moisture observations were collected with a neutron meter, in a radiating pattern of 10 access holes around a 6.5-ft (1.98 m) diameter, 300-ft (91-m) high tree. Measurements were taken in the entire 5-ft (1.5-m) profile at the beginning and end of the 24-hr period, and at the first and second foot depths every two hours in between. Soil moisture depletion was also noted to be greatest on the south side of the tree. Temperature and relative humidity observations were made at several levels in and outside the stand, and a device to measure the change in tree diameter was also installed.

The conclusions were that the water in the tree settled during the period as moisture tension at the canopy level decreased during a strong temperature inversion: evidence is in the higher soil moisture levels in percent by volume in the first foot at the end of the period than at the beginning; in the continuing increase in moisture content in the second foot; and in the short recorded period of steady diameter expansion. The transpiration draft on soil moisture shows a lag of about 4 hr. This is considerably more than is to be expected if one assumes that the maximum transpiration will occur at Local Solar Noon. That is when conditions of maximum light intensity and minimum relative humidity occur: transpiration rate is further influenced by the age of the leaves, moisture stress, and wind movement (Kramer and Kozlowksi 1979).

The observations were made on a rather unusual sunny day in the Redwood Region which included a strong temperature inversion, under which condition the flux of atmospheric moisture is downward.[12] The upward flux of moisture on the more typical foggy days when there is no temperature inversion would be restricted by the high relative humidity. Under either set of conditions, there is not a strong gradient along which the water is likely to move, and the Redwoods appear to "conserve" water, thriving in an area of only about 35 in. to 40 in. of precipitation per year (Black 1964).

The interrelationships between radiation and transpiration, especially when considering the individual leaf, are complex:

> When leaf and air temperature are equal, reradiation and transpiration dissipate the entire heat load. If stomatal closure stops transpiration, the heat load

---

[12] Fog prevails a great deal of the time, and has been, perhaps erroneously, identified as the limiting factor in Coast Redwood distribution (Fisher 1903): frost is more likely (Roy 1966).

must be dissipated by reradiation and sensible heat transfer, but usually there is a dynamic equilibrium in which all three mechanisms operate (Kramer and Kozlowski 1979).

The magnitude of the water stored in vegetation is, no doubt, considerably less in stands of other species. The principles would apply, however, even though instrumental detection and statistical verification might be difficult. Thus, it is important to conceptualize the more complete environmental moisture when investigating and managing the forest vegetation.

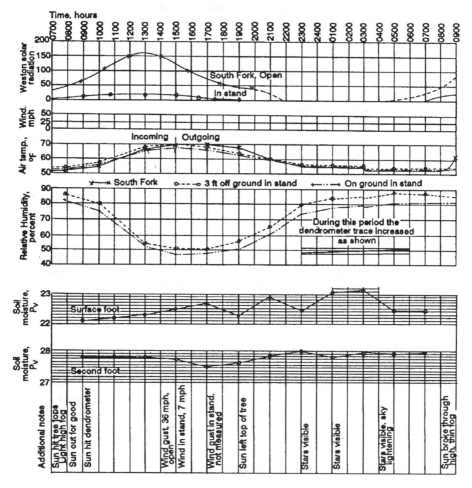

Figure 4-8 Data summary of the 24-hour study of coast redwood at Dyerville Flat, California

WATER STORED ON VEGETATION

This topic has already been introduced and partially discussed in the first portion of this chapter under the heading of the interception **process**, including reference to the volume of moisture temporarily stored. That moisture can be stored as snow, ice, or liquid water, and buffers the variability of precipitation, its "principal characteristic" (D. E. Miller 1977). Miller also points out that the "branching process of detention-film storage" can increase the variability under certain conditions. Typically, 1 to 3 kg/m^2 of liquid water can be stored on forest vegetation, and a like amount can be retained in the forest floor litter (Miller 1977). It normally takes about 10 minutes, dependent upon rainfall intensity, for the foliage to become wetted and, after that, additional moisture will be stored if water is evaporated or shaken off the foliage. When wind plasters snow on vegetative surfaces, cohesive and adhesive forces may be great enough to allow considerable buildup, and as much as 20 kg/m^2 have been reported. Snow storage on Western White Pine was observed at 2.5 kg/m^2, and on Douglas Fir at 3.8 kg/m^2.

Miller (1977) reports that of the amount of moisture retained on the foliage, two thirds either evaporates or is shaken off by wind; one third is retained by surface tension. It is this volume that is often reported as "interception loss."

WATER STORED IN HYDROSPHERIC SYSTEMS

Hydrospheric systems are represented by a wide range of wetlands. Wetlands normally are ecosystems that are in transition between open water and dry land, between aquatic and terrestrial systems (de la Cruz 1978). They are defined in several different ways in legal and government documents.

Wetland Definition

Definition of wetlands has traditionally and erroneously been based upon their classification. While some legislation persists in a vegetation-based definition, there is broad acceptance for a definition that includes water, soils, and vegetation combined. A wetland is an *area that is periodically inundated or saturated by surface or ground water on an annual or seasonal basis, that displays hydric soils, and that typically supports or is capable of supporting hydrophytic vegetation.*

Wetland Values

State laws have differentiated between *saltwater* (or coastal zone or estuarine) wetlands on the one hand and *freshwater* wetlands on the other. Like many others, New York's Freshwater Wetland Law (NYCRR 8-614) also attributes the following values to freshwater wetlands:

1. flood and storm control
2. wildlife habitat
3. protection of subsurface waters
4. recreation
5. pollution treatment
6. erosion control

Three of these are fanciful, at best: first, wetlands *do* hold back runoff waters that would contribute to flood peaks, but it is primarily owing to the *location* of the wetland adjacent to a stream. In fact, dry land usually can hold more water than can wet land. Thus, it is the *proximity* of these lands to the stream that is of value; that is, it is the floodplain that serves the flood control function; dry lands at the same location could undoubtedly do the job better.

Second, wetlands are usually wet because they have characteristically poor drainage: the bottoms of wetlands are usually coated with silts and clays that preclude downward movement of water into subsurface strata. More likely, the wetland is the result of subsurface water coming to the surface, or of undulations in the land surface intersecting a saturated zone. Carter et al. (1979) observe that:

> [w]etlands generally do not support growing season base flow because they lose water by evapotranspiration, at the expense of stream flow or ground water recharge. Some wetlands recharge the ground water system, but most do not. . . . Wetlands . . . are often discharge points for highly productive aquifers which could supply water for a community or municipality . . .

Verification of ground water interchange is best accomplished by calculation of water budgets, but these are nearly impossible to compute owing to the uncertainty of the location of wetland boundaries at any given point in time or even for demarcation on a map. In addition, the relative positions of the local water table, topographic surface, and wetland elevation (surface and bottom) are important in determining whether there is a flow of water between wetland and ground water reservoirs, or between the wetland and the subsurface flow reservoir, the aerated zone within the soil profile. In spite of the difficulties, and because estimates of evapotranspiration from wetland vegetation may be better than estimates for upland vegetation because water is rarely the limiting factor, water budgets have been calculated for a few selected wetlands (Mitsch and Gosselink 1986).

Third, since wetlands are flat, there is usually not much erosion to prevent. True, wetlands provide the function of catching sediment washed from upland slopes, but they do not produce sediment themselves. It is incorrect, therefore, to say that their presence next to a stream provides the function of "erosion control" (by reducing downstream flows, however, wetlands may reduce within-stream erosion). The root mat of vegetation in *tidal* wetlands does provide shoreline erosion control from the effect of both tide and waves (Carter et al. 1979).

Finally, a fourth value, recreation, is questionable: people do like to observe and photograph or study the wildlife and hunt and fish in wetland environments, but the impression is often given that these are favored locations for camping or picknicking, which is highly unlikely in light of the moist conditions and high mosquito and other undesirable wildlife populations.

Wetlands do provide attractive and productive areas for wildlife, furnish high-quality and often unique scientific and educational opportunities, and display noteworthy powers to remove pollutants from waters that enter them. They also provide open space. No doubt, with both the creation and destruction of wetlands in the United States, effort should be directed at protecting them: but they should not be protected for the wrong reasons (Black 1980a).

Wetland Classification

Wetlands are usually classified by the vegetative types that typically thrive under the adverse conditions found on them, primarily because that is the easiest way in which to identify the wetland area. In fact, most of the vegetation that grows on such sites is opportunistic in that the site is inimical to other species. Thus, many classification systems for wetlands are based on characteristics that have little to do with the hydrology of the area. For example, "wet meadow," "deciduous swamp," and "spruce bog" all suggest vegetation of a particular type. Other classifications by regulatory units of government are often on the arbitrary basis of size, anticipated vulnerability to direct or indirect impacts of development, or some "value," as discussed above.

Since "freshwater wetlands" obtain their name in both parts from "water," they ought to be defined and classified on the basis of their *hydrology*. Historically, this has not been the case because, without the appropriate hydrologic data, wetlands have been classified by what can be *seen*, namely the vegetation. In fact, many wetlands exist simply because water is not flowing well in one direction or another at the site. Mitsch and Gosselink (1986) point out the "importance of hydrologic flow-through for the maintenance and productivity of these ecosystems," but note that the only characteristic common to all wetlands is anoxic biochemical processes. Defining the local hydrology is difficult, at best, owing to the difficulty of and, therefore, lack of research on wetlands hydrology (Sather 1984).

Delineation of wetland boundaries has been and remains a constant problem for governments charged with the responsibility of wetland regulation, as well as for the researcher who wishes to define boundaries for the purpose of water budget determination. The resultant paucity of information on wetlands includes estimates of how much water is actually stored in them, for "analysis of hydrologic functions require a large number of detailed, precise measurements" (Sather 1984). Of greater importance than the hydrology is the **hydroperiod** (portion of the year that the land is actually under water), which appears to be more closely correlated with vegetative cover. Mitsch and Gosselink (1986) state that

> The hydroperiod is the seasonal pattern of the water level of a wetland and is like a hydrologic signature of each wetland type. It defines the rise and fall of a wetland's surface and subsurface water. It is unique to each type of wetland, and its constancy from year to year ensures a reasonable stability for that wetland. The hydroperiod is an integration of all inflows and outflows of water, but it is also influenced by physical features of the terrain and by proximity to other bodies of water.

The inundation is important ecologically because it generally prevents the lower portions of the aquatic fauna and flora from ever experiencing temperatures below $4^{\circ}C$ (Macan 1974). Most terrestrial ecosystems do experience such conditions annually, and the cold temperatures and frost affect both distribution of flora and hydrologic and water quality characteristics.

Wetlands that are connected with the ocean or other large body of water tend to assume at least some of the characteristics of that water body. The oceans are "the most constant of all environments" (Macan 1974) and, owing to the twice-daily inundation, tidal wetlands thereby will be most affected. Wetlands in the tundra are also subjected to relatively constant conditions, and therefore may not vary greatly. In between these two extremes, wetlands may be profoundly affected by the lakes or rivers with which they are in contact, and, if isolated, may be even more unique in terms of flora and fauna (Macan 1974). The same is likely true about the wetland's water quality characteristics.

The fundamental hydrologic characteristics of a wetland are the *source, velocity, renewal rate,* and *timing* of the inundating water (Gosselink and Turner 1978). The former two have a major influence on water quality, including sedimentation rate and amount. The latter two are closely related to hydroperiod.

An hydrologic classsification would separate coastal or saltwater wetlands from freshwater wetlands. The freshwater wetlands should then be separated into bogs and all other types. Bogs (and fens), which typically exhibit anaerobic conditions and acidic water associated with acidic vegetation and low net horizontal water movement in relation to rather high net vertical water movement, appear to be unique with regard to their water quality, known hydrologic characteristics, and vegetation. The remaining freshwater wetlands might be divided, again by hydrologic function, into two broad classes: (1) upland wetlands and (2) surface water-associated wetlands, including both riparian (streamside) and lacustrine (lakeside). The former are not connected to or associated with any (surface) water body. The reason for this breakdown, as noted above, is that the proximity of the wetland to a stream (lake or ocean) is of considerable importance in how the wetland functions hydrologically, that is, what role it plays in stream behavior and the water budget, if any, and its effect on water quality in the wetland. Further (secondary) classification of wetlands could be on the basis of ecosystem and vegetative characteristics. In fact, classification of wetlands ought to commence with the observation that, like the floodplain itself, *the streamside wetland is part of the stream, the lakeside wetland is part of the lake, and the saltwater wetland is part of the ocean.*

The Fish and Wildlife Service National Wetland Inventory (Cowardin et al. 1979) utilizes this philosophy of classification by categorizing wetlands first on the basis of five hydrologic classes: Marine and Estuarine in one branch, and Riverine, Lacustrine, and Palustrine in another. Further classification is on the basis of vegetation and other characteristics that affect water movement and storage, such as bottom characteristics, and water depth, color, and other water quality parameters.

Vegetation and storage of water in depth, not volume, units are intimately tied together in wetland ecosystems. A working demonstration, qualitative wetland hydrologic model was presented by Young, Klawitter, and Henderson (1972). The value of such models is the conceptualization of the effects of modification to the system and the opportunity to discuss scenarios from inroads of development. Recently, remote sensing of hydrologic information has been added to the wetland analyst's tool kit, and models have been developed for interpreting aerial photogrammetry based on ground truth information, as in Connecticut (Anderson, Lefor, and Kennard 1980). More detailed models of wetland hydrology and throughflow for water quality evaluation have also been developed. These are reviewed by LaBaugh (1986), with the following conclusions: "Long-term studies are often not done, despite the recognized importance of year-to-year variability in hydrologic conditions that can affect interpretation of chemical input-output budgets .... Only when all water budget components of [a] wetland are measured, will questions about the importance of hydrologic processes to wetland ecosystem structure and function be resolved."

An Hydrology and Water Quality-Based Classification

Clearly, wetlands have different water quality characteristics dependent upon their hydrology. Further, one of the most important values of wetlands is their capability to remove certain nutrients from the water column and sediments, thus "purifying" the water. (Kadlec and Kadlec (1978) point out that

> Every water quality parameter is altered by passage through a wetland ecosystem. Nutrients and other dissolved constitutents, heavy metals, suspended solids, and bacteria move into and out of the wetland with entering and leaving waters. Their concentrations can be altered by uptake, cycling, and dilution. ... *To interpret the performance of a wetland in altering any water quality parameter, a knowledge of the hydrology of the wetland is required*, along with data on concentrations, mass flow and storage of the constituent of interest. Wetland ecosystems are dynamic, with both stochastic and time varying effects being of great importance [emphasis added].

It would be logical, therefore, to classify wetlands on a combination of their hydrologic and water quality characteristics. An example is shown in Table 4-7. Generally, the cells are characterized based upon a review of current literature, and the differences between the different categories of wetland are apparent and, ultimately, useful in both understanding and managing the wetland. However, where the cells are not based upon precise research results, or the feature is completely unknown, a question mark is shown along with the estimate of the feature.

According to Good et al. (1978), there are "no meaningful average concentrations of nutrients in wetland ecosystems." "Some N and P may be translocated into belowground parts by several perennial macrophytes, but most is rapidly leached after death of vascular plants with up to 80% of the total N and even more of the P lost within 1 month .... It appears almost all habitats of freshwater tidal marshes may be sinks for inorganic N and PO_4-P during the growing season and that certain habitats may continually function as sinks" (Simpson 1978).

In a study of the influence of wetlands on water quality, Oberts (1981) reported that riverine wetlands are sinks but provide nutrients whenever there is a surface overflow, that is, flushing action. In contradiction, sub-alpine wetlands in Colorado were found to have no effect on water quality (Sundeen, Leaf, and Bostrum 1989), which may have to do with their low conductivity, N and P contents, and annual temperature, not to mention their resultant low decomposition rate. Obviously, some fundamental questions about natural wetlands, such as which nutrients are limiting, wetland role as sinks or exporters, and so forth, remain unanswered. In addition, the role of animals in nutrient cycling is uncertain (Good et al. 1978).

Kadlec and Alvord (1989) report that from a created wetland-water treatment system that handled 400,000 m^3/yr, (a) 96% of the total P was removed, with resultant concentration in the water less than 0.1 mg/l; (b) 97% of the total ammonium N was removed, to a water concentration of less than 0.3 mg/l; (c) there was a species shift to cattail and duckweed; (d) aboveground biomass increased by factor of 4; (e) root biomass decreased; (f) denitrification completely removed nitrate from water; and (g) sediment deposition and decomposition over ten years increased water levels by several centimeters.

Table 4-7 Hydrologic and Water Quality Characteristics of Wetlands

Characteristic	Tidal	Lacustrine	Riverine	Upland	Bogs/Fens	Tundra
Hydrologic Features						
Inundation						
Level	∿∿∿∿	~	⋀	⌣	—	—
Frequency	2x daily	irregular	irregular	seasonal/ irregular	annual (?)	annual (?)
Water movement	⟷	↔↕	↔↕	↔↕	↕	↕ (?)
Water Quality Features						
Temperature	regular: equal to ocean	regular: influenced by lake if connected	irregular: influenced by river when connected	seasonally variable	seasonally variable	seasonally variable
O_2	high	variable	high	medium	low	low
CO_2	Low	seasonal/ daily	variable	high	highest	high (?)
pH	equal to ocean	influenced by lake if connected	equal to river if connected	daily	lowest	high (?)
Nutrients	replenished 2 x daily	annual cycle	accumulate/ flushed	annual cycle	deficient	(?)

SUMMARY

Vegetation extends its influence downward into the soil as well as upward into the atmosphere. In both cases, it modifies the extremes of climate, and plays a major and accessible role in the hydrologic cycle. The primary hydrologic effect of vegetation is that variability of runoff may be either increased or decreased depending upon the circumstances, but natural peak runoff is often reduced owing to storage potential[13] and attenuation of abrupt changes in precipitation. These effects occur primarily through the collective evaporative processes of interception, evaporation, and transpiration, all of which increase the likelihood that the evaporative potential of the atmosphere will be attained.

---

[13] It is really rather presumptuous to make this observation, but it must be made, in spite of the fact that most of the attenuation of precipitation variability is caused by storage in the soil which, of course, must generally be present as the growing medium for vegetation.

PROBLEMS

1. Draw a networking diagram version of Table 4-3.
2. Consider pros and cons of whether precipitation should be evaluated on the slope rather than on the horizontal.
3. Using Table 4-5, what is the effect of doubling and halving the storage (STRGE) value to 50 mm and 200 mm, respectively?
4. Using Table 4-5, if this were the water budget for a watershed at the latitude shown that slopes south at about 40°N, how should the budget change and more accurately represent what occurs on the watershed?
5. Plot monthly (or, if computer plotting is available, daily) values of temperature, precipitation, and runoff for a nearby drainage on the same sheet of graph paper. Describe how well the expression $P - RO = E_t$ approximates E_t for the year and the plotting interval.

5 Water in the Terrasphere

In the soil,
water moves in response to gravity
when it is not responding to
tension gradients

This chapter focuses on the terrasphere following a brief discussion of storage in the lithosphere in order to provide relationship to and perspective on water in deep storage. Water in the terrasphere includes water in the aerated zone plus that ill-defined portion of ground water storage that plays an active role in storm runoff, especially on small watersheds. The lithospere consists of the solid earth, including the soil and its geologic substructure. The solid part of the earth that is beneath the water-bearing strata is not of primary concern in this book; indeed, that portion of the substrate that is in the realm of ground water geology (or geohydrology or hydrogeology) is not covered in great detail herein either.

STORAGE

There are three broad types of water storage in the lithosphere. The first, **geologically bound water**, is water which presumably has been in storage since the beginning of the earth (or at least a long time) and is not in circulation. It is not of concern in small watershed management or watershed hydrology; it is an uncertain quantity and is, in any event, largely inaccessible to man.

Second, **ground water** storage, which is covered only to a limited extent because, while an important resource, it is, by definition, not a factor in small watershed runoff. It is also very well covered in several excellent volumes (e.g., Chow 1964; Todd 1959; Freeze and Cherry 1979). Ground water is discussed here (and elsewhere) as necessary to provide perspective on watershed hydrology.

Third, in that portion of the terrasphere known as the soil, water occurs as **depression**, **retention**, and **detention storage**. The first is really an extension of interception storage, a category which indeterminately and inexorably grades into ponds and lakes and inland seas. Retention storage is water retained at high tensions in contrast to the nearly atmospheric tensions at which detention storage is detained: *the balance of and connections between depression, retention, and detention storage are at the core of the wildland water resource management system.*

GEOLOGICALLY-BOUND WATER

Juvenile water is newly formed from magma as it is brought to the surface of the earth by volcanic activity and discharged (Parker 1955). Much of the water emanating from hot springs and geysers may also be juvenile, but not all of it is. Some may be **connate**, defined as water which has been stored in sediments since the sediments were lain down. In both cases, the waters are likely to be highly mineralized and not generally useful (Thomas 1955). However, these two types of water may become economically important in the future as sources of minerals that are dissolved in them; they are also of considerable interest now in the petroleum exploration and development field owing to their use in forcing oil and gas resources to the surface and to their subsequent potential for polluting surface waters into which they are discharged. The quantity of juvenile and connate waters is not known.

GROUND WATER

Water in ground water storage is at or above atmosphere pressure, whereas water stored in the soil is under tension. The surface of this body of water is known as the **water table**. The water table is depicted clearly in many elementary (and some advanced) textbooks, and is often misleading, as noted in Chapter 1. In many locations, the exact position of the water table may be difficult to identify. In some locations, however, particularly where underlying strata are uniform and extensive, information on ground water may be readily available. This is typical in the Great Plains, along many alluvial bottomlands, and on moraines, such as Long Island. Ground water is "more than a resource" in that "it is an important feature of the natural environment; leads to environmental problems, and may in some cases offer a medium for environmental solutions" (Freeze and Cherry 1979).

The amount of water stored as "ground water" is shown in Table 1-1: 4.62 and 3.63×10^{12} AF in deep and shallow ground water, respectively. While the residence times as shown in Table 1-1 are 973 and 764 years for these quantities based upon the annual amount of precipitation, Freeze and Cherry (1979) point out that, in fact, residence times vary from 10s to 1,000s of years.

Ground water is important because:

1. it constitutes more than 80 times the amount of water available in the world's rivers and lakes combined (Table 1-1);
2. it supplies approximately 24 percent of the daily withdrawals of water in the United States, varying from a low of 1.8 percent in the Upper Colorado River Basin to over 68 percent in the Arkansas-Red-White River Basin (Water Resources Council 1978);
3. we use it at a far greater rate than that at which it is naturally replenished (Francko and Wetzel 1983),

4. it is subject to contamination that is a severe threat to health and "is a troubling environmental problem because, unlike the pollution of air or lakes, ground water is inaccessible, making cleanup virtually impossible" (Sun 1986),
5. it represents a huge natural storage sink in the hydrologic cycle itself; and
6. water that flows from small watersheds often recharges ground water reservoirs downstream where there is interchange between the river channel and ground water aquifers.

Over the long term, continual dilution and/or flushing action, along with chemical reactions between rock strata and ground water, help obscure any ground water contaminants.

Henry Darcy defined how water moves through a saturated porous medium with the analogy of a cylinder fitted with inflow and outflow pipes (Freeze and Cherry 1979). The law is derived from the basic relationship[1] that expresses the velocity of flow, v, in the pipe as a function of the discharge, Q, and its cross-sectional area, A:

$$v = Q/A$$

Darcy showed that velocity was a function of the difference in **head**, h (vertical distance between two levels) over a finite distance, l. From these observations, he defined the terms **hydraulic head**, dh, (change in head) and **hydraulic gradient**, dl, (change in head over change in distance) and combined them in what is known as Darcy's Law:

$$v = -K(dh/dl)$$

where K is the hydraulic conductivity. It has high values for porous materials such as sand and gravel and low values for clay (heavy) soils and rock. It applies in any direction. These basic concepts are of considerable importance in ground water investigations and in saturated soils as well.[2]

Ground water aquifers are classified on the bases of (1) water movement within the stratum, and (2) the relationship between the elevation of the atmospheric pressure surface and the earth's surface. A stratum that is completely impermeable to water is known as an **aquiclude**. A stratum in which water is held tightly, that is, by high tensions owing to adhesion between water and rock particles in the porous medium, is also in this category: it may hold large quantities of water, but does not transmit them readily and may be referred to as an **aquifuge**. An **aquifer** is a stratum that transmits water readily. If the water is under pressure in excess of atmospheric pressure, the aquifer is **artesian**. A flowing artesian well occurs if the pressure is sufficient to cause the water in a dug well to rise above the land surface (Soil Conservation Society of America 1976). Relationships and nomenclature are shown in Fig. 5-1.

The tremendous variety of scenarios for subterranean flow of infiltrated water on a watershed provides both complexity in describing runoff processes and, as a consequence, difficulty in modeling and predicting runoff. Thus, subsurface runoff on a small watershed may end up as ground water further downstream owing to interchange between the stream and

---

[1] This formula is used in streamflow determination in the form Q = AV (Chapter 6).

[2] For a full development of the basic concepts and applications, see Freeze and Cherry (1979).

riparian (bankside) exposure of ground water aquifers. Also, percolating water may contribute to deep ground water storage through deep seepage, often occurring along fissures in what otherwise appears to be impenetrable bedrock (Kirkby 1978).

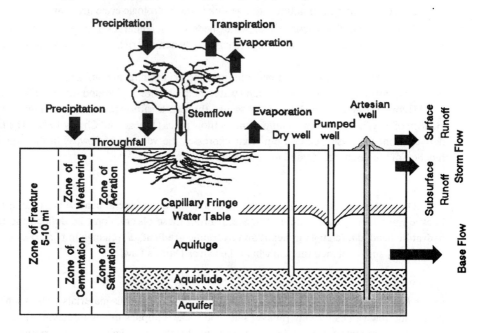

Figure 5-1 Water in the terrasphere

As noted in the discussion of the typical hydrograph in Fig. 1-4, runoff may consist of flow from the saturated zone beneath the water table. This discharge is known as **base flow** in contrast to **storm flow**, which is comprised of **surface runoff** (if any) and **subsurface runoff**, which comes from the unsaturated zone above the water table. Once in the stream, where they are collectively called **streamflow**, it may be difficult to separate the types of flow by origin.[3] The problem of hydrograph separation is of importance when analyzing runoff from large watersheds where storage includes ground water. On small watersheds, this problem is simplified because a standard definition of the small watershed depends upon there *not being any ground water storage at all* (Chow 1964). While this flexibility does not permit a finite determination by size of a small watershed, it typically restricts the definition of a "small watershed" to those of "a few to 1000 acres" (Chow 1964) or about 1 sq mi (259 ha). The remainder of this book is largely concerned with the storage and processes on watersheds of this size and smaller.

---

[3] Techniques of hydrograph separation based upon long-term records and water quality information may be used to help identify the origins of given waters.

SOIL WATER

The pore space in the soil is a giant reservoir that provides the primary buffering of precipitation that is delivered irregularly to the surface of the earth, and which is *attenuated* (delivered more smoothly) to the streams that drain it. In simplest terms, the soil is a porous medium. But it is not like a "sponge", as often referred to: a sponge is a material with holes of different sizes that (usually) are interconnected. The soil can be thought of as a negative sponge: it consists of particles that cannot fit completely in contact with one another because of their irregular shapes, and therefore includes **pores** (air spaces) which, if of the right size, distribution, and connectedness, behave hydrologically like a sponge. The soil is normally not at all uniform, although we often tend to think of it that way, especially if the surface is smooth, flat, and regular. Soil characteristics also vary wtih time (FitzPatrick 1971).

Porosity

Porosity is the percentage of the total soil volume that is not occupied by solid particles. Typically, the pore space is occupied by water and air, the relative distribution of each a function of the time available for drainage of the water and the ability of the pores to drain dependent, in turn, on their size, distribution, and connectedness. If all soil particles are the same size, shape, and packing, then the pore space will be the same regardless of particle size. Pore size is less in fine-textured soils, and is also reduced in size and total amount if some of the pores are filled with other, smaller soil particles.

In the soil pores, water is acted upon by forces of gravity and capillarity, often expressed as **tension** and reported in units of Pascals (atmospheres or bars) or as pF, the logarithm of the height of the water column, in centimeters[4] (the relationship among the units is shown in Table 5-1). The latter is the combined result of **cohesion** (attraction among water molecules) and **adhesion** (attraction between water molecules and soil particles). The tensions at which water is held in the soil pores against the force of gravity depend upon the **soil texture** and **soil structure**. The former is concerned with the size and relative proportions of mineral particles, while the latter is concerned with how those particles are arranged (Buckman and Brady 1960).

The amount of water that a soil can hold is, to use the greatly overworked word correctly, truly awesome: this is due, in part, to the mechanical structure of a porous medium and, in part, to the organic content and biological activity within the soil.

The mineral particles derived from rock that make up soil are classified as sand, silt, and clay. Fig. 5-2 documents relative sizes and several classifications of all particles; the "fine" fraction of soil particles are those that pass through a 2-mm sieve. Generally, sands are from 2.0 mm to 0.05 mm; silt, from 0.05 mm to 0.002 mm, and clay, less than 0.002 mm and are compared in Table 5-2. The percentage distribution of these three sizes of particles is used to classify the soil, as shown in the classical soil texture triangle (Fig. 5-3).

---

[4] It is a constant source of aggravation to the practicing hydrologist to have the soils terminology and units changed to accommodate soil classification and soils research which, although done in the name of clarity for and logic to the soil scientist, have considerably less relationship to the world of hydrology. Table 5-1 is an important case in point: the soils text version gives no explanation for including "15" and "31" atmospheres in a long list of one-over-factors-of-ten. They are there, in fact, because of their hydrologic significance.

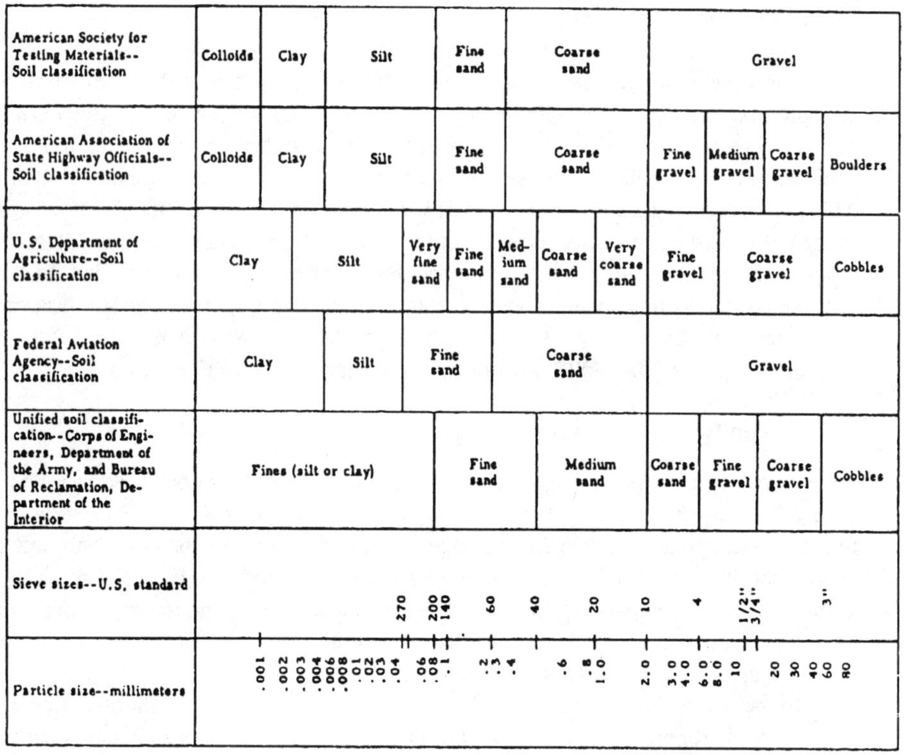

Classification System										
American Society for Testing Materials--Soil classification	Colloids	Clay	Silt	Fine sand	Coarse sand	Gravel				
American Association of State Highway Officials--Soil classification	Colloids	Clay	Silt	Fine sand	Coarse sand	Fine gravel	Medium gravel	Coarse gravel	Boulders	
U.S. Department of Agriculture--Soil classification	Clay		Silt	Very fine sand	Fine sand	Medium sand	Coarse sand	Very coarse sand	Fine gravel	Coarse gravel	Cobbles
Federal Aviation Agency--Soil classification	Clay		Silt	Fine sand	Coarse sand	Gravel				
Unified soil classification--Corps of Engineers, Department of the Army, and Bureau of Reclamation, Department of the Interior	Fines (silt or clay)			Fine sand	Medium sand	Coarse sand	Fine gravel	Coarse gravel	Cobbles	
Sieve sizes--U.S. standard				270, 200, 140	60, 40	20	10	4	1/2, 3/4	3"
Particle size--millimeters	.001 .002 .003 .004 .005 .01 .02 .03 .04			.06 .08 .1	.2 .3 .4	.6 .8 1.0	2.0 3.0 4.0 6.0 8.0 10	20 30 40 60 80		

Figure 5-2 Comparison of various systems of particle-size classes (from the Forest Service *Handbook on Soils*)

Soil texture affects **percolation**, the rate at which water moves downward through the soil, **permeability**, an expression of movement in any direction, and the **hygroscopic coefficient**, the water that is held at tensions beyond 31 atmospheres, essentially physically and chemically bound to the soil particles. The remainder of the soil particles are "coarse" fragments, and often of importance in soil-water relations dependent upon the abundance, size, and distribution in the profile.

The distribution of particle sizes and the pore spaces (also known as **interstices**) is of primary importance to the wildland watershed hydrologist. The simple geometric relationships involving volume and surface area that underscore this importance are dramatically demonstrated in Table 5-3. The basic assumptions behind this presentation are (1) as many spheres as possible are packed into a one-cubic-foot volume, (2) all the particles are represented by spheres of uniform size, and (3) the spheres in each case are uniformly and squarely packed.[5] The "Volume of all spheres" column shows that it doesn't matter what the

---

[5] The values for hexagonal packing are slightly different, but the principle illustrated is the same.

size of the particle is as long as the packing is uniform and the particles are all of the same size: of the 1728 cu in. available in the cube, slightly more than half, 904.78 cu in. (52 percent), are occupied by particles. The remainder, of course, is pore space (Fig. 5-4).

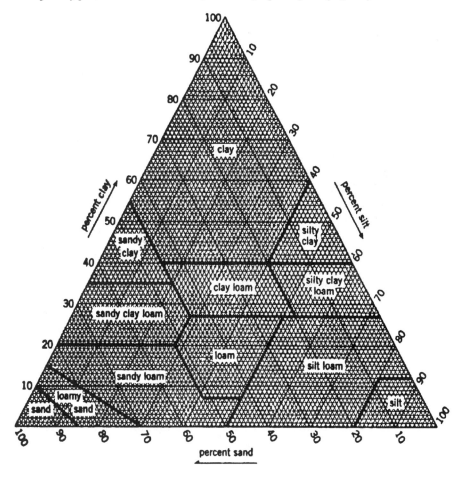

Figure 5-3 - Soil texture triangle (from the Forest Service *Handbook on Soils*)

In contrast to the volume, which remains constant with uniform packing, the total surface area increases exponentially as the diameter of the particle decreases. As a consequence, in the first line of Table 5-3, where the single sphere is 12 in. in diameter, very little of the pore space is close enough to the "soil" (sphere) surface to be held at very high tensions, as indicated by the (relatively) small surface area of the single sphere.

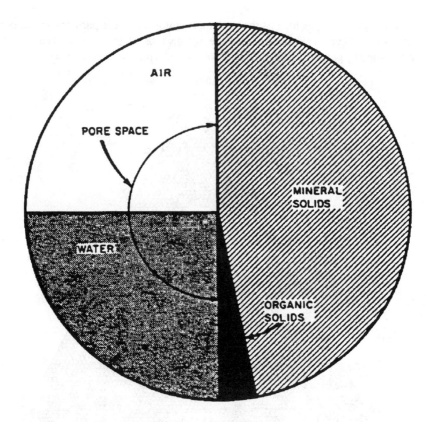

Figure 5-4 Volume composition of an average soil
(from the Forest Service *Handbook on Soils*)

Table 5-1 Energy Levels of Common Soil Moisture Constants

Soil Moisture Constant	Atmospheres (atm)	Height of Water Column (cm)	Soil Water Tension[a] (bars)	Soil Water Potential (bars)	pF value (log of water column height)
Field capacity, sandy soil	1/10	102	0.1	-0.1	2.00
Field capacity, clayey soil	1/3	306	0.3	-0.3	2.54
Wilting point	15	15,345	15.0	-15.0	4.20
Hygroscopic coefficient	31	31,713	31.0	-31.0	5.40

[a] The SI megapascal (MPa) unit is obtained by multiplying the tension in bars by 0.1
Source: Brady (1984).

Table 5-2 Size Limits of Soil Separates

Name of Separate	Range in particle diameter			
	in.		mm	
Sand				
very coarse	0.0787	0.0394	2.0	1.0
coarse	0.0394	0.0197	1.0	0.5
medium	0.0197	0.0098	0.5	0.25
fine	0.0098	0.0039	0.25	0.10
very fine	0.0039	0.00197	0.10	0.05
Silt	0.00197	0.00008	0.05	0.002
Clay	<0.00008		<0.002	

Source: Forest Service (1961).

For the 0.01-in. diameter spheres in the last line, however, 3770 sq ft of surface area are exposed in a single cubic foot of 1.728 *billion* (uniformly packed, spherical) particles. The last line is the approximate size of medium sand particles. Since silt particles are about 10 times smaller than medium sand, it may be calculated that *one acre of five-foot-deep silt includes about 188,500 acres of surface area*. This is a formidable amount of surface at which water is held in the pores close to the soil particles by very strong adhesive and cohesive forces.

Table 5-3 Effect of Particle Size on Space and Surface Area

Diameter (in.)	Number of Spheres in Cube (#)	Volume of		Surface Area of	
		one Sphere (in^3)	all Spheres in Cube (in^3)	one Sphere (in^2)	all Spheres in Cube (in^2)
12	1	904.78	904.78	452.39	452.39
6	8	113.10	904.78	113.10	904.78
2	216	4.19	904.78	12.57	2,714.34
1	1728	.52	904.78	3.14	5,428.67
0.1	1.728×10^6	.0005	904.78	.03	54,286.72
0.01	1.728×10^9	5.0×10^{-7}	904.78	3.0×10^{-4}	542,867.21

Thus, in fine-textured soils, more water may be held at higher tensions than in coarser-textured soils. Further, if there is a mixture of sizes (as is the case in most soils), whereby fine particles fit inside interstices between larger particles, more and more pore space will be taken up by solid material, resulting in (1) higher **bulk density** (weight per unit volume), (2) a smaller amount of pore space, (3) less potential for holding water at all, and (4) water held at higher tensions.

The amount, nature, and distribution of the particles and pores clearly determine the water relations of the soil and, as a consequence, stream behavior. Some of the differences in hydrologic behavior are shown in Table 5-4. The measure of pore space is called **porosity**, n, defined as V_v/V_t, and ranges from about 0.25 to 0.7 in natural soils dependent upon packing, distribution, and uniformity of soil particles, and from 0.1 to 0.3 in rock (Dingman 1984).

Table 5-4 Textural Classes of Soils and Permeability

General Texture	Textural Classes	Permeability
Clayey (heavy)	clay silty clay sandy clay	very slowly
Moderately clayey	silty clay loam clay loam	slowly
Loamy (medium)	sandy clay loam silt loam loam very fine sandy loam	moderately
Sandy (light)	fine sandy loam sandy loam loamy fine sand	rapidly
Very sandy	loamy sand fine sand sand	very rapidly

Large quantities of water can readily infiltrate sandy soils as there are considerable large, interconnected pores. However, the tensions at which much of the water is held in sands is quite low, and water will permeate the profile rapidly and drain out; there is not sufficient tension to hold the water back against the force of gravity. In contrast, at the other end of the curve, clays hold more of the water at very high tensions; more water may be drawn into a clay soil rapidly by capillarity, and a greater proportion of the water in the pore space will be held at higher tensions than in the sand. The result is that clays will not give up drainage water rapidly (if at all) since the force of gravity is too weak to remove the water from its high-tension bond with the small clay particles. Water may be held at even higher tensions and in response to chemical bonds, too, in colloidal suspensions within the soil. Viscosity plays a major role here, its influence in retarding flow being greater in the fine-textured soils (Dingman 1984).

Just as important as particle/pore size in the movement through and storage of water in the profile is the *distribution* of pore space. This is because as long as the pores are reasonably similar in size and connected to one another, water may flow from one pore to another: if there is a break in the tension gradient, that is, in the connectedness of the pores, the water may not flow at all. The basis for this is illustrated by Fig. 5-5, which shows how fast water can move through the various particle size classes and the tensions at which that movement takes place, expressed as **capillary conductivity**. It is important to note, too, that "the rate of flow between two pores of different size is controlled by the pore of the smallest size" (Brutsaert 1963).

Soil material characteristics are governed by the initial conditions of particle size, density, shape, distribution, and packing. Under the influence of the other soil-forming factors, those characteristics become modified. The ranges of *hydraulic conductivity* of the different soil particle textures overlap so that water will move smoothly from one texture class to another if there is physical continuity of sizes. If that continuity is broken, water may not flow at all. Water relations will reflect the combination of parent material characteristics and modifying factors. An example is demonstrated in time-lapse sequences in the film "Water Movement in Soil" created at Washington State University: the results are partially described by Gardner (1968). Here, sharply defined boundaries between strata of markedly different particle sizes within the profiles of alluvial soils of the Columbia Basin are common, and can effectively prevent downward movement of water by virtue of a "tension barrier." Tension barriers occur where there is an abrupt change in soil pore size. For example, water in a saturated layer of clay overlying a layer of nearly pure sand could not percolate to the sand because there was no gradient along which the water could move (cf., Fig. 5-5). In the reverse situation, with the sand above, only a little water would move from the sand to the clay, even though the water was under the influence of strong tension forces that would pull the water into the clay layer; there was a clear difference in tension between the two layers, but without the gradient, the water did not move.[6]

A practical example of this phenomenon occurs following extreme drought in the southwest: even if water is piled up on or is running across the surface, infiltration may not occur owing to a strong tension barrier between the atmosphere and the soil. This phenomenon, referred to as "soil wettability," has also been observed after wildfires "associated with an organic coating on soil particles that makes them hydrophobic" (Krammes and Debano 1965). Hussain, Skau, and Meeuwig (1968) report that pine litter may also be a cause of low infiltration. Unwettable soils have also been observed on unburned sites, and appear to be changeable over time (DeBano 1969).

The difficulties of understanding and modeling water flow in macropores in the soil was reported by Beven and Germann (1982). They point out that, even in the large, noncapillary pores, "flow rates will be controlled by the void of the smallest size in any single continuous flow path." They add that macropore flow may be a significant component of local hydrology in certain areas. Finally, the complex interactions between the many soil and water

---

[6] A classic example of this occurs when tiles are laid under heavy soils in an attempt to drain a field: without providing some gradient of particle sizes from the fine particles of the soil to the open flow in the drain, the water may not move into the drain tile at all.

characteristics and, in certain instances, atmospheric characteristics, are ramified in water *quality* issues as well (Nielsen, van Genuchten, and Biggar 1986): these are discussed in more detail in Chapter 6.

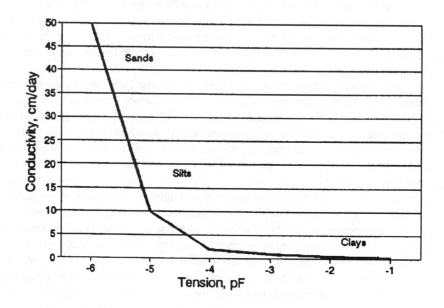

Figure 5-5 Capillary conductivity, tension, and soil particle sizes (after Buckman and Brady 1960)

The organic content of the soil includes burrowing insects and animals as well as decaying litter and living and decaying root systems. The rich organic activity that is associated with all this material plays an important role in the water relations of the soil: in addition to maintaining infiltration capacity and porosity discussed below under "Biological Agents," the organic fraction represents a portion of the complex soil ecosystem, affecting soil fertility, nutrient cycling, and vegetation growth, survival, and reproduction. Obviously, all of these considerations are important in a discussion on soil moisture which, therefore, cannot be considered in a vacuum; however, biological activity in detail is taken up in order, below, as a soil-forming factor. Suffice it to say that, as the most dynamic of the soil-forming factors, the biologic activity has a profound effect on the soil's development, characteristics, and appearance as well, obviously, as on its hydrologic behavior.

The particles that make up the solid, inorganic portion of the soil are derived from parent material and undergo change as the soil forms by action of the soil-forming factors. The inclusion in the soil of organic material from the vegetated zone further modifies the basic inorganic soil material and its hydrologic characteristics.

Typical rate and extent of capillary rise in three different soil textures are shown in Fig. 5-6. The value of a mixture of soil particle sizes and organic content of the loam is apparent. Note, too, that the sand and clay exhibit roughly similar upward limits to capillary rise, but the *rates* are quite different.

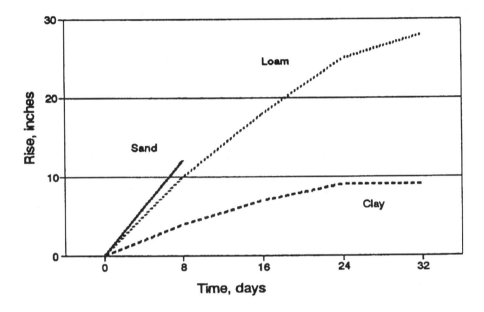

Figure 5-6 Capillary rise in soils (after Buckman and Brady 1960)

Soil-forming Factors

The soil-forming factors are (1) parent material, (2) time, (3) climate, (4) geology and topography, and (5) biological agents (Jenny 1941). They provide a logical outline and sequence within which to understand water movement and storage in soil.

Parent Material

The parent material refers to the rock type from which soil material and, eventually, soil, is developed. The basic rock types are igneous, sedimentary, and metamorphic.

Igneous rocks form through solidification of molten magma by fusion of two or more unaltered minerals. These include granites, usually feldspar, or quartz, which tend to be coarse-grained, along with included mica; fine-grained basalt, rock containing plagioclase or augite; and porous materials such as pumice from volcanic eruptions. With the exception of pumice, igneous rocks weather and erode slowly, and often are found as dikes or resistant parts of the landscape where more erodible rock and soil has left them exposed.

Sedimentary rock is formed either by chemical cementation of deposited minerals, rock, or shell particles, or by precipitation of salts. They include the large groups of sandstones and limestones. The former range in particle size from very fine to coarse dependent upon the deposits from which they were derived, whereas the latter are at least 50 percent carbonates and often calcareous (sandstones may also be calcareous). Water movement through both types is likely to be very rapid, although by different routes: sandstones are usually very porous and water moves between cemented sand grains, whereas the movement through limestones tends to be along solution channels where soluble minerals have been dissolved by the water with which they are in contact.

Metamorphic rock is formed when either igneous or sedimentary rock is acted upon by heat and/or pressure, or by waters charged with gas or dissolved salts. Thus, granite becomes gneiss, shale becomes slate, and limestone becomes marble. These rocks are even more resistant to erosion and solution than are the igneous or sedimentary rock, and often form aquicludes.

Erosion is the gradual alteration, breaking up, and movement of the parent material caused by climate, geostrophic, and biospheric forces working over time (the other four soil-forming factors). The particles derived from erosional processes, or particles available directly from wind and water deposits, are known as **soil material**, and may develop into soils under the further influence of the soil-forming factors. A **soil** is distinguished from other soils and from the **soil material** (and parent material) beneath it by horizons, layers of variable thickness that are differentiated by chemical composition, color, function, organic content, structure, and texture. It may be further distinguished from other soils that are nearby by degree of horizonation, depth, and other features.

Time

The length of time that the several soil-forming factors have had available to act upon the parent material may or may not determine which of those factors is dominant. For example, if time has not given any factors a sufficient period in which to act, the soil will display the characteristics of the parent material. If there has been sufficient passage of time, one of the other soil-forming factors may have acted long enough to modify the parent material sufficiently to allow the soil to show some characteristic feature, usually in one or more of the horizons. Soil may be formed in terms of years or centuries: it depends upon how effective the soil-forming factors are, given the time within which they act on the parent material.

The length of time has no particular correlation with "age" of the soil, which is an expression of its degree of horizonation caused largely by climatic factors. Soils that have been buried by glacial and/or alluvial activity and become exposed may, in fact, be very old in terms of years, yet young in terms of development. Thus, it is the degree of morphological development that is important in terms of soil age.

Climate

The unrestrained influence of climate on the soil material is to develop a series of horizontal zones within the soil called **horizons**. The characteristics and nomenclature of the theoretical horizon system for a forest soil are shown in Fig. 5-7.

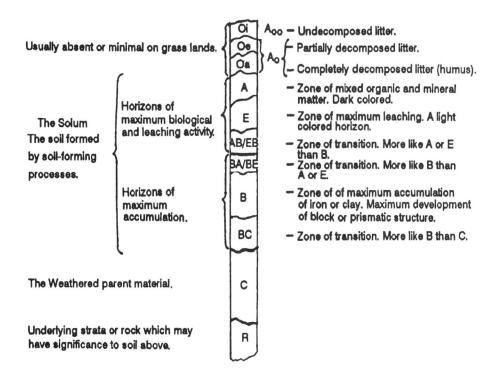

Figure 5-7 Soil profile showing theoretical horizonation (adapted from the Forest Service *Handbook on Soils*)

Generally, five broad categories of horizons are identified:

1. the *organic surface layers* characterized by the designated Oi, Oe, and Oa horizons for the litter, fermentation, and humus layers (formerly L, F, and H layers), respectively;
2. the *zone of eluviation* characterized by (a) organic matter input from the Oa to the A horizon (formerly the $A1$), (b) by nutrient removal by recycling through vegetation, (c) by colloid action (humus and clay removal), and (d) leaching (by percolating water) which leaves behind the usually more sandy and nutrient-deficient E horizon (formerly the $A2$). A transitional AB (if no E horizon is present) or AE (if the E is present) may occur;
3. the *zone of illuviation*, consisting of the B horizon, sometimes with differentiated sub-horizons designated Bs (if iron and aluminum hydrous oxides are accumulated), Bg (if waterlogged during part of the year — "gleyed"), Bx (if a dense layer called a "fragipan" that restricts water movement is present), and Bk (alkaline, with carbonate accrual, most as $CaCO_3$ or $MgCO_3$);
4. the *soil material* (C), and
5. the unweathered *parent material*, or R horizon (formerly the D or D_R horizon).

In the Oi (litter) layer, plant parts are readily identifiable; in the Oe (fermentation) layer, plant parts are actively undergoing change by chemical and biological (such as fungi and insects) agents and may be barely recognizable as plant material; in contrast, in the Oa (humus) layer, the organic material is a dark, amorphous mass of indistinguishable organic origin. The material in this layer mixes with mineral (inorganic) soil particles to form the hydrologically important A horizon.

If the litter is acidic in nature, there may not be a great deal of microbial activity and, as a consequence, most of the litter will remain undecomposed on the surface; there will be little humus, and the mull layer will be slow to develop. Infiltration rates are likely to be high and the resultant sandy nature of the A horizons will not provide a great amount of retention storage. In contrast, if the conditions are favorable for chemical and bacterial activity to break down the organic matter, the Oe, Oa, and A layers will be relatively thick, while the Oi layer will be considerably thinner (dependent upon season and leaf fall). If the surface is undisturbed, infiltration rates may still be high, and the layer with a combination of organic and inorganic material will provide high retention storage potential.

The occurrence of frost is an important climatic influence on hydrologic properties of soil (Buckman and Brady 1960; Thorud and Anderson 1969). It maintains the high infiltration and percolation rates associated with low bulk density and may, owing to the earlier-observed fact that the vapor pressure over ice is less than that over water, be responsible for increasing soil moisture in the surface horizons, dependent upon the form of frost. **Honeycomb** frost is common in the litter layer, and **stalactite** frost "is unimportant to infiltration as it is obviously of an open porous nature, transient in time and of very limited areal occurrence (Trimble, Sartz, and Pierce 1958). A third type, **granular** frost, was found to have higher infiltration rates than where there was no frost (Trimble, Sartz, and Pierce 1958), while **concrete** frost forms at greater depths than other types (Striffler 1959) and appears to influence infiltration only if the soil above it is saturated or the frozen conditions extend all the way to the surface. Prévost et al. (1990) reported that "the natural topographical drainage network was augmented in the spring by superficial concrete frost." Soil water may also be brought to the surface by ice formation, as well as be limited by excessive soil moisture withdrawals owing to complex heat balance relations and changes therein by the heat of condensation (Soons and Greenland 1970) during ice formation and melt.

The E horizon is the zone of maximum leaching and, in the extreme, may appear to be bleached nearly white by acidic leachate, typically enhanced by naturally acidic precipitation and by infiltrating water made more acidic by contact with the acidic litter from pioneer forest species. The E horizon is usually quite sandy, devoid of substantial amounts of plant nutrients, and with high percolation rates. The AB (or EB) and BA (or BE) horizons are transition zones, as noted in Fig. 5-7. The B is the primary zone of deposition within the soil profile, and may be differentiated by sub-designations showing dominant element(s) deposited. This is the zone from which the feeding roots of plants recycle nutrients. Together, the A and B horizons are known as the **solum**: the **soil** consists of the surface organic, A, B, and C horizons; and the parent material is the R.

All of these horizons will not necessarily be found in one soil: they are the theoretical potential. However, with different climates scattered around the globe that typically affect vegetation distribution and seasons of water availability, evapotranspiration, heat, and drought, they can be used to classify soils into broad groups. This was done in midcentury in the form of the **Great Soil Groups**, a climate-based classification now largely replaced by the less hydrologically applicable if scientifically more accurate Comprehensive Soil Classification System (CSCS), also known as the "Seventh Approximation" and published as "Soil Taxonomy" by the Soil Conservation Service (1975). This development was in spite of a call by Musgrave (1956) for "*other and more efficient and expeditious methods of evaluating soils hydrologically*" [emphasis in original]; "grouping according to reponse under intense rainfall" was a suggested alternative. FitzPatrick (1971) points out that "the most striking feature [of the CSCS] is the overall appearance of or orderliness," but that the groupings are not clearcut or mutually exclusive. The new classification clearly serves the need for a neat, pure abstruse classification scheme, not for better understanding of the soils, which are not necessarily orderly, neat, or abstruse. More importantly, from a management standpoint, the groupings under the CSCS do not relate well to climate, the driving element of the hydrologic system, whereas the Great Soil Groups do. The remaining discussion in this section, then, focuses on the Great Soil Groups, with the CSCS presented where the relationship can be provided.

Soils and vegetation are further related to the two important climatic characteristics of Precipitation Effectiveness and Temperature Efficiency, as shown in the greatly simplified, diagrammatic representation of the areal distribution of soils and vegetation (Figs. 5-8 and 5-9). The two climatic factors were defined by complex mathematical formulas by C. W. Thornthwaite as he sought to model the water budget and use it for climatic classification; they are scaled along the horizontal and vertical dimensions of the diagram, and are generalized into the descriptive comments shown at the four corners (Buckman and Brady 1960). The relationship has been usefully exploited in the soil-vegetation survey system in use throughout California (Department of Natural Resources 1955).

The natural distribution of soils and vegetation for the continuous United States are obviously related, as will be noted by comparing the two maps in Figs. 5-10 and 5-11. Like Figs. 5-8 and 5-9, "the right to left [and north to south] succession of soils and vegetation shown in the diagrams [and maps] approximates those that will be encountered" east of the Cascade mountains in the United States (Buckman and Brady 1960).

The current classification into the CSCS groups represent recognizable profiles of (formerly) **zonal** (a climate-based classification) soils where the unrestrained effects of climate such as temperature and precipitation are ramified in the soil profile by horizons that are characteristic of the climatic zone. The Great Soil Groups have no direct translation to the current CSCS groups; however, the current soils map of the United States (Fig. 5-10) may be visually compared with the Great Soils Group layout (Fig. 5-8), and rough comparisons may be made in Table 5-5.

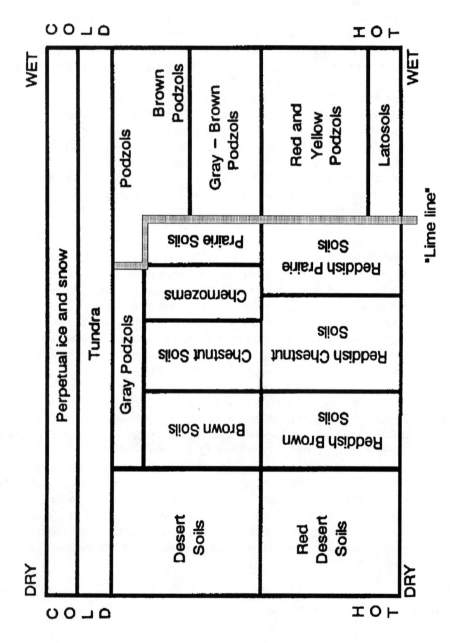

Figure 5-8 Great Soil Group distribution

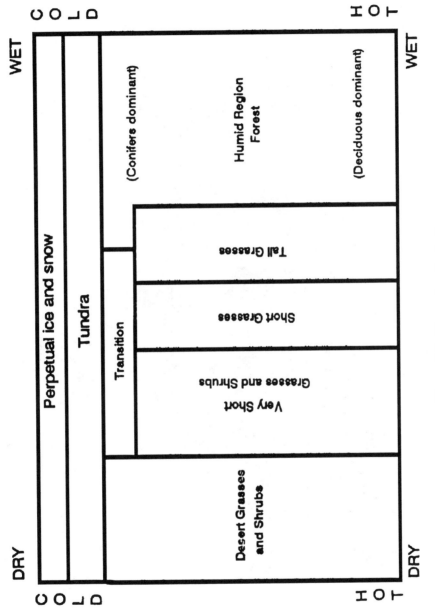

Figure 5-9 Vegetation distribution

157

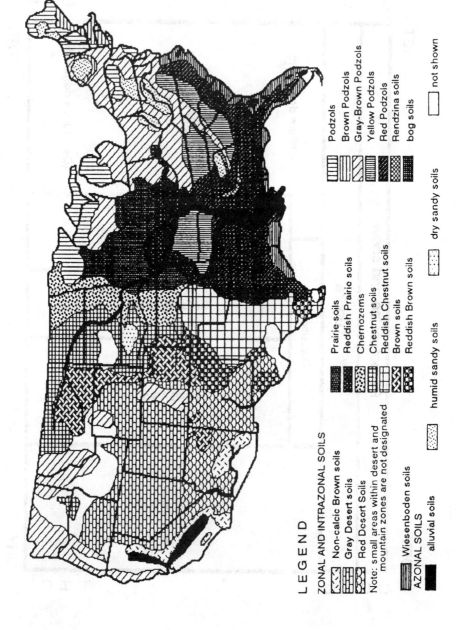

Figure 5-10 Distribution of Great Soil Groups in the United States

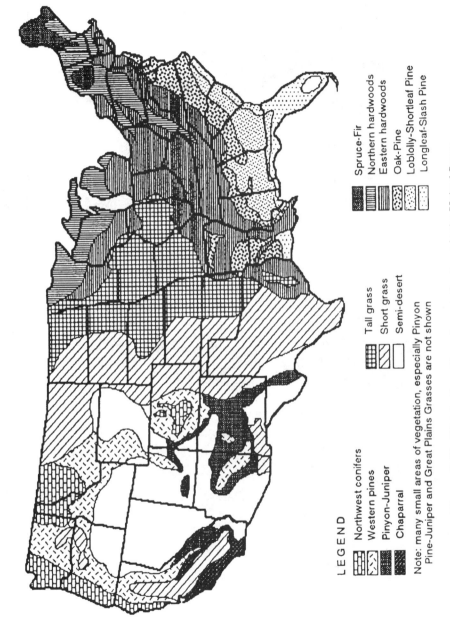

Figure 5-11 Distribution of broad vegetative types in the United States

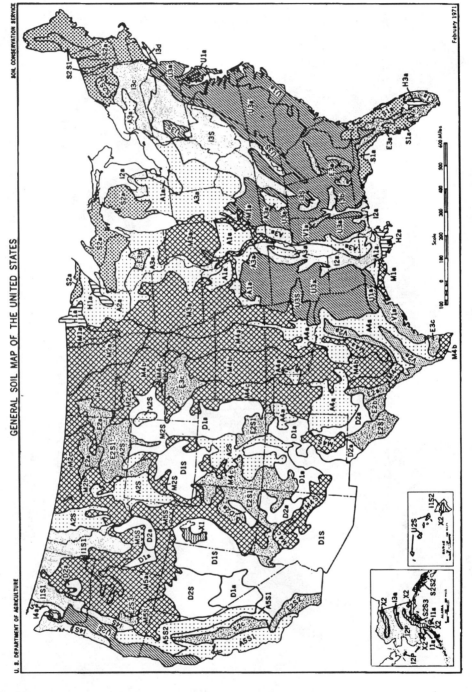

Figure 5-12 Distribution of soils under the Comprehensive Soil Classification System (Soil Conservation Service 1975)

ALFISOLS

AQUALFS
A1a—Aqualfs with Udalfs, Haplaquepts, Udolls; gently sloping.

BORALFS
A2a—Boralfs with Udipsamments and Histosols; gently and moderately sloping.
A2S—Cryoboralfs with Borolls, Cryochrepts, Cryorthods, and rock outcrops; steep.

UDALFS
A3a—Udalfs with Aqualfs, Aquolls, Rendolls, Udolls, and Udults; gently or moderately sloping.

USTALFS
A4a—Ustalfs with Ustochrepts, Ustolls, Usterts, Ustipsamments, and Ustorthents; gently or moderately sloping.

XERALFS
A5S1—Xeralfs with Xerolls, Xerorthents, and Xererts; moderately sloping to steep.
A5S2—Ultic and lithic subgroups of Haploxeralfs with Andepts, Xerults, Xerolls, and Xerochrepts; steep.

ARIDISOLS

ARGIDS
D1S—Argids with Orthids, Orthents, Psamments, and Ustolls; gently and moderately sloping.
D1S—Argids with Orthids, gently sloping; and Torriorthents, gently sloping to steep.

ORTHIDS
D2a—Orthids with Argids, Orthents, and Xerolls; gently or moderately sloping.
D2S—Orthids, gently sloping to steep, with Argids, gently sloping; lithic subgroups of Torriorthents and Xerorthents, both steep.

ENTISOLS

AQUENTS
E1s—Aquents with Quartzipsamments, Aquepts, Aquolls, and Aquods; gently sloping.

ORTHENTS
E2a—Torriorthents, steep, with borollic subgroups of Aridisols; gently or moderately sloping.
E2b—Torriorthents with Torrerts; gently or moderately sloping.
E2c—Xerorthents with Xeralfs, Orthids, and Argids; gently sloping.
E2S1—Torriorthents; steep, and Argids, Torrifluvents, Ustolls, and Borolls; gently sloping.
E2S2—Xerorthents with Xeralfs and Xerolls; steep.
E2S3—Cryorthents with Cryosamments and Cryandepts; gently sloping to steep.

PSAMMENTS
E3a—Quartzipsamments with Aquults and Udults; gently or moderately sloping.
E3b—Udipsamments with Aquolls and Udalfs; gently or moderately sloping.
E3c—Ustipsamments with Ustalfs and Aquolls; gently or moderately sloping.

HISTOSOLS

HISTOSOLS
H1a—Hemists with Psammaquents and Udipsammants; gently sloping.
H2a—Hemists and Saprists with Fluvaquents
H3a—Fibrists, Hemists, and Saprists with Psammaquents; gently sloping.

INCEPTISOLS

ANDEPTS
I1a—Cryandepts with Cryaquepts, Histosols, and rock land; gently or moderately sloping.
I1S1—Cryandepts with Cryochrepts, Cryumbrepts, and Cryorthods; steep.
I1S2—Andepts with Tropepts, Ustolls, and Tropofolists; moderately sloping to steep.

AQUEPTS
I2a—Haplaquepts with Aqualfs, Aquolls, Udalfs, and Fluvaquents; gently sloping.
I2P—Cryaquepts with cryic great groups of Orthents, Histosols, and Ochrepts; gently sloping to steep.

OCHREPTS
I3a—Cryochrepts with cryic great groups of Aquepts, Histosols, and Orthods; gently or moderately sloping.
I3b—Eutrochrepts with Uderts; gently sloping.
I3c—Fragiochrepts with Fragiaquepts, gently or moderately sloping; and Dystrochrepts, steep.
I3d—Dystrochrepts with Udipsamments and Haplorthods; gently sloping.
I3S—Dystrochrepts, steep, with Udalfs and Udults; gently or moderately sloping.

UMBREPTS
I4a—Haplumbrepts with Aquepts and Orthods; gently or moderately sloping.
I4S—Haplumbrepts and Orthods; steep, with Xerolls and Andepts; gently sloping.

MOLLISOLS

AQUOLLS
M1a—Aquolls with Udalfs, Fluvents, Udipsamments, Ustipsamments, Aquepts, Eutrochrepts, and Borolls; gently sloping.

BOROLLS
M2a—Udic subgroups of Borolls with Aquolls and Ustorthents; gently sloping.
M2b—Typic subgroups of Borolls with Ustipsamments, Ustorthents, and Boralfs; gently sloping.
M2c—Aridic subgroups of Borolls with Borollic subgroups of Argids and Orthids, and Torriorthents; gently sloping.
M2S—Borolls with Boralfs, Argids, Torriorthents, and Udolls; moderately sloping or steep.

UDOLLS
M3a—Udolls, with Aquolls, Udalfs, Aquepts, Fluvents, Psamments, Ustorthents, Aquepts, and Albolls; gently or moderately sloping.

USTOLLS
M4a—Udic subgroups of Ustolls with Orthents, Ustochrepts, Usterts, Aquents, Fluvents, and Udolls; gently or moderately sloping.
M4b—Typic subgroups of Ustolls with Ustalfs, Ustipsamments, Ustorthents, Ustochrepts, Aquolls, and Usterts; gently or moderately sloping.
M4c—Aridic subgroups of Ustolls with Ustalfs, Orthids, Ustipsamments, Borolls, Ustochrepts, Torriorthents, Aquolls, Salorthids, and Ustocryepts; gently or moderately sloping or steep.
M4S—Ustolls with Argids and Torriorthents; moderately sloping or steep.

XEROLLS
M5a—Xerolls with Argids, Orthids, Fluvents, Cryoboralfs, Cryoborolls, and Xerorthents; gently or moderately sloping.
M5S—Xerolls with Cryoboralfs, Xeralfs, Xerorthents, and Xererts; gently sloping or steep.

SPODOSOLS

AQUODS
S1a—Aquods with Psammaquents, Aquolls, Humods, and Aqualfs; gently sloping.

ORTHODS
S2a—Orthods with Boralfs, Aquents, Orthents, Psammants, Histosols, Aquepts, Fragiochrepts, and Dystrochrepts; gently or moderately sloping.
S2S1—Orthods with Histosols, Aquents, and Aquepts; moderately sloping or steep.
S2S2—Cryorthods with Histosols; moderately sloping or steep.
S2S8—Cryorthods with Histosols, Andepts and Aquepts; gently sloping to steep.

ULTISOLS

AQUULTS
U1a—Aquults with Aquents, Histosols, Quartzipsamments, and Udults; gently sloping.

HUMULTS
U2S—Humults with Andepts, Tropepts, Xerolls, Ustolls, Orthox, Torrox, and rock land; gently sloping to steep.

UDULTS
U3a—Udults with Udalfs, Fluvents, Aquents, Quartzipsamments, Aquepts, Dystrochrepts, and Aquults; gently or moderately sloping.
U3S—Udults with Dystrochrepts; moderately sloping or steep.

VERTISOLS

UDERTS
V1a—Uderts with Aqualfs, Eutrochrepts, Aquolls, and Udolls; gently sloping.

USTERTS
V2a—Usterts with Aqualfs, Orthids, Udifluvents, Aquolls, Ustolls, and Torrerts; gently sloping.

Areas with little soil
X1—Salt flats.
X2—Rock land (plus permanent snow fields and glaciers).

Slope classes

Gently sloping—Slopes mainly less than 10 percent, including nearly level.
Moderately sloping—Slopes mainly between 10 and 25 percent.
Steep—Slopes mainly steeper than 25 percent.

Figure 5-12 Continued

Table 5-5 Comparison of old and new soil classification system
(after Brady 1974)

Name[b]	Formative Element[a]		Approximate Equivalents in the Old System
	Derivation	Pronunciation	
Entisol	Nonsense symbol	re*cent*	Azonal, some Low-Humic Gley soils
Vertisol	L. *verto*, turn	in*vert*	Grumusols
Inceptisol	L. *inceptum*, beginning	in*cept*ion	Ando, Sol Brun Acide, some Brown Forest, Low-Humic Gley, and Humic Gley soils
Aridisol	L. *aridus*, dry	ar*id*	Desert, Reddish Desert, Sierozem, Solonchak, some Brown and Reddish Brown soils and associated Solonetz
Mollisol	L. *mollis*, soft	mo*ll*ify	Chestnut, Chernozem, Brunizem (Prairie), Rendzinas, some Brown, Brown Forest, and associated Solonetz, and Humic Gley soils
Spodosol	Gk. *Spodos*, wood ash	Po*d*zol; odd	Podzols, Brown Podzolic soils, and Groundwater Podzols
Alfisol	Nonsense symbol	Ped*alf*er	Gray-Brown Podzolic, Gray Wooded, and Non-Calcic Brown soils, Degraded Chernozems, and associated Planosols and some Half-Bog soils
Ultisol	L. *ultimus*, last	u*lt*imate	Red-Yellow Podzolic soils, Reddish-Brown Lateritic soils of the U. S., and associated Planosols and Half-Bog soils
Oxisol	F. *oxide*, oxide	*ox*ide	Laterite soils, Latosols
Histosol	Gk. *histos*, tissue	*hist*ology	Bog soils

[a] The italicized letters in the pronunciation column are used in the suborder and great group categories to identify the order to which they belong.
[b] Note that all orders end in sol (Latin *solum*, soil).

A simple and older classification of zonal soils was based upon the degree to which elements have been leached out of all or some of the soil profile and is relate to precipitation effectiveness. West of the line representing about 30 in. of annual precipitation in the United States (not including the Complex western mountain and coastal areas), there is insufficient leaching to remove the calcium (Ca) from the soil, and the soils are broadly classified as **Pedocals**; east of the line there is sufficient leaching to remove the calcium and leave behind the now-dominant aluminum (Al) and iron (Fe) in soils, broadly classified as **Pedalfers**.[7] Pedocals support the grassland and desert vegetation of the United States prairie and desert

---

[7] While the terms are not currently in use, they appear in the literature and represent two broad classes of soil which, in a rather simple manner, reflect relationships between climate, soil, and water.

regions, while pedalfers tend to support forests and to exhibit soils with strongly developed E horizons and thick surface litter layers. While quite crude, the classification into these two categories is helpful in understanding relationships between several of the climatic factors of soil formation. The line separating the pedocals from the pedalfers is known as the "lime" line. The pattern is repeated in less well-defined areas of the country, but is not as readily modeled as shown in Fig. 5-8. Again, the relationship aids in understanding the connection between soils, vegetation, and climate but, with the GSG classification no longer in use and no one-to-one correlation between the "Seventh Approximation" and the GSG, translation of this figure to modern terminology is not possible, useful as it might be. Hydrologists will probably continue to use the Great Soil Groups to whatever extent it serves the hydrologic purpose and linkage with climate; the new CSCS system does not.

Where there has not been sufficient time for climatic factors to develop any horizonation in the soil material, the soils are classified as **azonal**; where the typical zonal pattern for a region has been modified by geologic or topographic influences, especially drainage, the soils are classified as **intrazonal**.

Geology and Topography

The primary effect of geologic and topographic factors in soil formation are the distribution of texture classes within the soil profile and drainage. The factors are ramified in (1) the geologic processes through which soil material might be accumulated, and (2) the modification of long-term effects of precipitation and climate on horizonation by local topographic features. Historical geologic processes affect the creation of soil material; topography affects soil development.

Geologic processes that produce or redistribute soil material may be classified in two ways: (1) by whether the soil material develops from the parent material below it or is moved into the site of soil development from another location, and (2) whether the soil is stratified by particle size.

    A. Soils that develop from soil material derived from underlying parent material developed *in situ*, particle size variable, unstratified...**Residual Soils**
    B. Soils developed from soil material transported in from another location
        1. by gravitational forces:
            unstratified, talus, rock falls, avalanches..**Colluvial Soils**
        2. by water
             particle size variable, stratified, not all rounded, calcareous (oceans) **Marine Soils**
            rounded, usually very fine particles in lakes (still waters) .... **Lacustrine Soils**
            rounded particles, well stratified by rivers (running waters)................**Alluvial Soils**
            angular particles, well stratified, (water-from-ice).....................**Fluvioglacial Soils**
        3. by ice
            coarse, angular particles, poorly stratified, if at all................................ **Glacial Soils**
        4. by wind
            very fine particles, well stratified (dunes, Loess deposits)....................**Aeolian Soils**

The hydrologic significance of this classification lies in the occurrence and extent of stratification, and in the characteristics of the soil that develops from the soil material. The results of the geologic history of the soil material on the soil that ultimately develops therefrom may also be further influenced by the soil-forming factors, including local topography.

The effect of local topography on soil development is to alter the typical amount of precipitation that percolates through the developing profile. This is accomplished by the process of leaching, in which soluble materials on and in the surface layers are dissolved and transported downward to where they are either deposited in lower horizons (perhaps to be recycled by vegetation) or are moved completely out of the soil into a water body. Increased or decreased opportunity for infiltration is ramified in how the nutrients are distributed within the soil profile and, consequently, in the degree of horizonation that develops. For example, if there is a local concavity in the land surface, more water will be available for infiltration, percolation, and leaching; the soil that develops will be different in characteristics from an immediately adjacent one that has a continuously sloping or a convex surface. Specifically, it is this process that is influenced by topography.

Another topographic consideration is aspect, which affects evapotranspiration, snowmelt, and erosion. Evapotranspiration and snowmelt influence the amount of water available for percolation and attendent leaching, consequently affecting nutrient availability and distribution and, over the years, erosion rates and extent will in turn be affected by water available for plant growth and soil protection.

The result of the foregoing along a hillside, for example, gives rise to a **catena** (or topographic sequence). A catena is a series of soils that develop from the same or similar soil or parent material, under common climatic conditions, but that differ with respect to characteristics of the solum because of differences in relief or drainage (Fig. 5-13).

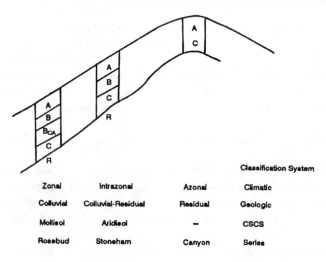

Figure 5-13 A typical catena of the high plains

The conditions under which the catena develops are influenced by aspect and slope. This is because infiltration, percolation, and basic soil water relations are profoundly affected by solar radiation, evapotranspiration and decomposition rates, freeze/thaw cycles, and depth of soil development, all of which are influenced by the temperature regime which, as was shown in Chapter 2, is greatly dependent upon latitude, aspect, and slope. Essentially, the catena is condensing the change one would normally observe over several hundred miles of latitude into a relatively short geographic distance (perhaps one hundred yards). A typical sequence of soils in a catena will exhibit poorly- to moderately-drained soils at each end, and well-drained soils near the middle. The catena might serve as a basis for further classification of soils to assist hydrologic interpretation (Chiang and Petersen 1970). The catena is important hydrologically because it involves the processes of infiltration, percolation, and deep seepage as well as storage of soil water.

Biological Agents

Factors in the biotic community influence soil fertility, organic content, and humus type. Soil fauna, including earthworms and burrowing insects, increase and maintain porosity and permeability in the soil. Many of the pore spaces created are re-filled with waste materials, but these tend to be organic and lower in density than the soil that is displaced, thus enhancing soil storage potential. This activity within the surface horizons is especially important, along with fermentation of litter and mixing of inorganic and organic material, in establishing and maintaining a well-developed A horizon. Although some breakdown of organic matter is accomplished by faunal activity, especially earthworms and insects, bacteria are responsible for most of the decomposition, which affects pH.[8]

Organic content of the soil reflects and is affected, too, by root development and decay. Root development, along with burrowing insects and animals, displaces soil particles and, in the process, affects porosity and water relations. Rotted root channels provide free water drainage, sometimes to great depths: this occurs if some of the rotted material remains and aids directly in water movement through capillary action, or if it is removed and leaves behind an opening that is maintained by the soil structure that surrounded the former root. Bacteria further the recycling of organic material and nutrients. The combination loosens and aerates the soil, lowering bulk density, increasing water-holding capacity, infiltration, and percolation.

It is the combined effect of soil biotic agents that determine humus type. The humus type is described by which of the two horizons, **mor** (the *Oi* layer) or **mull** (the *A* layer), is better developed (Fig. 5-14). The mull type is identifed by a well-developed *A* layer as the result of active decomposition of the litter (*Oi*) and its mixing with inorganic material in the *A* (mull) horizon. The lack of decomposition in the mor type results in litter remaining on the surface (a thick *Oi*), and small or nonexistent *Oe*, *Oa*, and *A* horizons. The resultant acidic leachate (percolating water) leaches all but the iron and aluminum out of the upper soil, leaving a well-defined *E* horizon.

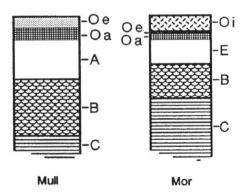

Figure 5-14 Mull and mor humus types

---

[8] Log of the hydrogen ion activity: a measure of the degree of soil acidity or alkalinity along a scale of 0 to 14, 7.0 being neutral. For further explanation, see the portion of Chapter 6 dealing with water quality.

There are endless possibilities between the two humus extremes: the mor humus type generally exhibits high infiltration and percolation rates, high amounts of detention storage, and fairly low retention. In contrast, the mull humus type will exhibit low detention and high retention storage, and infiltration will be high if the surface is undisturbed. Mull humus types tend to develop in concert with later successional stages of vegetation where reproduction will survive under the dense shade of climax vegetation and nutrients will be recycled from the litter to the *Oi*, to the *B*, and back to the plants. Mor humus types tend to be associated with early plant successional stages, where nutrients are retained on the surface of the forest floor and not recycled, and the reproduction that survives is of a new, subsequent successional stage, which utilizes nutrients that are available.

It is possible to manage the humus type, as illustrated in Fig. 5-15, showing the upper soil profile under a 12-year-old Eastern White Pine stand on one of the Coweeta watersheds: prior to cutting the hardwoods on the site, the humus type was classed as a mull; after the white pine was established, the acidic litter inhibited microbial activity in the *Oe* and *Oa* horizons, and the *Oi* layer built up so that now the humus type is distinctly a mor type.

Clearly, the management of soil and vegetation resources must be founded upon a straightforward understanding of the hydrology of the area which, in turn, requires these fundamentals of soil formation.

Types of Soil Moisture Storage

Water is stored on and in the soil in depression, detention, and retention storage. The former is on the surface of the soil, whereas the latter two are within the soil profile. The three are interconnected (cf., Fig. 1-2): water can move between depression and detention storage, although it generally is from the former to the latter. Water can also move from either depression or detention storage directly into retention storage, but not vice versa. Water moves out of retention storage only by evaporative processes which have sufficient tension to overcome the combined adhesive and cohesive forces that keep water in the capillary pores.

Depression Storage

This is water stored, usually temporarily, on the surface of the soil in small puddles. It may be quite high following a rain or snowmelt event, especially if the soil is saturated. Eventually, climatic conditions following a rainfall event will either permit evaporation of the depression storage, or it may gradually seep into the soil itself. It is logical, therefore, to think of depression storage as part of interception storage, or to consider both as water temporarily stored on the soil surface and interception storage as depression storage.

Detention Storage

Detention storage is water temporarily detained in the soil in the noncapillary pores. It is subject to drainage, and is often referred to as "drainage water," responding to the dominating force of gravity, as illustrated in Fig. 5-16. Detention storage is that water which is stored below 1/3 atmosphere of tension, and generally will flow out of the soil within 24 hr. Detention water is in the soil for too short a time, and often under conditions not particularly favorable to evaporative processes to permit much of it being returned directly to the atmosphere by transpiration. This is due to the rather obvious facts that: (1) weather is likely to be cloudy in the short number of daylight hours following a rain, (2) the relative humidity will be high,

Figure 5-15 Humus type under an eastern white pine stand

which inhibits any evaporative process, and (3) typically, half of the 24-hr period will be nighttime when the energy for transpiration is either not available or is greatly reduced.

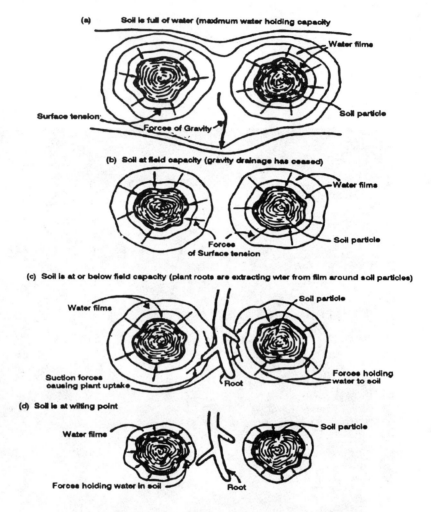

Figure 5-16 Schematic diagrams of soil-moisture relationships (from Forest Service *Handbook on Soils*)

The dividing point between detention and retention storage is **Field Capacity**, defined as that amount of water left in the soil after it has been saturated and allowed to drain for 24 hr. The sequence of drainage/evapotranspiration/retention and the different types of storage are shown in Fig. 5-17. The thinnest water films retained around each soil particle may be at tensions as high as 10,000 atmospheres (Buckman and Brady 1960). Since soils of different textures drain at different rates, field capacity is identified as being approximately 1/3 atmosphere (-0.33 bar[9]) for clayey soils, and -0.1 bar to -0.2 bar for sandy soils.

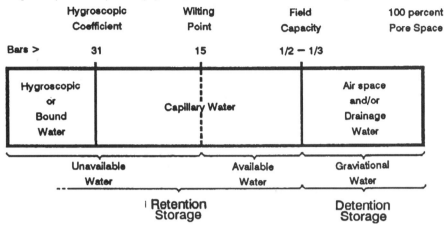

Figure 5-17 Soil moisture storage relations

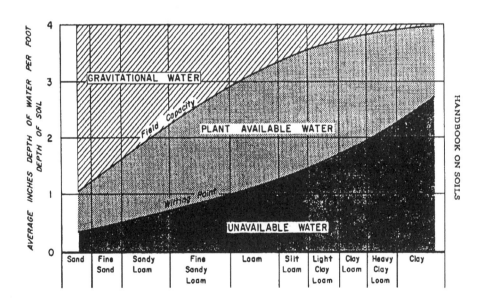

Figure 5-18 Water Relations and Soil Texture
(from Forest Service *Handbook on Soils*)

---

[9] Negative bars (-bars) are used to indicate water held at forces greater than gravity; positive bars (+bars) used to designate water under pressure.

Retention Storage

Retention storage is water stored in the capillary pores, held at tensions between 1/3 and 31 bars against the force of gravity. (Water held at tensions beyond 31 bars is "bound" or hygroscopic water.) That part of retention water that is between the wilting point, or 15 bars, and the hygroscopic coefficient of 31 bars is not available to plants. The greater the percentage of the total soil pore space in capillary size ranges, therefore, the greater may be the amount of retention storage[10] and the greater the amount of water available for transpiration and plant growth. The relationships are shown in Fig. 5-17. The relationship between the relative portion of the pore space occupied by retention water and soil texture is shown in Fig. 5-18. The greatest amount of (and percentage of pore space occupied by) gravitational water is found in sandy soils; the greatest unavailable water in clays; and the largest amount of available water is found in loams.

MEASUREMENT-INSTRUMENTS-LIMITATIONS

There are five general methods (plus a miscellaneous category) based upon several different types of devices for monitoring soil moisture in the aerated zone of the lithosphere. The categories are gravimetric, tensiometric, electrical resistivity, neutron scattering, and remote sensing.

Gravimetric Method

The oldest and most widely used method of soil moisture determination, gravimetric sampling, is accomplished by removing a cylinder of soil of known dimensions, weighing it, drying it, reweighing it to determine the percent water weight loss, and then multiplying by the bulk density to convert to a percent volume loss figure, which can be translated to an inches-depth figure.

For example, a soil sample of bulk density (BD) 1.19 and a water content percent by weight (P_w) of 21, will have a percent by volume (P_v) of:

$$P_v = P_w \times BD = 21 \times 1.19 = 25$$

Thus, a cu ft of soil contains

$$\approx \times 1728 \text{ in}^3/\text{ft}^3 = 432 \text{ in}^3$$

of water. More generally,

$$\approx \times 12 \text{ in.} = 3$$

that is, a foot (depth) of soil contains 3 in. of water. By extension, an acre of soil 5 feet deep would contain 15 in. of water, or 1¼ AF per acre.

A variety of sampling tubes for extracting small soil volumes were developed at the several Agricultural Research Service, Forest Service, and Soil Conservation Service

---

[10] If carried to extremes, this statement implies that clay soils would provide more water for plant growth than, say, loams, an obviously false conclusion.

experiment stations (Lull and Reinhart 1955). One of the most popular was the King (or Veihmeyer) Tube, made of heavy steel in variable lengths up to 20 ft, with an inside diameter of 1 in. and an outside diameter of 1-1/8 in., as described originally by Veihmeyer (1929). The tube was driven into the soil either from the natural surface or within a stepped pit that provided sampling at different levels with less disturbance to the sample (but more to the site). Driving was accomplished with a 15-lb weight welded on a rod that fit into the open upper end of the tube. The tube could be jacked out of the soil, or a cross-ways keyhole within the weight allowed it to be turned sideways and used to help remove the tube from the soil by upward swings against "ears" at the upper end of the tube while holding the weight by the ends of the rod to which it was secured. Needless to say, this was strenuous exercise for the soil sampling crew. Further, excessive time on the site and trampling caused considerable compaction of the soil, and if a pit were dug, disturbance would be so great that the site was rendered of questionable value for further study of soil moisture, which eliminated the opportunity for obtaining reliable long-term trends of soil moisture.

The soil sample obtained while driving through some desired depth of soil (usually 6 in. at a time) was removed and deposited in a numbered 4-oz metal can, the tare of which had been recorded. Back in the laboratory, the can and contents were weighed, dried at $105^{\circ}C$, reweighed, and the percent by weight of the water in the soil was calculated (Olson and Hoover 1954). Devices to assist in the preparation and handling of samples are described by Ferguson and Duke (1954).

Bulk density measurements may be obtained by the block method or by removing a core (Broadfoot 1954). In the former, a pit is dug isolating a block of soil which can be removed with minimum disturbance for measurement, weighing, and drying to drive off moisture. This is not only an energy- and time-consuming procedure, it requires a rather large soil pit, which disturbs the site where the research is being conducted. Several different types and sizes of corers, usually metal cylinders on the end of a T-handle were developed by researchers at the several soils experiment stations. In order to obtain an undisturbed sample, some of these were elaborate in design: a thin-walled cylinder might be inserted in the thicker outer cylinder which could be used to penetrate the soil and thereby minimize error in the volume sampled. Bulk density samples are most reliable when tensions are between 0 and 1 atmosphere (Reinhart 1954).

Overlooked in early gravimetric soil moisture determinations of percent by volume was inclusion of an error[11] term. Usually, several percent-by-weight samples were taken in order to obtain an average soil moisture content for a given site and date; an error term could be (and occasionally was) calculated for this component. The value of P_w may be quite accurately determined; however, only one bulk density sample was usually taken owing to site disturbance and tediousness of sample preparation and removal; no error term could be calculated. The formula for determination of percent by volume, P_v, from percent by weight, P_w, actually is:

$$P_v = P_w \times BD \pm e$$

---

[11] An extensive report on statistical methods commonly used in soil data analysis was prepared by the Forest Service (Blaney, Ponce, and Warrington 1984).

where *BD* is the bulk density and *e* is the error. Hewlett and Douglass (1961) point out that this error term is potentially large since both the P_v and P_w terms do in fact have errors associated with them. As a consequence, the true error term was often greater than the variation of soil moisture encountered and most of the gravimetric soil moisture data for the soils studied were not at all reliable for long-term trends or for comparing moisture levels between different soils, vegetative covers, or different treatments.

A further difficulty with the gravimetric method is that during the time-consuming process of sample collection (sometimes a day-long affair), hydrologic processes continue and evaporation or rainfall may interfere with obtaining data from more than one area under comparable conditions. "Despite the disadvantages ..., the gravimetric method of soil moisture measurement is the only one capable of direct measure of soil water quantitatively ..." (Olson and Hoover 1954). In addition, precision aside, it is the method that is used to calibrate all the other methods.

Tensiometric Methods

These methods rely upon the tension at which water is retained in the soil. A porous medium at one end of a tube, usually 6 in. to 4 ft (but up to 15 ft) in length, is introduced into the soil and is allowed to remain in place over several weeks so that its water content is the same as that of the soil. The upper end of the device is a mercury or vacuum-type manometer which indicates the tension at which the water is held in the porous medium. The devices monitor soil moisture over a rather limited range of 0 to 0.84 atmospheres which, in fine-textured soils, is only about half the range from field capacity to wilting point; in coarser soils, it may be as much as 90 percent of the range (Lull and Reinhart 1955). At tensions greater than -1 bar, air bubbles may develop in the tensiometer's water column, destroying its ability to monitor tensions.

The principal value of the tensiometer is that there is little disturbance to the site. However, water may run down the tube itself, and water relations at the tube-soil interface may therefore not be identical to those in the soil being sampled. As a consequence, good relative measurements over time may be easier to obtain than determination of concurrent differences in moisture content. Another major disadvantage is that the site must be very well protected, for the instruments are left in situ. Finally, the devices must also be calibrated by some other method of soil moisture determination.

Ground water, that is, subterranean water at atmospheric pressure or greater, can be monitored with a **piezometer**. An open-ended, solid-walled pipe inserted in a water-bearing stratum holds water at the height to which the water pressure in the stratum holds it. The upper surface defined by these levels is called the *piezometric surface*. If the well is *artesian*, the piezometric surface is above the land surface.

Electrical Resistivity Methods

Since water is a good conductor of electricity and soil is not, the resistance to an electrical current of a block of moist soil may be used to indicate the amount of moisture in the soil. Direct use of electrodes is not possible since electrical resistance is also influenced by temperature and dissolved salts in the soil. The result is that small fluctuations of salts, which actually conduct the current, were found to have a greater effect on resistivity than fluctuations in soil moisture (Olson and Hoover 1954).

As with tensiometers, a porous medium is introduced into the soil and allowed to come into equilibrium moisture content. Initially, these devices were constructed of plaster-of-Paris with embedded electrodes; later, a variety of materials were used. Theoretically, a properly constructed sensing unit is not affected by salt concentrations, but must be capable of monitoring temperature. Since the electrical current moves through water because salts are present, the desired accuracy and stability are not practicably attainable. The device designed by E. A. Colman (Colman and Hendrix 1949) is a sandwich of fiberglass and monel screens that serve as inner electrodes and outer shells measuring 1.5 in. x 1.0 in. x 0.12 in. The electrode package, which includes a thermistor, is placed in the soil at the desired level with minimal disturbance and the wires connected to the electrode are deflected downward for a few inches (to preclude water traveling down the wires and into the sensor) before being extended to the surface where, whenever desired, they may be connected to a meter that monitors the resistance of the buried unit. The principal monitoring device is known as the Colman Meter, consisting of a power source and an alternating current ohmeter with readout on a microammeter.

Again, as with tensiometers, site disturbance is at a minimum, and good temporal data may be obtained. Cost is less than with tensiometers, and there is less chance in the field of disturbance to the sensing devices, which are also inexpensive. The meter itself is only taken into the field when actual measurements are being made, and only one is needed to monitor soil moisture over a large area. Calibration for each soil must be done in order to evaluate volumetric soil moisture data. However, for most soils there is a relatively constant relationship between tension and resistivity and "a calibration curve can be constructed so that moisture tensions can be related to resistance measurements" (Olson and Hoover 1954). At high water content, that is, at low tensions, the accuracy of the resistivity method falls off.

Neutron Scattering Methods

About the time of perfection of the foregoing methods, it was discovered that introduction of a high-energy neutron source into the soil could be used for rapid, repetitive, and highly accurate soil moisture measurements. The neutrons are slowed only by collisions with the H^+ ion, the presence of which is largely due to soil water content. (In highly organic soils, a significant percentage of the H^+ ions would be organic; thus, the method is of limited use in these situations.)

The neutron scattering equipment consists of a portable scaler for counting the returning (slowed) neutrons and for calibration, including battery power; a 1.5-in.-diameter probe (15 in. long) containing a radium-beryllium source of fast neutrons with a half-life of 1620 yr, a slow-neutron detector, and amplifier; and a cable connecting the two units. The probe is contained in a lead shield which, when mounted on top of a 1.55-in.-diameter metal (usually aluminum) access tube of any length that has been previously and permanently inserted in the soil, allows the probe to be lowered without radiation danger to the operator. The source can be positioned quite precisely in the tube by the length-calibrated cable. Timed measurements or counts during one to two minutes are obtained, dependent upon soil moisture content

(Merriam and Knoerr 1961). An entire 6-ft profile can be monitored within 10 minutes, including notetaking time.

The radiation is nondestructive and actually establishes a sphere of influence, a volume of about a cubic foot when the soil is at field capacity, larger when the soil is drier, and vice versa. Consequently, the counts obtained on the scaler are directly proportional to the percent by volume of soil moisture. Accuracies to ±1 percent are possible, especially if care is taken to stratify sampling by soil texture (Douglass 1962a). Calibration of the equipment is readily accomplished (Sartz and Curtis 1961; Douglass 1962b), and minor problems with use of the equipment relate to sampling near the surface, where the sphere of influence is partially out of the soil (Ziemer, Goldberg, and MacGillivray 1967), installation of tubes in rocky soils (Cline and Jeffers 1975; Koshi 1966; B. Z. Richardson 1966), and sealing of access tubes (Mace 1966). A caution is noted by Cotecchia, Inzaghi, Pirastru, and Ricchena (1968) in that certain elements in soils may have an influence on the neutron capture rate. The errors noted are small but significant, and may be controlled by monitoring density. In addition to the frequently used depth moisture probe, both depth and surface density probes are available.

Remote Sensing Methods

Photogrammetry from either aircraft or spacecraft is a relatively new and promising method of obtaining information about moisture on and near the surface of the soil. Modern equipment can simultaneously collect images at several different wavelengths that can provide a variety of information about the soil. Currently, infrared and visible wavelengths are used to evaluate a variety of hydrologic parameters that are useful in modeling water yield and flood forecasting. The parameters include extent of snow cover and moisture content of snow and are combined with models of the energy and water balances and suitable ground truth (Rango 1985). Thermal and microwave spectra are expected to be particularly useful for evaporation estimates (Miller and Rango 1985) and for soil moisture determinations, but may be affected by illumination geometry, organic content, atmospheric interference, surface roughness and reflectance, and soil texture (Schmugge, Jackson, and McKim 1980). Reporting a benefit-cost ratio of 75:1 for satellite determination of snow cover data, Rango (1985) declares that the future for remote sensing is bright:

> It can be expected that if remote sensing data are found to be useable in hydrologic models and the ability to directly input the data is improved, a significant cost savings over conventional methods will result. Additionally, remote sensing may be the only practical source of data for many remote areas of the world.

Rango (1987) predicts that "in the future, remote sensing should play a dominant role in hydrological research and operations." The process of refinement of hydrologic models will be enhanced by remote sensing data, and the availability of such data will make the models increasingly useful.

Miscellaneous Methods

Soil moisture may be determined by the Immersion Method, as described by Wilde and Spyridakis (1961). As the title suggests, the water content of a soil sample is determined by immersion in water. The difficulty with the method lies, in part, with the fact that the results are, like the gravimetric methods, still reported in percent by weight and must be converted to percent by volume to be related to other water balance components, which introduces further error. Wilde and Spyridakis (1961) report that the results of weighing in an immersion procedure are within 1 percent of the gravimetric determination. A major drawback is that water must be carried into the field for immediate determination on the site.

D. E. Miller (1964) reported a method for estimating soil moisture in soil layers from flow out of coarser materials below, but noted that the method was not applicable when sharp textural boundary between the layers was not present or if the lower layer contained "appreciable" soil-sized material.

Mount (1972) described a method for determination of a soil dryness index based upon runoff data. The purpose of the research was to evaluate forest fire danger from runoff data. The premise was that surficial organic material that is subject to conflagration dried out at a rate similar to the soil which, being more uniform than the burnable materials, would be easier to monitor. The method predicts the amount of rainfall necessary to bring the soil back to field capacity. It was developed in Australia where such indices have proved useful. The index can also be used for predicting water yields and for flood forecasting.

Lysimeters may be used to monitor soil moisture in conjunction with measurement of other hydrologic cycle components, as described in Chapter 4. Similarly, water balance models may be employed to estimate soil moisture along with evapotranspiration. On a larger scale, soil moisture may be estimated from the water balance on experimental watersheds, but on-site information is often necessary and advisable for calibration purposes.

Finally, for permafrost, a probe has been developed that withdraws interstitial water for analysis, and also can be used to measure hydraulic conductivity, temperature, and water pressure (Harrison and Osterkamp 1981).

LINKING STORAGE AND MOVEMENT FROM SOIL TO STREAM

Water that is introduced to a partially dried soil will move to the area of greatest tension, to the zone that is driest, to the zone with the smallest and greatest number of small pores. Thus, following rain or snowmelt, water will move first into retention storage and will continue to do so until that type of storage is full or satisfied prior to providing any surplus water to fill detention storage. When the latter is full, the soil is saturated and, if no further infiltration takes place, water runs off the surface and is **Surface Runoff**. Water which flows out of the aerated zone, that is from detention storage, is **Subsurface Runoff**, and water that is routed to a stream or river via the ground water reservoir is **Base Flow**, as shown in Fig. 5-1. As noted at the beginning of this chapter, the distinction is important in hydrograph analysis, understanding, stream behavior, and even small watershed characterization and definition. Even on such a well-studied area as the Coweeta Hydrologic Laboratory, Hewlett (1961) notes that "despite progress in studies of the hydrology of small watersheds, there remains a gap in our practical knowledge of the behavior of water during [the] passage [of water through the soil mantle]."

Recognition that different types of precipitation events typically occur at different seasons and in typically different forms, and that a typical sequence of events surrounding a precipitation or runoff event may exhibit characteristic antecedent moisture conditions (dependent upon soil and vegetation), provides the foundation of watershed hydrology.

PROCESSES

The movement of water from one type of storage to another within the soil is given several different names, dependent upon the discipline of the namer or the direction of flow. The hydrologist tends to classify the processes into three functional categories: infiltration, percolation, and transmission.

INFILTRATION

As a relatively recent term (Chow 1964), "infiltration" has been defined differently, from "surface entry, transmission through the soil, and depletion of storage capacity in the soil" (Chow 1964) to "A soil characteristic determining or describing the maximum rate at which water can enter the soil under specified conditions, including the presence of an excess of water" (infiltration rate) and "The actual rate at which water is entering the soil at any given time" (infiltration velocity) (Soil Conservation Society of America 1976). As used herein, the term infiltration rate is synonymous with infiltration capacity; and infiltration velocity with infiltration.

Infiltration is here defined as the movement of water from the atmosphere to the soil across some definable but intangible interface. It is reported in units of depth per hour. As noted in Chapter 1, infiltration might best be regarded as a concept because one cannot see or directly measure it without influencing its value. However, it may be approximated by a variety of different methods, and the concept itself is useful in understanding this critical zone where precipitation or snowmelt first encounters the porous medium, the properties of which will determine how the water ultimately arrives at the stream.

The maximum rate that the air-soil interface will "take" is known as the infiltration capacity. This attribute of the soil, is expressed as a rate in inches or centimeters per hour, and is a function of soil surface conditions (e.g., occurrence of litter) and surface horizon characteristics (texture and structure). It is further influenced by the rate at which the water is supplied to the soil and by the percolation rate (next section), which in turn is dependent upon the amount of water in the soil at the start of the event and the time since the event began. Stallings (1952) notes that as long ago as 1877 raindrops were observed to seal the soil surface and decrease infiltration by inwashing of fine particles.

The rate at which infiltration actually occurs varies, and is always equal to or less than the infiltration capacity. Generally, both infiltration capacity and infiltration are influenced by the rate at which water is supplied to the soil surface, for instance, rainfall intensity. Infiltration and infiltration capacity are coincident if the rate at which water (precipitation, runoff, or snowmelt) supplied to the soil surface exceeds the infiltration capacity. They decline during a runoff event and tend to equilibrate after about an hour (Kane and Stein 1983).

From a distance, the interface between air and soil is rather clear; however, upon close examination it is almost impossible to identify where each sphere begins and ends. There is, indeed, atmosphere in the soil pores when they are not filled with water. Air permeates the soil pores near the surface, a process whereby gases, including water vapor, readily move between atmosphere and soil by diffusion (Hillel 1980): relative humidity is likely to be high within the soil pores, as is carbon dioxide, which may be as much as one hundred times higher than the normal atmospheric content of 0.03 percent (Brady 1984). Both of these factors may influence gaseous exchange and water movement in the pores. As noted earlier, the occurrence of a light frost at the soil surface (also affected by water vapor content) may enhance infiltration (Striffler 1959); Kane and Stein (1983) reported soil that had gone through a freeze/thaw cycle exhibits a higher infiltration capacity.

Under natural, undisturbed conditions in the forest, infiltration capacity is almost always high enough to preclude natural overland flow (surface runoff, in general). Occasionally, under particularly intense storms, on thin saturated soils, surface runoff for short distances may be observed, but such conditions are rare. If the soil surface is severely disturbed, infiltration rates may suffer drastic reductions, sometimes to as low as zero. With varying degrees of disturbance and protection, infiltration may recover to pre-disturbance rates, especially with adequate protection and occurrence of frost which restores permeability.

On natural rangelands, where there often is insufficient annual precipitation to support forest growth and attendant faunal activity within the soil, recovery of infiltration will need assistance from man, including removal of the cause of the reduced infiltration (e.g., reduction of compaction by removal of cattle which may be overgrazing the area) and mechanical breakup of the surface. In contrast, infiltration rates that have been decreased by logging activities on forest lands will tend to recover without assistance other than cessation of logging: good frost action, plant growth, faunal activity, and development of the A horizon will all work to restore the infiltration rate on cutover areas. On severely compacted areas, such as logging roads and skid trails, however, vegetation and soil fauna may be slow in becoming reestablished.

Quantification of infiltration and infiltration capacities has been the subject of a large number of studies, with most of the successful measurements being made on disturbed, nonforested lands, especially crop and rangelands. Summarizing a large number of studies with relatively modern infiltrometer equipment, Kirkby (1978) reported infiltration capacities up to 8.4 cm/hr (3.31 in./hr) in grazed western rangelands, and 8 cm/hr (3.15 in./hr) on Vermont pasture, formerly pine woodland soils. The sprinkling infiltrometers (see below) appear to be most successful in these situations but cannot be used on forested lands. Virtually the only way in which to obtain reliable infiltration capacity data for forested soils is to combine precipitation and runoff records with the field observation that, as is usually the case, there was no surface runoff during a particular storm. Even then, one only knows that the infiltration capacity of the soil was not exceeded, but not what it is, in fact.

MEASUREMENTS-INSTRUMENTS-LIMITATIONS

Measuring infiltration and infiltration capacity is difficult, since both are influenced by the rate of application. The generic term used for the device used to "measure" infiltration is an infiltrometer. There are several types, including variations of the sprinkling, and flooding ring infiltrometers. A gaged watershed or plot may also be used as an infiltrometer. Difficulties include: (1) edge influences, that is, the effect of a dry area around the instrument into which water may move laterally beneath the surface or across the receiving surface area; and (2) effects on the surface layer(s) of the soil being tested by the artificial application of water.

Sprinkling Infiltrometers

The North Fork Infiltrometer was developed in 1934 in California. "It is sufficiently portable to permit random and comparably rapid sampling and can be installed with very little disturbance to the soil" (Rowe 1940). The device works on slopes from 0 percent to 100 percent, and provides rainfall at 3/4 in. to 10 in. per hr over an area 12 by 30 (horizontal) in. The equipment for the unit can be carried in a pickup truck; thus, application is obviously limited by access. The second of the major difficulties identified above is minimized by providing a wetted border around the plot. The artificiality of the rainfall, however, means that, despite claims to the contrary, it can only measure infiltration capacity if the rate of water sprinkled is great enough to provide surface runoff. That rate must be reached initially, since the infiltration capacity decreases with time since the start of precipitation. The device actually measures infiltration.

The Rocky Mountain infiltrometer is a modification of the Soil Conservation Service's FA type (Dortignac 1951). It is a much more cumbersome device, and requires a larger vehicle for transport, from which it also must be operated. It also measures infiltration on a 12 in. by 30 in. plot, but can only be used on slopes less than 50 percent. The device was calibrated so that "on gentle slopes 1 cubic centimeter of runoff per minute equals 0.01 in. per hour" (Dortignac 1951). The instrument does not provide a wetted buffer zone around the plot, and has limited rainfall intensities of from about 2.5 in. to 5.5 in. per hour, but can also be used to monitor rates of erosion.

Ring Infiltrometers

A metal ring of about 12 in. in diameter is inserted in the soil and flooded with an inch of water. The time it takes for that water to infiltrate may be converted to the infiltration rate in inches per hour. The water in the ring may be maintained at a constant depth from a convenient reservoir, which is then monitored to provide the rate of infiltration.

To avoid movement of water horizontally outward from the ring, a second, larger ring is used to surround the ring in which the measurements are made and is flooded at the same time. The size of the device has not been standardized, and varies from 9 in. to 36 in. in diameter (Sherman and Musgrave 1942). Flooding has also been employed on plots to evaluate infiltration rates.

Major disadvantages of this type of infiltrometer are that (1) air may be entrapped, thus precluding normal movement of water downward into and through the soil profile (Adrian and Franzini 1966); (2) in sandy soils the water cannot be applied rapidly enough to maintain a saturated surface without also washing fine particles into the interstices; and (3) the introduction of the ring itself disturbs the soil and provides for a potentially rapid travel pathway for the water at the interface between the metal and the soil.

The ring infiltrometer may be used repeatedly, however, to illustrate the depletion of infiltration capacity over time. It is also useful for demonstrating radically different infiltration capacities and rates under different soils and land uses (Dunne and Leopold 1978). A simple recording version of a concentric ring infiltrometer was described by Cox (1952).

The Watershed as an Infiltrometer

Infiltration may be determined from precipitation and streamflow records over a runoff plot or a watershed (Sherman and Musgrave 1942). If the area is forested, canopy interception must be evaluated and any variation in soil type must be accounted for as well. "The precision of the determination depends largely upon the adequacy of the records" (Sherman and Musgrave 1942).

The concept is demonstrated in Fig. 5-19 where a curve of infiltration capacity is shown superimposed upon a histogram of a hypothetical rainstorm. Both the line and bars are in inches per hour, so it is possible to determine, knowing the length of time by which each bar of rainfall intensity exceeds the infiltration curve value, just how much rainfall will not infiltrate the soil. Working backwards from the hydrograph, then, it is possible, though tedious, to determine what the infiltration curve is during a particular storm (Horton 1939).

PERCOLATION

Percolation is the rate at which water moves downward through the soil profile. The water that thus moves is gravitational water (detention storage), since retention storage is held against the force of gravity by tensions in excess of 1/3 bar. The percolation rate is dependent upon the characteristics of the horizons through which the water moves. The list, by now, should be familiar: soil texture, structure, and moisture and organic content. The latter may include biotic agents such as roots or residuals thereof, and faunal burrows, as well as decomposed litter material incorporated into the A horizon. Field analysis shows that "the rate of redistribution within the soil profiles [in three different agricultural soils] depended upon the initial depth of wetting and the soil water content-soil pressure-capillary conductivity relations" (Biswas, Nielsen, and Biggar 1966), as one might expect. The percolation rate cannot be greater than the rate at which the water is supplied, that is, the infiltration rate or, if less, the precipitation intensity. It is also reported in units of depth per unit time, for example, inches per hour.

The downward movement of gravitational water may be limited by excess moisture in the horizon below that through which the water is moving, by the water table, or by impervious layers such as a fragipan, an impermeable silt-dominated horizon often found under natural eastern forest soils.

Theoretically, percolation does not apply to water in retention storage: at levels of moisture content below saturation, other factors may come into play. For example, retention storage water in coarse-textured soils:

> Apparently, at high tensions the water in sands is held only at points of contact between the relatively large sand particles. Under these conditions, there is

no continuous water film and thus no opportunity for liquid movement. Water transfer, if it occurs must take place in the vapor state (Buckman and Brady 1960). For this reason, as noted earlier, a dry soil layer may actually act as a barrier to water movement. If one wishes to drain a soil, it is important to provide a gradient of soil particle sizes from the area of saturation to the point at which the water is to enter the drain tile. Such a gradient will provide good drainage, whereas dumping gravel around slotted tile pipe in a trench dug in clayey soil may provide no drainage at all.

Measurement-Instruments-Limitations

Lysimeters have been used to monitor percolation. Like the ring infiltrometer, the effect of artificially influenced water movement owing to interaction between the soil and the instrument wall is likely to be biased toward high readings. Further, the lysimeter puts a lower physical limit on the downward movement of water in the system: the water must either enter a measuring tank, or a runoff trough.

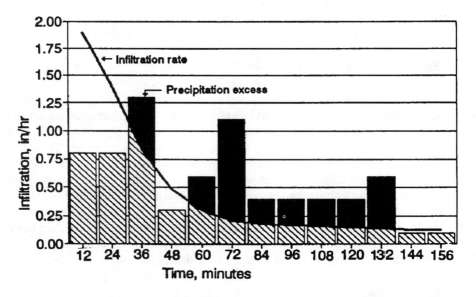

Figure 5-19 Infiltration rate and precipitation excess

The neutron meter may be used to evaluate percolation within the sphere of influence of the instrument. Successive readings at closely spaced time intervals, with the radioactive source located as carefully as possible at the same location for each reading, allows monitoring the progress of both the wetting front and the recession curve of moisture in the soil. A well-placed stack of Colman units may also be used to monitor the percolation rate.

When saturated, percolation may be modeled. Thus, soil pressure measurements at 5-cm intervals were used to corroborate percolation predictions based upon Darcy's Law with reportedly good results (Srinilta, Nielsen, and Kirkham 1969).

TRANSMISSION

On occasion, water may move through the soil under an hydraulic head or pressures greater than -1/3 bar, much like it does through a pipe or a confined aquifer as described earlier in this chapter. This is saturated flow to which the term "percolation" generally applies. Where a highly permeable surface is flooded, for example, water will rapidly enter the soil and diffuse outward from the source, exhibiting rapidly reduced pressure with distance as the energy of the head is dissipated.

Transmission, indeed all saturated flow, is dependent upon permeability (and the rate of water added to the soil), the rate of water movement through a porous medium regardless of direction. The coefficient of permeability is standardized as the maximum possible rate of flow in gallons per day through a 1-sq ft cross section, at a temperature of $60°F$ and an hydraulic gradient of 100 percent. Laboratory tests by the Geological Survey indicate a range of some 450 million times from the lowest to the highest coefficients of permeability (Meinzer 1942). Meinzer classifies the methods for determination as direct and indirect laboratory analyses, field-velocity measurements, and field-discharge methods. These are well described in current ground water geology texts (e.g., Freeze and Cherry 1979).

Measurement-Instruments-Limitations

In addition to the methods noted for percolation, a great deal of interest has been shown recently in measuring the rate of movement (in any direction) in soils owing to the increased interest in the movement of nutrients, salts, contaminants, and various dissolved substances in the soil water. Studies have been carried out in natural ecosystems to understand normal nutrient cycles: one elaborate scheme involved lysimeters with controlled flow rates, acidity, water flow, and conductivity[12] measurements (Cole 1968). Both pH and conductivity were found to change more readily under conditions of warmer temperatures, and have significance to transport of iron:

> Such studies are directly relevant in understanding soil development, in quantitizing [sic!] the cycle of minerals between the soil and vegetative components, and in evaluating the impact of land management practices, including the effects of fertilizing, burning, and harvesting on elemental losses from the soil profile (Cole 1968).

SOIL/VEGETATION/WATER CLASSIFICATION AND HYDROLOGY

A massive cooperative federal-state effort in California in the early 1960s focused on creating detailed soil/vegetation maps that were particularly useful to land managers (Zinke 1958). These maps recognized that there were natural associations of flora and substrate and, further, that these associations exhibited consistent hydrologic behavior on characteristic sites. The

---

[12] See Chapter 6 for description of conductivity, also identified as "specific conductivity" or "specific conductance."

soil/vegetation complexes described, in the highly variegated California countryside, run the gamut from true Podzols (along the coast between Eureka and San Francisco under Monterey Pine) to Prairie soils on some of the grass-covered coastal hills. In between, one may observe almost every Great Soil Group with the exception of a Chernozem.

It is not surprising that this approach to wildland management is in concert with basic concepts of watershed hydrology. Understanding the occurrence, nature, and extent of soil moisture storage, humus type, and distribution of water in the wildland environment is the cornerstone of effective watershed management. That the California soil/vegetation survey system is not easily applied elsewhere is testimony to extensive disturbance of the natural environment by man's activities. However, the fact that the specific soil/vegetation associations did exist prior to widespread disturbance constitutes the basis for recognition of manageable ecological niches. The primary principle derived from this idea is the Principle of Watershed Equilibrium, discussed in detail in Chapter 7.

SUMMARY

The soil buffers the irregular and often "lumpy" delivery of precipitation to the surface of the watershed. It attenuates that delivery into a smooth outflow hydrograph that reflects the characteristics of the porous medium through which the precipitation passed. The physical properties of the soil, based upon the parent material, provide the buffering, in turn modified by climate, time, geologic history and processes, and biotic factors. The soil/vegetation complex is the foundation of watershed management.

PROBLEMS

1. What are some simplifying assumptions made in the presentation of Fig. 5-1?
2. Verify the 188,500 acre-figure given for the surface area of silt particles in one acre of 5-ft-deep soil
3. What is the volume of water that can be retained in the upper 3 ft of soil per acre if the retention capacity is 30 percent?
4. Diagram the italicized statement on page 175-6, that starts: "Recognition that different types of precipitation events typically . . ."
5. Modify the diagram quantitatively in Fig. 5-17 to illustrate a very sandy or a very clayey soil.

6 Water in the Hydrosphere

*In the stream
runoff is the integrator of all of the factors
which affect its quantity, quality, and regimen,
and is one of the factors of watershed ecology itself*

The hydrosphere consists of the waters of the earth. The concern herein is primarily with the earth's surface waters, but ground- and subsurface waters (including soil moisture) are discussed where appropriate. Like the atmosphere, the hdyrosphere permeates the terrasphere and biosphere to greater or lesser degrees depending upon local conditions, and the circumstances and purposes of the observer.

This chapter primarily treats the *water body* and *precipitation-runoff relations*, while Chapter 7 treats *the watershed*. The previously used division into *storage* and *process* is still used in the first section of this chapter initially, but is inapplicable thereafter as it becomes necessary to consider relationships between storage and process, and the watershed as an integrated whole: the transition actually begins in this chapter.

WATER YIELD

Water Yield implies consideration of both water quantity and regimen (the rate of delivery or quantity per unit time). Since water on the surface of the earth may be found in a variety of circumstances, some expansion on the presentation of data in Table 1-1 is necessary. Specifically, there must be concern for the tremendous amounts of water in and the disproportionate distribution of the water in the lakes of the world, the fact that some of the lakes are connected to flowing streams, some of which do not flow to an ocean, and the streams and rivers, that at any point in time, contain no small amount of storage themselves.

Water body is a generic term for any collection of water, whether it be on the surface or below the water table. Usually, the term is reserved for surficial bodies of water: water below the water table is referred to as *ground water storage*, and surficial water bodies are referred to collectively as *depression storage*, as noted in Chapter 1. The amount of water stored in depressions is about 10 times that in the rivers of the earth. While streams and rivers are actively flowing, it is difficult to consider them as storage, yet it is clear that volumetrically they must be included in the figures in Table 1-1. In fact, of course, the water in the stream is in transit, a process known as *runoff*, and therefore considered separately. Lakes and ponds are not all storage: they are in many ways involved in hydrologic processes; and streams and rivers are not *all* process: they frequently sustain storage.

STORAGE IN LAKES AND PONDS

According to Table 1-1, the 0.3 percent of the earth's water in storage in lakes (and ponds)[1] amounts to approximately 100 billion acre-feet, with an average residence time of 21 years. Many surface water bodies have considerably shorter residence times, however, especially if they are part of a stream system. Flushing of such a lake may occur several times per year.

Bathymetric surveys are used in order to evaluate lake volumes. The science of *limnology* deals with the characteristics and behavior of lakes. Lake investigations also include, and often are focused on, detailed studies of biological elements and chemical characteristics of the water body. The topic is the subject of one or more college courses on its own, and is also well documented in several references (cf., Hutchinson 1957; Stumm and Morgan 1970; Faust and Aly 1981).

While the water in depression storage is technically not included in the runoff process, any water stored in a lake connected to a stream is, in fact, a part of the runoff of the entire drainage basin. Lakes do have profound effects on the runoff characteristics as measured in the stream or river system, not in the lake.

In general, just as the soil reservoir attenuates the abrupt precipitation pattern of a storm or snowmelt event into a smooth hydrograph, so does a lake through which a stream flows further attenuate the hydrograph, lowering the peak flow and increasing the time it takes for the runoff waters to exit the watershed. In short, and other things being equal, a watershed with an included lake will exhibit lower flood peaks, higher minimum flows, and probably less erosion and sedimentation than a similar watershed without a lake. Of particular importance in such an observation is the relative position of the lake on the watershed: one that is near the watershed outlet will be more effective in attenuating the flood peak than one further upstream, for example. On a watershed where a large proportion of the drainage area is occupied by lake surfaces, attenuation of peak runoff may be partially offset by virtue of the fact that a large proportion of the precipitation that falls on the drainage may go more directly to runoff, rather than being routed through the terrasphere.

Lakes have been classified on the basis of size (Winter 1977) but, more frequently, on the basis of chemical or biological variables. Two trophic extremes are recognized: *oligotrophic* (clear, relatively low productivity, high oxygen content, water bodies), and *eutrophic* (usually the reverse). Classification on the basis of hydrologic characteristics would be of benefit to being able to generalize about them and to provide insights into their natural physical, biological, and chemical cycles, responses to intrusions, and management. Some

---

[1] Subsequently, the use of the term "lake" will imply ponds as well. Wetlands are considered in Chapter 4, and may be included here by inference.

important hydrologic characteristics include whether the lake is an open (connected with a flowing stream) or closed (isolated) system;[2] water budget; ratio of drainage basin area to lake area; residence time (related to the water budget and area ratio); and relation to ground water bodies. According to Winter (1977):

> ... it is clear that a classification of lakes based on all factors related to lake hydrology has not been developed. Such classification is needed if results of detailed studies are going to be applied with confidence to other lakes. It is evident also that the factors related to lake hydrology are not generally known or understood. Much work needs to be done to identify and evaluate the relative importance of factors that control lake hydrology, especially those that control the interaction of lakes and groundwater.

In Winter's (1977) multi-variate study of hydrologic characteristics of 150 lakes in the upper Midwest, he concluded that the precipitation-evaporation balance and water quality factors were the most important factors, followed by inflow and outflow, geologic and ground water characteristics and, finally, the drainage basin/lake area ratio. The original variables that seemed best for characterizing the lakes were statistically independent and "clearly related to one another." Until further research on natural, modified, and man-made lakes is accomplished, much of the concern over lakes will remain in the bailiwick of the limnologist.

RUNOFF - STREAMS AND RIVERS

According to Table 1-1, the 0.003 percent of the earth's water in rivers (and streams)[3] amounts to approximately 10 billion AF, with an average residence time of 2.1 years. While this water is flowing and, therefore, technically not in storage, it is for some specific time in the channel and often referred to as *channel storage*, as noted in Chapter 1. The water may be temporarily detained within a channel, for example, under an overhanging streambank, in the recesses of a deep pool, in a floodplain while the stream is in flood, or in a water body through which the river or stream passes. The consideration here is of water in motion, hence as a process, moving the fluid from one storage location to another: as runoff.

The distribution of runoff over time is generally referred to as **regimen**. A wide variety of attributes have been identified for the purpose of characterizing runoff, with the ultimate goal being either managing or predicting streamflow for human benefit. An entire science of flow in open and contained channels, known as *hydraulics*, is covered in several excellent references (e.g., King et al. 1960; Chow 1964; Dingman 1984), and, like limnology, is the subject of one or more advanced college courses.

Runoff is water in motion. At this point in the presentation of watershed hydrology, it is considered as being in the stream. In chapter 5, three general sources of runoff within the terrasphere were identified, namely, surface runoff[4], subsurface runoff, and base flow. Once

---

[2] This is obviously of importance to consideration of the lake in its storage or process function.

[3] The generic term "streams" will be used unless reference to a specific river is desired.

[4] Surface runoff is used herein to designate water running over the surface (of a watershed), interchangeable with overland flow. In certain cases, the term is used to indicate all runoff in creeks, streams, and rivers, as in the Geological Survey's "Surface Water-Supply Papers."

collected in the stream, however, runoff also consists of water that falls directly in the stream, known as **channel interception**. On a small watershed, where the area of the stream may constitute 1 percent to 4 percent of the total watershed area at the start of a storm event, this may be a disproportionately large percent of the total water in the stream, and is often detectable on the storm hydrograph as a small pre-storm-peak peak. Channel interception can be expected to expand as the channel itself extends and expands during the storm. On large watersheds, channel interception is generally a less significant percent of the total.

Measurement-Instruments-Limitations

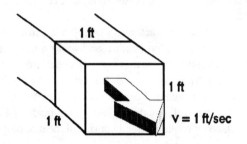

Figure 6-1 Open channel streamflow relations: the cubic foot per second

The basic unit of streamflow (also referred to as **discharge**) in the United States is the *cubic foot per second*[5], abbreviated cfs and sometimes referred to as a second-foot. The concept of the unit is shown in Fig. 6-1. In simple terms, and for visualization, it is a volume of 1 cu ft of water flowing through a one-square-foot (vertical) plane in one second (the cubic foot refers to volume, and can be any dimensions). The formula for the unit of flow, as noted in Chapter 5, is:

$$Q = A \times V$$

where Q is the volume of flow in cubic feet per second, A is the area through which the water is flowing in sq ft, and V is the velocity in feet per second. For example, if the rate of flow in a stream is measured to be 2.5 ft per second, the width is 6 ft, and the depth is 1.5 ft, the discharge is simply

$$2.5 \times 6 \times 1.5 = 22.5 \text{ cfs}$$

Measurement of discharge may be accomplished under a variety of conditions: rapid approximations may be made along a reasonably stable and uniform stretch of stream by timing the rate of movement of some twigs or leaves thrown into the stream and by wading, or range-finder or eyeball estimates of width and depth. At the other end of the spectrum, precise, continuous recording stream gage installations for research purposes may cost upwards of $50,000 and require, of course, annual maintenance, calibration, and record-keeping. Making good field estimates of streamflow depends upon an understanding of how the flow is distributed in the stream, for it is not at all uniform.

Determining Velocity

Velocity of water in a channel is a predictable function of slope and roughness. The relationship is known as Manning's Formula:

$$V = (1.5/n) R^{2/3} S^{1/2}$$

---

[5] In the metric system, the unit is the cubic meter per second. Translation within and between the English and SI units may be accomplished with the aid of the conversion tables in Appendix A.

where V is the velocity in feet per second, n is the dimensionless Manning roughness coefficient, R is the hydraulic radius (width plus two depths) in feet, and S is the slope as a decimal. The formula is not constant over the range of possible values (Chow 1964), but it is a useful estimator of velocity where measurements are not available. Values of n are given in Table 6-1:

Table 6-1 Values of Manning's Roughness Coefficient, n

Channel Material	n
Cast-iron pipe	0.015
Rubble masonry	0.017
Smooth earth	0.018
Firm gravel	0.020
Corrugated metal pipe	0.022
Natural channels in good condition	0.025
Natural channels with stones and weeds	0.035
Very poor natural channels	0.060

Source: Linsley and Franzini (1964).

Velocity may be measured by a current meter as described above, or with a **velocity head rod**. This device is simply a wooden rod about 6 ft in length, with a triangular cross section and the distance from the lower end marked in feet and tenths. The rod is held vertically on the bottom of the stream, with the sharp end pointing upstream: the surface waters divide smoothly around the rod providing a measure of the depth of the stream at that point. Then, the rod is rotated 180 degrees so that the blunt edge points upstream: the non-streamlined side causes a **hydraulic jump** (head), the height of which is read on the scale and is proportional to the velocity, which may be looked up in a table for the rod, as calibrated.

Research investigations in open channels have shown that velocity varies across the width of the stream and from top to bottom. In a uniform cross section, velocity will be greatest near the middle of the channel and about one-third of the distance from the top of the water (**stage**) to the bottom of the channel, as shown in Fig. 6-2. The method is used by the U.S. Geological Survey, the federal agency with the prime responsibility for monitoring streamflow in the United States.

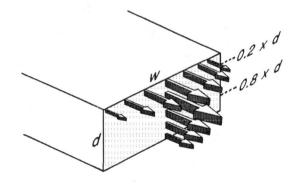

Figure 6-2 Theoretical horizontal and vertical distribution of flow velocities in a uniform channel

Sample pages from USGS Water-Supply Papers are shown in Figs. 6-3 and 6-4. A typical page from a USGS surface water supply-paper is shown in Fig. 6-3. Peak flows are shown in Fig. 6-4.

SUSQUEHANNA RIVER BASIN

01503000 SUSQUEHANNA RIVER AT CONKLIN, N.Y.

LOCATION.--Lat 42°02'07", long 75°48'12", Broome County, on left bank at abutment of former highway bridge, 500 ft upstream from bridge on County Highway 304 at Conklin, 0.7 mile downstream from Little Snake Creek, and 3.5 miles downstream from Pennsylvania-New York State line.

DRAINAGE AREA.--2,232 sq mi (revised).

PERIOD OF RECORD.--November 1912 to September 1970.

GAGE.--Water-stage recorder. Datum of gage is 840.95 ft above mean sea level (levels by Corps of Engineers). Prior to Oct. 4, 1914, nonrecording gage at same site and datum.

AVERAGE DISCHARGE.--57 years (1913-70), 3,526 cfs (21.45 inches per year).

EXTREMES.--Maximums and minimums (discharge in cubic feet per second, gage height in feet).

Annual maximum discharge (*) and peak discharges above base (18,000 cfs), water years 1966-70

Date	Time	Disch.	G.H.	Date	Time	Disch.	G.H.	Date	Time	Disch.	G.H.
Mar. 6, 1966	1200	*18,000	10.68	Nov. 19, 1968	-	*24,000	-	Nov. 9, 1969	2200	19,100	11.00
				Mar. 26, 1969	-	23,000	-	Feb. 4, 1970	0100	-	a13.28
Mar. 30, 1967	1145	*16,800	10.38	Apr. 6, 1969	-	23,000	-	Apr. 3, 1970	2000	*25,300	12.54
								Apr. 10, 1970	1200	23,700	12.09
Mar. 23, 1968	2000	*21,200	11.63								

a Ice jam.

Annual minimum discharge, water years 1966-70

Wtr yr	Date	Disch.	G.H.	Wtr yr	Date	Disch.	G.H.
1966	Sept. 2, 1966	157	1.64	1969	Sept.30, 1969	242	1.89
1967	Nov. 1, 1966	294	2.02	1970	Aug. 22, 1970	224	1.84
1968	Sept. 1, 2, 1968	a270	-				

a Minimum daily.

Period of record: Maximum discharge, 61,600 cfs Mar. 18, 1936 (gage height, 20.14 ft); maximum gage height, 20.83 ft Mar. 22, 1948; minimum discharge, 85 cfs Oct. 14, 1964 (gage height, 1.30 ft).

REMARKS.--Records fair except those for water year 1969, which are poor. Diurnal fluctuation at low flow caused by mill several miles above station. Minor regulation by upstream lakes and reservoirs.

REVISIONS.--WSP 696: Drainage area.

DISCHARGE, IN CUBIC FEET PER SECOND, WATER YEAR OCTOBER 1965 TO SEPTEMBER 1966

DAY	OCT	NOV	DEC	JAN	FEB	MAR	APR	MAY	JUN	JUL	AUG	SEP
1	464	763	2,760	3,510	1,000	4,530	6,260	6,070	1,980	700	280	175
2	530	763	2,610	3,870	920	7,910	6,280	5,720	1,790	637	300	163
3	609	763	2,360	3,760	900	7,960	6,140	4,870	1,690	577	280	163
4	700	728	2,340	3,600	920	7,200	5,740	4,280	1,440	480	270	210
5	679	742	2,610	3,630	900	11,900	5,330	3,960	1,310	469	260	253
6	595	742	2,780	3,460	800	16,800	5,120	3,810	1,230	480	250	261
7	630	756	2,600	4,320	740	13,000	4,750	3,850	1,110	577	230	665
8	707	763	2,300	4,770	700	9,130	4,870	3,360	1,100	700	220	749
9	829	845	2,060	3,870	690	7,150	4,790	4,050	1,110	798	210	486
10	1,020	1,310	1,960	2,600	680	5,960	4,510	4,970	4,750	637	210	375
11	1,200	1,610	1,980	2,300	1,800	6,260	4,380	4,830	4,870	480	210	352
12	1,020	1,460	2,340	2,150	7,000	6,530	4,130	5,100	4,190	430	220	285
13	953	1,270	2,730	2,100	11,000	9,520	3,800	7,350	2,760	380	220	242
14	970	1,230	3,030	2,100	16,000	9,570	3,670	8,820	2,060	400	210	234
15	910	1,260	3,500	2,200	15,900	8,330	3,530	7,420	1,920	340	210	249
16	910	1,230	3,550	1,800	12,200	7,450	3,390	6,210	1,880	320	210	245
17	910	1,500	3,220	1,700	9,600	6,320	3,110	5,540	2,200	310	220	273
18	877	2,580	2,920	1,600	6,800	6,210	2,970	5,200	2,640	300	220	245
19	919	3,210	2,820	1,550	6,000	7,380	2,790	6,300	2,370	310	220	220
20	861	3,020	2,530	1,550	4,500	9,810	2,730	8,510	1,880	370	230	206
21	837	2,680	2,190	1,500	3,100	10,200	3,380	7,810	1,680	560	210	214
22	791	2,790	1,700	1,450	3,300	9,210	4,400	6,390	1,480	400	200	242
23	798	3,080	1,900	1,400	3,800	9,000	3,920	5,520	1,300	420	196	339
24	869	3,310	2,120	1,250	3,800	10,100	4,450	4,630	1,200	360	196	290
25	970	3,190	2,400	1,200	3,390	13,900	8,250	4,210	1,100	310	192	281
26	1,060	2,760	4,200	1,050	3,300	15,100	8,560	3,670	997	280	196	321
27	970	2,670	4,200	1,000	3,300	12,500	6,440	3,030	962	270	203	420
28	885	3,260	2,800	960	3,100	9,760	5,410	2,740	928	270	206	452
29	869	3,870	2,700	960	------	7,680	5,100	2,660	861	390	210	480
30	829	3,240	2,950	960	------	7,010	4,610	2,410	798	360	217	380
31	784	------	3,060	1,000	------	6,790	------	2,180	------	310	196	------
TOTAL	25,955	57,395	83,220	69,170	126,140	280,170	142,810	155,470	55,586	13,625	6,902	9,470
MEAN	837	1,913	2,685	2,231	4,505	9,038	4,760	5,015	1,853	440	223	316
MAX	1,200	3,870	4,200	4,770	16,000	16,800	8,560	8,820	4,970	798	300	749
MIN	464	728	1,700	960	680	4,530	2,730	2,180	798	270	192	163
CFSM	.38	.86	1.20	1.00	2.02	4.05	2.13	2.25	.83	.20	.10	.14
IN.	.43	.96	1.39	1.15	2.10	4.67	2.38	2.59	.93	.23	.12	.16

CAL YR 1965 TOTAL 753,966 MEAN 2,066 MAX 14,000 MIN 167 CFSM .93 IN 12.57
WTR YR 1966 TOTAL 1,025,913 MEAN 2,811 MAX 16,800 MIN 163 CFSM 1.26 IN 17.10

Figure 6-3 Typical year-by-days streamflow data page from Geological Survey (1976)

Distribution of flow velocities is detected with a current meter (Fig. 6-5). These anemometer-like devices are suspended (or, as shown in the figure, attached to a wading rod) in the channel at various depths (often 2- and 8-tenths of the depth) to obtain a true average in the vertical. The formulation of discharge from a large, nonuniform cross-section of a stream is accomplished by establishing a number of "verticals" across the channel, creating a series of trapezoids (Fig. 6-6). The average velocity in each vertical is, in turn, averaged with (a) the

Chapter 6 Water in the Hydrosphere

4770. Eel River at Scotia, Calif.

<u>Location</u>.--Lat 40°29'30", long 124°05'55", in SW¼ sec.5, T.1 N., R.1 E., near center of span in left pier of bridge on U.S. Highway 101, 0.5 mile north of Scotia and 6 miles upstream from Van Duzen River.

<u>Drainage area</u>.--3,113 sq mi.

<u>Gage</u>.--Nonrecording prior to Dec. 12, 1940; recording thereafter. Datum of gage is 36.15 ft above mean sea level, datum of 1929.

<u>Stage-discharge relation</u>.--Defined by current-meter measurements below 300,000 cfs.

<u>Bankfull stage</u>.--Not subject to overflow.

<u>Remarks</u>.--Peaks slightly regulated by Lake Pillsbury (capacity, 86,400 acre-ft) beginning in December 1921. Peaks are affected by diversion through Potter Valley powerhouse (capacity, about 350 cfs) beginning prior to 1911. Peaks for the years 1911, 1915, 1918, 1920-26, 1929, 1930 and 1932 are maximum observed. Only annual peaks are shown prior to Oct. 1, 1945. Base for partial-duration series, 72,000 cfs.

Peak stages and discharges

Water year	Date	Gage height (feet)	Discharge (cfs)	Water year	Date	Gage height (feet)	Discharge (cfs)
1911	Jan. 20, 1911	-	136,000	1952	Dec. 1, 1951	35.4	140,000
1912	Jan. 26, 1912	-	170,000		Dec. 27, 1951	46.50	262,000
1913	Jan. 18, 1913	-	150,000		Feb. 2, 1952	43.6	223,000
1914	Jan. 22, 1914	52.5	309,000				
1915	Feb. 2, 1915	55.5	351,000	1953	Dec. 7, 1952	39.95	177,000
1917	Feb. 25, 1917	51.25	292,000		Jan. 9, 1953	42.98	215,000
1918	Feb. 7, 1918	27.7	78,800		Jan. 18, 1953	37.36	151,000
1919	Mar. 17, 1919	38.3	149,000		Mar. 20, 1953	30.20	94,400
1920	Apr. 16, 1920	25.0	82,000	1954	Jan. 17, 1954	45.20	245,000
1921	Nov. 19, 1920	38.2	148,000		Jan. 23, 1954	32.10	109,000
1922	Feb. 19, 1922	34.50	123,000		Jan. 28, 1954	39.18	169,000
1923	Dec. 28, 1922	26.9	75,400		Feb. 13, 1954	30.35	95,400
1924	Feb. 8, 1924	26.9	75,400		Apr. 6, 1954	28.00	80,000
1925	Feb. 6, 1925	35.20	127,000	1955	Dec. 31, 1954	23.29	52,400
1926	Feb. 4, 1926	42.20	176,000	1956	Dec. 6, 1955	27.65	77,800
1927	Feb. 21, 1927	45.2	221,000		Dec. 20, 1955	43.15	217,000
1928	Mar. 27, 1928	46.3	235,000		Dec. 22, 1955	61.90	541,000
1929	Feb. 4, 1929	21.30	41,000		Jan. 7, 1956	27.75	82,500
1930	Dec. 15, 1929	34.10	120,000		Jan. 16, 1956	42.20	205,000
					Feb. 22, 1956	43.45	221,000
1931	Jan. 23, 1931	29.0	87,000				
1932	Dec. 27, 1931	36.10	127,000	1957	Feb. 25, 1957	36.11	153,000
1933	Mar. 17, 1933	26.10	58,100				
1934	Mar. 29, 1934	24.8	50,900	1958	Nov. 14, 1957	29.45	98,200
1935	Apr. 8, 1935	29.62	79,900		Dec. 22, 1957	28.53	91,900
					Jan. 30, 1958	35.30	145,000
1936	Jan. 16, 1936	44.7	216,000		Feb. 10, 1958	27.34	84,400
1937	Feb. 5, 1937	37.0	134,000		Feb. 12, 1958	38.16	174,000
1938	Dec. 11, 1937	55.1	345,000		Feb. 15, 1958	31.64	118,000
1939	Dec. 3, 1938	35.90	133,000		Feb. 19, 1958	38.27	161,000
1940	Feb. 28, 1940	52.25	305,000		Feb. 25, 1958	40.55	202,000
1941	Dec. 24, 1940	36.4	150,000		Apr. 3, 1958	32.23	124,000
1942	Feb. 6, 1942	42.2	209,000	1959	Jan. 9, 1959	29.59	102,000
1943	Jan. 21, 1943	50.75	315,000		Jan. 12, 1959	34.58	145,000
1944	Mar. 4, 1944	24.60	57,800		Feb. 15, 1959	30.55	110,000
1945	Feb. 3, 1945	30.55	99,100				
				1960	Feb. 8, 1960	51.45	343,000
1946	Dec. 4, 1945	-	103,000		Mar. 8, 1960	27.16	83,300
	Dec. 23, 1945	-	98,000				
	Dec. 27, 1945	44.60	239,000	1961	Dec. 1, 1960	28.23	90,000
	Jan. 5, 1946	-	78,300		Dec. 17, 1960	28.90	94,400
					Jan. 31, 1961	29.42	97,900
1947	Feb. 12, 1947	29.02	86,100		Feb. 11, 1961	31.45	113,000
	Mar. 10, 1947	-	72,100				
				1962	Feb. 14, 1962	29.92	107,000
1948	Jan. 8, 1948	32.6	114,000		Feb. 16, 1962	27.10	84,800
1949	Mar. 15, 1949	35.4	140,000	1963	Oct. 13, 1962	32.48	128,000
					Dec. 3, 1962	32.89	132,000
1950	Jan. 18, 1950	32.85	117,000		Feb. 1, 1963	47.00	252,000
					Mar. 28, 1963	26.55	90,800
1951	Oct. 29, 1950	33.33	121,000		Apr. 6, 1963	28.12	87,800
	Dec. 4, 1950	37.37	160,000		Apr. 15, 1963	26.50	76,800
	Jan. 18, 1951	35.5	141,000				
	Jan. 22, 1951	45.39	249,000	1964	Jan. 21, 1964	39.40	178,000
	Feb. 5, 1951	40.55	193,000	1965	Dec. 23, 1964	72.0	752,000
					Jan. 6, 1965	36.04	148,000

Figure 6-4 Typical peak runoff data page from Geological Survey (1967). The data are used in Table 6-4

average from the adjacent vertical, (b) the trapezoid's width, and (c) the average of the two depths (height of the verticals themselves) to determine the discharge for the trapezoid (if only one depth can be measured in a given vertical, it is at $0.6d$). All the discharge figures are then added to obtain the discharge for the entire cross section. To preclude repeated revision of the cross section geometry, gage site should be selected to be stable at a wide variety of discharges. Thus, in general,

$$Q_s = Q_{s_1} + Q_{s_2} + Q_{s_3} + \ldots Q_{s_n}$$

where Q_s equals the total discharge, Q_{s_i} equals the discharge in the *ith* trapezoid. Then,

$$Q_{s_i} = V_m \times W \times (d_7 + d_8)/2$$

where d equals the depth of the vertical, W equals the horizontal distance between the verticals, and V_m equals the average velocity, $(V_{.2} + V_{.8})/2$.

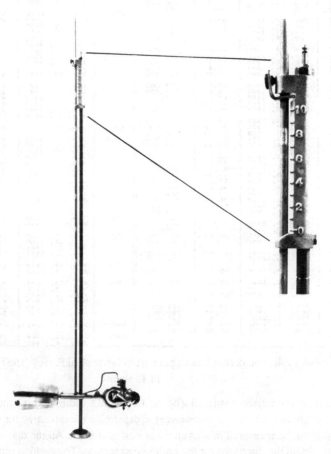

Figure 6-5 Current meter attached to top-setting wading rod (Buchanan and Somers 1969)

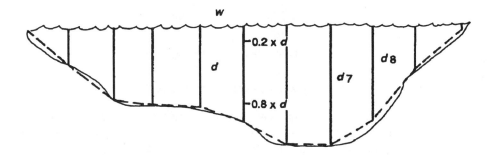

Figure 6-6 Stream divided into trapezoids for measurement of discharge

Determining Discharge

Discharge is calculated from velocity and depth measurements. Special field note paper makes this calculation job easier (Buchanan and Somers 1969), as does computer software. Careful determinations of streamflow by this method are used for calibration of continuous-recording devices: calibration must be performed at various stages of flow in order to establish a reliable **rating curve** (relationship between stage or height of water above some reliable datum and discharge) for the gage site, and verified after unusually high runoff events to be sure that the channel has not changed, or that any changes are documented.

At heavily instrumented research installations, streamflow may be measured by a **weir**. A weir consists of a ponding basin in which the streamflow is temporarily ponded prior to exiting over some sort of controlled outlet. Under such conditions, the pond may be made secure against leakage around it so that all the flow is actually measured, and the pond also serves to still the water so that the height or *head* of water over the bottom of the weir blade may be accurately measured (Figs. 6-7 and 6-8). The blade may be in any of a variety of shapes, classified as **sharp-crested** (typically $90°$ or $120°$ V-notches, or rectangular in cross section) and **broad-crested weirs**: The former are more accurate, but are subject to distortion by debris floating in the stream; the latter are not adversely affected by debris, but are not as accurate. Where large amounts of sediment and/or debris are anticipated (for example, if the installation is on an agricultural watershed), a flume may be used (Fig. 6-9).

Like the standard USGS gaging station, the research weir installation requires a stable cross-section. It also has a stilling well that is connected to the level of the water in the pond or uniform stretch of channel, and a shelter of some sort (Fig. 6-8) over the well that houses the **water level recorder** (Fig. 6-10). A float in the stilling well is attached to a perforated steel tape that passes over a pegged pulley: the other end is connected to a weight that keeps the tape taut. The pulley is on the same shaft as an eccentric that activates a pen which leaves a trace on a chart on a revolving drum. It is this trace that, during the period of a runoff event, is referred to as the **hydrograph**.

The cubic foot per second may be readily converted to other units with two simple conversions and one definition:

1 cu ft holds 7.48 (US) gallons (gal);

1 acre contains 43,560 sq ft; and

1 AF is the volume of water covering 1 acre to a depth of 1 ft (43,560 cu ft).

Since there are 86,400 seconds in a day, a flow of one cfs for one day equals 86,400 cu ft, very close to twice 43,560 or 2 AF (af) per day. And one AF contains approximately 325,800 gal. With these conversions, the three modern units of water measurement, cfs, AF, and gal, may be used as needed. The hydrologist should be able to make these conversions readily and to be able to analyze and report in the units appropriate to the water use. Given a watershed from which a known volume of discharge has been measured over a period of time, it is possible to convert that volume to an inches-depth figure over the watershed. For example, knowing that the average flow of 6 cfs is observed from a 2-sq-mi watershed for a full day produces:

$$\frac{6 \text{ ft}^3/\text{sec} \times 60 \text{ sec/min} \times 60 \text{ min/hr} \times 24 \text{ hr/day} \times 12 \text{ in/ft}}{2 \text{ mi}^2 \times (5280 \text{ ft/mi})^2} = 0.11 \text{ in. of runoff for the day}$$

An important unit of runoff is the *cubic foot per second per square mile* (of watershed area), abbreviated "csm." In the previous example, the runoff would be

$$6 \text{ cfs}/2 \text{ sq mi} = 3 \text{ csm}$$

The csm permits valid comparison of runoff from watersheds of different sizes. In discussions that follow where comparative statements are made, it is presumed that the data are being compared in units of csm unless otherwise stated. Runoff may also be reported in inches of runoff (or cfs or csm) per inch of precipitation.

Note that the permanently installed stream gage measures 100 percent of the runoff: there is no sampling of the flow. Thus, in the water budget (Chapter 4) computations, there is need for caution over the relative accuracy and precision with which the various components are monitored. For example (from Chapter 3), with about one gage for every 350 sq mi, precipitation in the United States is sampled at a rate of about 1 in 9.9×10^{12}; evaporation is measured in somewhat larger devices than the standard rain and snow gage, but with an even more sparse distribution of gages; soil storage is assumed to be the same at the beginning and the end of each Water Year; transpiration is estimated, and 100 percent of the runoff is measured. Even in simple regressions or other statistical methods of comparing two or more water budget components, *it must be kept in mind that the two components were probably measured with drastically different degrees of accuracy and precision*: such a discrepancy may affect conclusions.

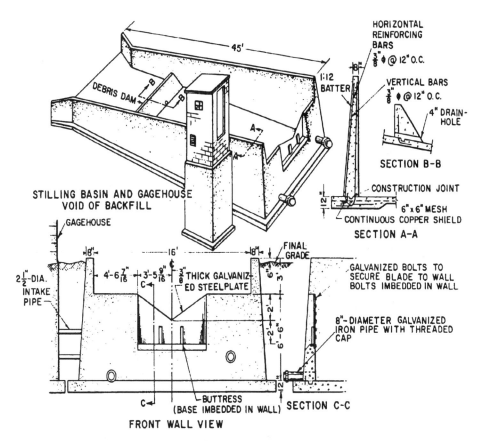

Plans for concrete weir of watertight-box design constructed at Hubbard Brook Experimental Forest, West Thornton, N.H.

Figure 6-7 Details of research weir (Reinhart and Pierce 1964)

Patterns of Flow

The regimen of a stream is, in most cases, not at all uniform. While there are some streams in the southern United States and some tropical locations where streamflow is nearly identical from month to month, that is the exception. Streams typically exhibit storm, monthly, and seasonal patterns that are characteristic of the watershed from which the runoff emanates. These patterns are also quite predictable, often with a high degree of confidence.

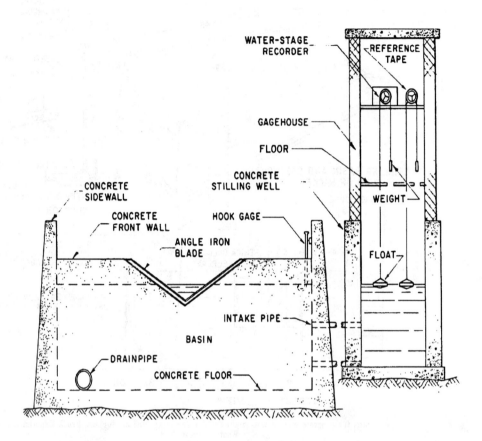

Figure 6-8 Schematic diagram of 120° V-notch weir (Reinhart and Pierce 1964)

Average runoff for the United States is shown in Fig. 6-11, using lines of equal runoff called **isopleths**. Similar maps of runoff for any selected period may be drawn for any area. The pattern, of course, reflects the amount of water available for runoff and thus closely resembles an isohyetal map of annual precipitation.

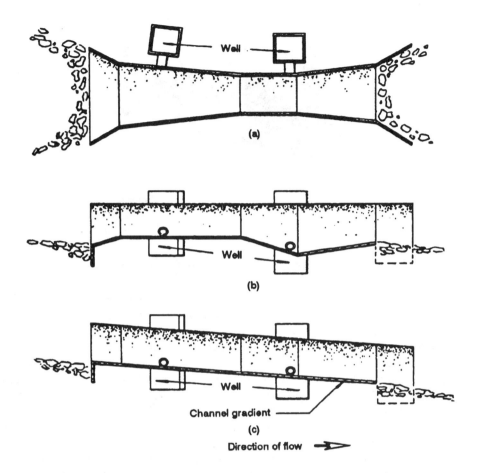

Figure 6-9 Schematic diagram of Venturi flume. It requires measurement of stage at two points. Floor is same grade as stream channel: (a) plan view of Venturi flume; (b) cross section of Parshall flume; (c) cross section of Venturi flume (Reinhart and Pierce 1964).

The normal distribution of streamflow by months for the United States is shown in Fig. 6-12. It illustrates the highly variable amount and type of precipitation that occurs across the nation, although there are some similarities as well. Details and further reference to Fig. 6-12 are discussed as appropriate in subsequent sections.

Figure 6-10 Typical streamflow recorder (courtesy Leupold & Stevens, Inc., Beaverton, OR)

Hidore (1971) identified four categories of temporal changes of streamflow: (1) regular periodic diurnal, monthly, or seasonal changes; (2) "irregular fluctuations that repeat themselves but at irregular intervals"; (3) long-term trends; and (4) random occurrences, such as floods or droughts. Predictability of such changes is not possible with the present state of our knowledge since we do not even know with certainty what category of change we observe: "What is difficult to determine is whether a given change in the flow of a stream is a long-term trend, part of a cyclic or periodic change with time, or several different types of changes superimposed upon one another" (Hidore 1971).

Hydrologic behavior for longer time periods (seasons and years) of a large number of locations has been the subject of investigation for purposes of basic research or management. As a consequence, the hydrology of many areas is well known. For example, long-term records from forested watersheds in the Sierra Nevada mountains have yielded a solid understanding of the long-term hydrologic behavior of these deep-soil drainages, which are characterized as follows:

> Winter snow accumulation and spring snowmelt are the primary influences on the annual hydrograph. However, groundwater release keeps stream flow relatively high well into summer, and all peak flow events have resulted from mid-winter rain-on-snow events. These rain-on-snow peaks have been up to five times greater than spring snowmelt peak flows. Annual runoff has varied from less than 12 cm to more than 150 cm over the period of record. Evapotranspiration appears to be relatively constant from year to year with a mean of about 60 cm and accounts for about half of the annual average precipitation (Kattelmann 1989a).

In contrast, in the alpine zone of the Sierra Nevada mountains, runoff from an unvegetated shallow-soil cirque exhibits radically different behavior:

> Groundwater storage and release account for only a small portion of the total quantity of water in the annual water balance of the basin. However, subsurface water is very important in the temporal distribution of water. Releases from

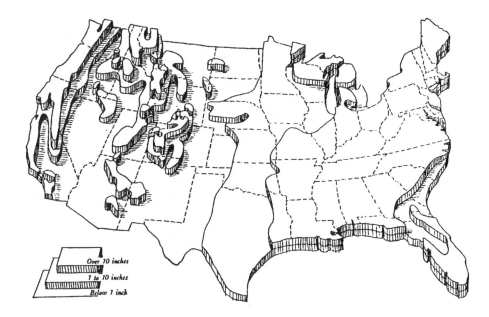

Figure 6-11 Average annual runoff in the United States (Department of Agriculture 1955)

Figure 6-12 Normal distribution of runoff in the United States by months (Department of Agriculture 1955)

subsurface storage are the primary water input to the stream and lake system for eight to nine months of the year. Although the quantity of this water is small compared to snowmelt runoff, groundwater discharged from various deposits and fractures has the potential to control lake chemistry for more than two-thirds of the year (Kattelmann 1989b).

These are examples of general hydrologic characterization; at the other extreme from the broad, areal, and monthly patterns of flow is the runoff from a single storm. Two fundamentally different approaches have been developed and utilized to describe, analyze, and provide the basis of management of stream behavior. These are the unit hydrograph and the variable source area concepts.

The classic approach to evaluating runoff in the short term is the engineering-oriented **unit hydrograph** (discussed in detail following discussion of the variable source area concept) based on the relationship between precipitation intensity and infiltration during a storm.[6] The unit hydrograph is underlain by Hortonian hydrology, focusing on the observation that the unit hydrograph is produced by surface runoff or overland flow that occurs because precipitation intensity exceeds infiltration capacity. Introduced first, the unit hydrograph and its attendant methods for hydrograph separation (into storm flow and base flow, primarily) dominate the engineering approach to hydrology. Based on several important assumptions, the unit hydrograph and its associated analytical methods have considerable utility in providing a means for rather precisely and in reliable replicated fashion analyzing runoff events in considerable detail. Another value is analysis of the underlying assumptions themselves: it provides insight into the nature of the runoff process, as well as a means of evaluating and predicting stream behavior.

Wildland hydrologists, on the other hand, have had difficulty with the unit hydrograph approach, however, because infiltration capacity always seems to be great enough to accommodate any precipitation intensity in all but unusual situations of very thin soils, extremely steep slopes, or unusually heavy rainfall events, yet a storm hydrograph still occurs. That is, the watershed responds to a water input, often in some predictable pattern. Thus, on undisturbed, vegetated watersheds "Hortonian overland flow [is] a rare occurrence" (Freeze 1972). More recently, this ecological approach to understanding short-term runoff behavior has evolved.

Variable Source Area

Here is where the distinction between storage and process begins to break down: this concept embraces both elements. Runoff is the result of interaction of a rainfall (or snowmelt) event and numerous different types of storage over the entire watershed. This gives rise to the **variable source area** concept, which recognizes the three-dimensional, dynamic nature of the runoff process, along with the acknowledgment that that process is in no way a simple one. The concept was initially named and presented by Hewlett and Hibbert (1967) who, after pointing out that "hydrograph separation is one of the most desperate analysis techniques in use in hydrology," noted that:

---

[6] The landmark paper is Horton's 1933 work entitled "The role of Infiltration in the Hydrologic Cycle."

stream flow is generated chiefly by processes operating beyond perennial stream channels, [that] the yielding proportion of the watershed shrinks and expands depending on the rainfall amount and antecedent wetness of the soil, [and] the concept that stream flow from a small watershed is due to a shrinking and expanding source area - the variable source area concept - grew out of studies of the drainage of sloping soil models at the Coweeta Hydrologic Laboratory.

Prior to that, Betson (1964) had reported that "runoff originates from a small but relatively consistent, part of the watershed," but that, in apparent contradiction thereof, there seemed to be variable portions of the watershed that contributed runoff at different times during storms. In a subsequent study, Betson and Marius (1969) had reported that the area contributing runoff was definitely *not* constant and that "variation in the depth of the topsoil caused a heterogeneous runoff pattern." A theoretical example of the variable source area concept is illustrated in Fig. 6-13.

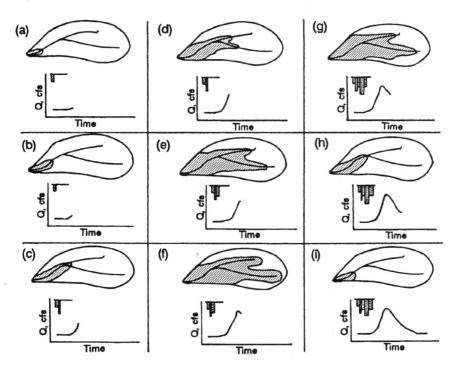

Figure 6-13 A theoretical example illustrating the variable source area concept. The several stages of the variable source area (shaded) are coordinated with a storm hydrograph. Precipitation ends between (e) and (f).

Observation of both watersheds and models reveals zones of the drainage that appear darker (owing to higher moisture content) in the immediate vicinity of the stream. These zones may, in fact, be saturated. During a runoff-causing event such as rainfall or snowmelt, the zone expands to include area further and further from the live stream. Streamflow increases during this expansion, and conversely, contracts during recession. The size limits of this variable source area are, by definition, erratic and normally incalculable. Runoff from this zone may occur from ground water in the form of localized base flow (Abdul and Gillham 1984), or from other portions of the watershed, mostly subsurface flow and channel interception. Thus, during the runoff-causing event, storm flow (often equated with **quickflow**) is actually a combination of surface runoff (often equated with **overland flow**), subsurface runoff (often equated with **interflow**), base flow and channel interception.

Evidence supporting and amplifying the variable source area concept has been the subject of both real watersheds and laboratory model studies. Working on laboratory models, Abdul and Gillham (1984) reported that "if the zone of tension saturation extends to, or near, ground surface, the application of a small amount of water can cause an immediate rise in the water table." Black (1970a) found that the storm hydrograph up until the peak was generated on the lower half of the watershed model, and that the saturated variable contributing area could be readily seen on the light-colored sponge used to simulate soil. Lane, Diskin, Wallace, and Dixon (1978) similarly reported that 45 percent to 60 percent of the watershed they studied was contributing to runoff at the peak. In a controlled field experiment, Dunne (1978) documented the variable source area, and noted:

> During a storm the rate of runoff depends upon the properties of the soil, pattern of subsurface fluid pressures, the area over which water is emerging from the ground and the intensity of rainfall. Consequently, for any single storm there is probably a hysteretic relationship between the runoff rate and the extent of the saturated area (or its length). Within a few minutes of the end of rainfall, runoff from direct precipitation onto the saturated area has ceased, return flow is also declining rapidly, but the extent of the saturated area diminishes slowly.

DeWalle and Rango (1972) reported that mean annual runoff appeared to be correlated with channel width, a suggestion that variable source area, also related to channel width, was capable of being monitored in the field. Indeed, Anderson and Burt (1978) found that an automatic tensiometer system could be used to identify and document the variable source area. It would seem logical that direct precipitation in the channel, plus overland flow (surface runoff), if it occurs, contribute disproportionately to the rising limb of the hydrograph.[7] Runoff from banks, near-channel ground water, or temporarily saturated riverine zones are added to produce the peak. Subsurface flow is more likely to contribute to the recession limb (Dunne 1978). Dunne and Black (1971) reported that runoff from snowmelt also varied diurnally, noting that:

> Snowmelt runoff is strongly influenced by nonuniform characteristics of snow accumulation, concrete frost, saturation of soils, and radiation as modified by topography and cover. The combination of these factors causes the area contributing quick runoff to be dynamic in the sense that it varies during and between days. These facts suggest that the 'partial area' concept of runoff production during rainstorms in Vermont may be a useful conceptual framework . . . .

---

[7] The term is used here to indicate the generic rise and fall of stream discharge in response to a runoff-causing event, not the unit hydrograph.

Reporting the results of creating artificial rainstorms on a real watershed, Corbett and Lynch (1989) observed:

Under dry antecedent soil moisture conditions the rising limb and hydrograph peak are produced by storm flow contributions from the channel and base slope zones, primarily in the front 30 percent of the watershed. The percent of rainfall converted into quickflow ranged from 21.9 percent for the channel-base slope application to 9.8 percent for the total watershed application. Under wet antecedent soil moisture conditions the percent of rainfall converted into quickflow ranged from 55.9 percent for the channel application to 82.6 percent for the channel-lower slope application.

Proportion of storm flow from different sources, lag times, and peak flow, all seem to be correlated with variable source area, but the limited number and extent of studies, and some differences with definitions among researchers, has not yet led to a uniform model or theory (Dunne 1978).

Variable source area is, in many ways, more difficult to comprehend than is the unit hydrograph. It demands a conceptualization of the entire watershed. Ultimately, therefore, it demands synoptic, critical analysis of all the relevant factors affecting runoff from the drainage basin. Of especial importance is consideration of the watershed's response to water input under a given set of antecedent moisture conditions. Essentially, *all* of the factors that affect the movement and storage of water to this point in the book must be within the vision of the watershed hydrologist. They are the underpinning of an ecological approach to hydrology, to management of the watershed as a resource.

Further comprehension of the variable source area concept is inherent in the presentation of management concepts in Chapter 7.

The Unit Hydrograph

Introduced by Sherman (1932), the unit graph or unit hydrograph represents on paper the combined surface and subsurface runoff ("storm flow," as shown in Fig. 5-1) from each separable segment of a watershed. It is a specialized case of the storm hydrograph, the pulse response of the watershed to the water input. Wisler and Brater (1959) provide a succinct statement of the principles of unit hydrograph theory:

1. A unit hydrograph is a hydrograph of surface runoff resulting from a relatively short, intense rain, called a unit storm.
2. A unit storm is defined as a rain of such duration that the period of surface runoff is not appreciably less for any rain of shorter duration. Its duration is equal to or less than the *period of rise* of a unit hydrograph, that is, the time from the beginning of surface runoff to the peak. For all unit storms, regardless of their intensity, the period of surface runoff is approximately the same.
3. A distribution graph is a graph having the same time scale as a unit hydrograph and ordinates, which are the percent of the total surface runoff that occurred during successive, arbitrarily chosen, uniform time increments. Alternative and interchangeable units for the ordinates are cubic feet per second per sq mi per inch of surface runoff. The most important concept involved in the unit hydrograph theory is

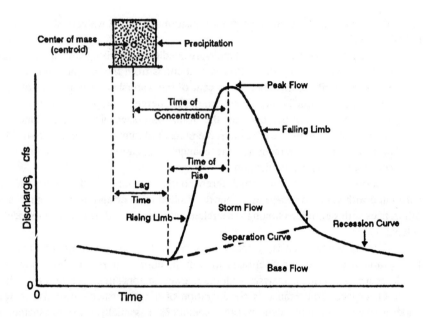

Figure 6-14 A typical storm hydrograph

that all unit storms, regardless of their magnitudes, produce nearly identical distribution graphs.

Although it can be easily proved theoretically that the foregoing relationships cannot possibly hold true rigidly, the error is so trivial that from a practical viewpoint they may be considered as being correct.

The basic assumptions underlying the unit hydrograph theory, namely that (1) the contribution of each watershed segment does not interfere with the runoff from other segments, and (2) that the runoff from all the units are additive, are not necessarily consistent and are the subject of much controversy (Singh 1976). Nevertheless, the unit hydrograph is a valuable analytical and educational tool. Its analytical value is particularly useful in determining storm-designed facilities such as culverts, reservoirs, and flood control works (Dunne and Leopold 1978). Engineering texts and handbooks detail the analytical utility of the unit hydrograph (e.g., Chow 1964; Hydrology Committee 1957), and computer programs are available (Bethlahmy 1972) that not only provide rapid calculation, but also permit easy study of the modelled runoff process by providing for multiple iterations with readily changeable input parameters. Its educational value is in its integrative and descriptive nature: it depicts the net output of the watershed following a storm or snowmelt event.

Linsley, Kohler, and Paulhus (1949) point out that consideration of the unit hydrograph "leads naturally to the hypothesis that identical storms with the same antecedent moisture conditions produce identical hydrographs." Proportionality exists between various measurable parameters of the hydrograph (e.g., height, length) and rainfall durations and, since the recession or falling limb is asymptotic to zero, and its rate of fall and duration are functions of its initial value (related or equal to the peak flow), the integration of the area under the hydrograph, which is volume of flow (cubic feet per second times time in seconds) will also be proportional to the storm's parameters.

Research has shown that the ratio of storm hydrograph height to length is a constant, that peak flow is a function of rainfall excess, that the recession or falling limb has a characteristic and constant shape, and that the unit hydrograph may be used for separating storm flow from base flow in order to achieve the foregoing measurements. In the event of accretion to ground water during the storm, previous knowledge concerning the isolated runoff-causing event's unit hydrograph may be useful in separating storm flow and base flow during these more complex periods as well (next section).

The unit hydrograph works best for relatively compact watersheds with no major channel or ground water storage, and hence may be used for watersheds under about 2000 sq mi; it is best if the rainfall duration is approximately one-fourth the **basin lag** (the time between the centroid of precipitation and the occurrence of the peak discharge) (Linsley, Kohler, and Paulhus 1949). On occasion, application of the unit hydrograph theory has been extended to larger and more complex watersheds and even to ground water hydrographs.

Smoothed, the plot of discharge (or head) over time is an attractive and oversimplified representation of a single event in a stream's history. In fact, a stream gage does trace such a curve as that shown in Fig. 6-14, and such curves (1) are of characteristic shape for a given watershed, (2) may be used to further understanding of the runoff processes on that watershed, and (3) will change in the event that land use or other runoff-affecting factors are altered.

The hydrograph is really a very complex integration (Hall 1968) of runoff from each portion of the watershed that contributes to the peak flow,[8] as well as an integrator of all the factors that affect it (American Society of Civil Engineers 1949). Violation of the assumptions underlying the method provide the range of limitations of its use: the most common violation is that the storm does not uniformly, instantaneously, and completely cover a watershed; if the storm moves across the watershed, or up or down its main axis, it will have a considerable impact on the shape of the resultant storm hydrograph. Probably the most commonly occurring natural violation is that the outflow from one watershed unit does not interfere with the outflow from another.

Given, then, that the measurements associated with the hydrograph are empirical or incomplete approximations of the true and full relationship between influencing parameters (Dingman 1984), there remains some useful application of unit hydrograph theory.

A primary purpose of using the unit hydrograph method is to predict peak flows, and a variety of simplified methods that make use of the unit hydrograph theory or short cuts based upon its theory are available.

Peak Flow Determination

Because of the drama and importance of flood peaks, a disproportionate percentage of programmatic monies for flood control and relief, and for hydrologic research, has been

---

[8] Not all of the watershed necessarily contributes to the flow at the time of the peak; however, all parts usually do contribute to the total storm runoff. See the discussion under the variable source area concept.

directed at finding ways to accurately predict the magnitude of flood peaks from storm or snowmelt events.

On an impervious watershed surface with constant slope, area, and roughness (really minute depression storage as well as resistances to surficial laminar flow) the peak flow will be a function of precipitation intensity, roughness, and watershed area, and is relatively easy to predict. Even in this simplified case, as was the situation with the unit hydrograph, one must make assumptions concerning the areal extent of the storm and the time-distribution of the precipitation. Usually, the assumption is that the watershed is instantly, uniformly, and completely covered by precipitation (rainfall) that has a constant rate from start to finish. This makes the solution considerably easier; and concurrently, any deviation from such assumed uniformity drastically complicates the solution.

The time necessary for the runoff from an impervious watershed to equal the precipitation rate is defined as the **time of concentration**. This is the period that is necessary for saturation of the surface to occur. The **rational formula** uses the time of concentration concept to determine peak flow:

$$Q_p = ciA_d$$

where Q_p is the peak flow in cfs, c is a runoff coefficient (Table 6-2), i is the rainfall intensity in in. per hr, and A_d is the area of the drainage, in acres. If a rainfall rate of 1 in. per hr is applied to a 1-acre watershed for the time of concentration, t_c, of the watershed, then the runoff will be very closely equal to 1 cfs. The use of the formula is restricted to small, developed watersheds, such as urban drainages or highways (Viessman et al. 1977), and thus c-values are not given for typical wildland uses.

Table 6-2 Runoff Coefficients for Use in the Rational Formula

Surface	Value of c
Level terrane not affected by snow	0.200
Rolling farmland, long narrow valleys	0.333
Uneven terrane, wide valleys	0.500
Rough, hilly country, moderate slopes	0.667
Steep, rocky ground, abrupt slopes	1.000

Source: Linsley, Kohler, and Paulhus (1949); Linsley and Franzini (1964).

Although the intent was to apply the rational formula to small watersheds or even to plots, the inclusion in engineering texts of c-values for "hilly country," "wide valleys," and "rolling farmland" suggests wider application. Linsley, Kohler, and Paulhus (1949) present an adaptation of the rational formula for larger areas and for longer periods of rainfall, as well as different return periods: the method developed by Bernard involves a two-stepped formulation: first, a determination of the c value:

$$c = c_{max}(T_p/100)^x$$

where c is as above, c_{max} is the value associated with a return period of 100 years and, along with the regional constants in the second step, are given in a series of maps (Figs. 6-15 to 6-20).

Inclusion of frequency of occurrence of the precipitation intensity used in the peak flow determination is possible with a general formula (Linsley, Kohler, and Paulhus 1949):

$$i = (kT_p^x)/t^n$$

where i is the rainfall intensity in inches per hour, T_p is the return period, t the is duration in minutes, and $k, x,$ and n are regional constants. Two different maps are required for the values of k dependent upon whether the duration of rainfall is between 5 min and 60 min or 60 min and 1440 min. The results for the 60-minute duration by the two formulas are not identical.

Additional complicating factors include presence of ground water storage; varying subsurface runoff; length of time between storms; nonuniformity of watershed; temperature, aspect, slope, type of vegetation, and season, all of which impact energy available for evapotranspiration and, as a consequence, the hydrologic season and thus the response of runoff to existing hydrologic conditions. The high degree of variability demands (1) acceptance of uncertain predictions, and (2) the need for estimation by more than one technique. Several have been identified.

Of primary importance is the presence or absence of ground water storage. This is ramified in minimum or sustained flow levels, and the difficulty of ascertaining streamflow-precipitation relations (see base flow recession, following). For this reason, much of the early peak flow determination work was done with "small" watersheds, those that have, by definition (Chow 1964), a drainage area of less than 100 sq mi. Unfortunately, the formulations thus developed have too often been erroneously applied to watersheds with ground water storage and/or areas in excess of that limit.

Predictions of peak flows at forest road culverts may be improved by consideration of drainage basin size, mean basin elevation, and mean annual precipitation (Campbell and Sidle 1984). For detailed computation of peak flows at culverts and other flow devices, see Bodhaine (1968). This idea becomes even more useful in flow duration and flood frequency analysis, as discussed below.

Base Flow Recession

Fundamentally, the base flow recession portion of the hydrograph (Fig. 6-14) comes from the ground water storage component on the watershed. It may be likened to an electrical capacitor that has a characteristic "die-away" constant once the electrical loading is removed. A water storage component exhibits a similar exponential decrease of output with time, expressed as

$$Q_t = Q_o K^{-t}$$

where Q_t is discharge at time t, Q_o is the initial discharge, and K is the recession constant (Hydrology Committee 1957). The next time you wash dishes (or the car), soak the sponge with water and, without squeezing it, let it drain, watching the reducing stream of water flowing from it. You are watching a recession curve.

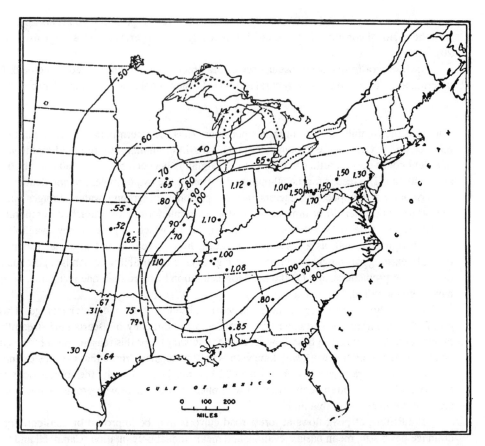

Figure 6-15 Values of the runoff coefficient, c_{max} (reproduced with permission from Linsley, Kohler, and Paulhus 1949)

Given that the inherent nature of the drainage system is what provides base flow recession characteristics, it is presumed that repetitive recessions will superimpose perfectly. However, too many independently varying factors affect base flow, such as soil moisture content, evapotranspiration, channel storage, recharge during recession, and antecedent moisture conditions in general. Methods of base flow recession determination have varying degrees of success. Different slopes for recession curves on the same watershed have been found to vary seasonally, a phenomenon that is probably the combined result of storm type and antecedent moisture conditions as ramified in soil storage potential and potential evapotranspiration.

Most of the mathematical development work on the falling limb of the hydrograph was accomplished by the beginning of this century. Interpretation of the curve still lacks high correlation with field data owing to (1) the complexity of the stream/watershed system, (2) the fact that base flow (also termed simply **delayed flow**) "can come from numerous sources

Chapter 6 Water in the Hydrosphere 207

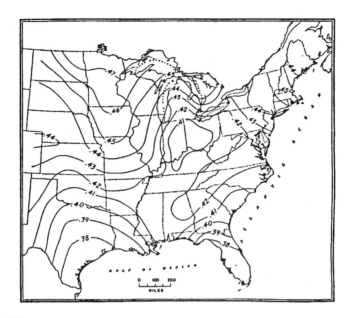

Figure 6-16 Map of n values for durations between 5 min and 60 min (reproduced with permission from Linsley, Kohler, and Paulhus 1949)

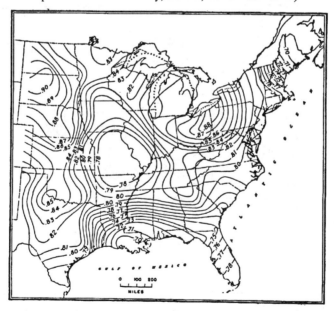

Figure 6-17 Map of n values for durations of between 60 min and 1440 min (reproduced with permission from Linsley, Kohler, and Paulhus 1949)

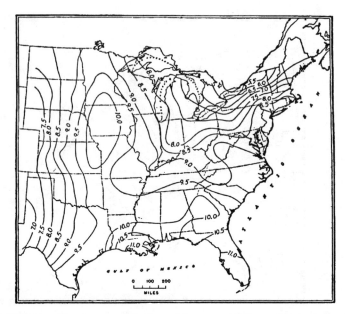

Figure 6-18 Map of k values for durations of between 5 min to 60 min (reproduced with permission from Linsley, Kohler, and Paulhus 1949)

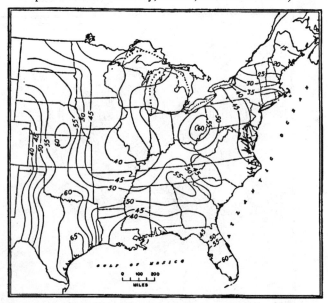

Figure 6-19 Map of k values for durations between 60 min and 1440 min (reproduced with permission from Linsley, Kohler, and Paulhus 1949)

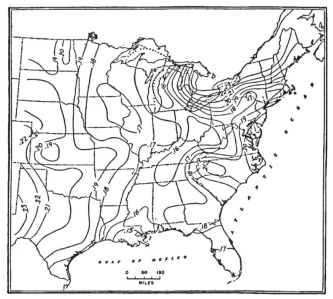

Figure 6-20 Map of exponent x values (reproduced with permission from Linsley, Kohler, and Paulhus 1949)

besides ground water" (Hall 1968), and (3) the fact that the three methods of analysis do not always produce duplicate results.

The three primary methods are (1) *graphic* or *matching strip*, making use of transparent overlays; (2) *mathematical* or *tabulation* method, where multiple observations are simply averaged for common time intervals to arrive at the base flow recession curve; and (3) *statistical* or *correlation method*, involving serial correlation of discharge over time. "The matching curve method is probably the easiest and quickest, but it is hard to do objectively without lengthening or shortening the true length of time of recession. Also, the coordinates must be chosen so that the curves are not too flat. The tabulating method eliminates some of these problems but does not allow adjustment for irregularities in the recession curves. The correlation method is potentially the most useful, but in practice several problems arise. If a straight line can be fitted to the data points, then [some direct equations] can be used. However, nonlinear curves cannot be handled so easily" (Hall 1968).

Ultimately, the value of a reliable base flow recession curve is (1) to estimate the total ground water flow and, thereby, to evaluate ground water budgets, and (2) to separate storm flow from base flow, thereby to evaluate single storm event relations between precipitation and storm flow.

If, during the runoff event, there is an accretion to ground water, or there are more than one storm pulses, then a complex hydrograph will result. Much of the base flow recession analysis is of necessity carried out on clearly separated storms but, as a consequence, interpretation of the probably more common complex multiple-pulse storm event is not clean: hopefully, as the processes are better understood, a more complete ability to predict the base flow recession and storm flow will result.

Certainly, obtaining a good hydrograph for peak flows and recession curves is not the major problem. However, interpreting and evaluating streamflow or discharge records, especially in light of how we measure other hydrologic cycle components, is a matter of some concern.

Stream Behavior

Given the general comments on runoff sources, measurement, and limitations in the preceding section, it is possible to continue with more detailed considerations of how runoff varies in time and space.

There is no single method of analyzing or characterizing runoff. Several have been developed and some are still being refined. In any given hydrologic investigation, one or more analytical methods may be useful aids to understanding, predicting, and manipulating the hydrologic cycle's prime output — runoff.

Fixed Time Periods

Patterns of streamflow that differ from the rather uniform streams of the southern United States are evident in Fig. 6-12. Precipitation in the form of snow, and attendant cold weather that preserves the snow in frozen condition until late spring or even early summer, causes disproportionate distribution of runoff during the typical year. Noticeably high peaks in May and June are natural in the mountainous areas of the western states and in parts of New England. In lower-lying areas of the northeast, for example, the peak that typically occurs in March or April is lower, and the percentage of the total annual runoff that occurs during the peak runoff months is less than for the high-snowpack runoff, mountainous regions.

More formal methods of analyzing temporal distribution of runoff include percent of total occurring in each month or season, such as is shown in Fig. 6-12; half- and quarter-flow intervals; flow duration analysis; and flood frequency analysis.

Half- and Quarter-Flow Intervals

The half-flow interval was initially defined as the length of time from the start of the record-keeping (calendar or water) year that it takes for one half of the annual runoff to exit the watershed. It is now defined as the shortest time it takes for one half of the annual runoff volume to exit the watershed. This parameter seems to be more consistent from year to year and within hydrographic regions. The quarter-flow interval is defined similarly. One or the other may be useful in any particular situation for detailed description and analysis of a stream-watershed system.

The unit, of course, is a unit of time, either days, weeks, or months. The half- and quarter-flow intervals seem to be especially sensitive to land use changes: any change in the infiltration capacity of the air-soil interface, or in one or more of the major storage components on a watershed — especially a small watershed — can be expected to produce a dramatic change in the flow interval. For example, between 1929 and 1961 on the 181-sq mi Wappingers Creek watershed in southeastern New York, the one-quarter flow interval increased by 20 percent (from 27.7 days during the first 21 years to 33.3 days during the second

Chapter 6 Water in the Hydrosphere

21 years) as the land use changed from 80 percent farmland to 80 percent forest cover (Black 1968).

Flow Duration

The **flow duration curve** expresses the relationship between magnitude of daily flow (in cfs per day, for example) and the number of days[9] during which that flow is likely to be equaled or exceeded. It is derived by tallying the number of days for a year (or over the period of record) in each size class of runoff based upon the range of observed values. The resultant relationship between the two (number of days and size class) plots as nearly a straight line on semi-log or other specialized graph paper.

The curves are part of a low flow analysis system (Riggs 1968a) that includes several additional analysis techniques. Good low flow analysis is essential for the long range planning of water supplies and community growth that can be promoted only by reliable runoff prediction. Flow duration curves may also be used to analyze the distribution of the high end of the curve, providing information on how well flows of any specified magnitude might be expected to be sustained.

A typical flow duration analysis is shown in Table 6-3; the corresponding flow duration curve is shown in Fig. 6-21. Note that:

- there need not be much more than ten classes in order to obtain good data for the curve;
- the size of the classes does not affect the curve derived;
- the curve is there, and one need only evaluate its position; the classes do not need to be the same size, nor continuous (as shown);
- the lower class limit is used because the proper way to "read" the table and, consequently, the curve, is "X-percent of the observations are greater than S-size";
- the specialized graph paper used exhibits a logarithmic vertical axis, and a normal distribution of percentages (the X-axis on commercially available paper is actually in "probits"), that is, for a perfectly normally distributed set of observations, the accumulated percentages should plot as a straight line;
- the extent to which the curve deviates from the straight line is a measure of skew: if the extreme values from Table 6-3 were included, (1) the paper would have to be very much larger, and (2) the skew would be more apparent (numerous techniques, especially in flood frequency analysis, have been developed to evaluate and account for the extreme values).

(This analysis and figure should be compared with the parallel analysis and figure showing flood frequency analysis in Table 6-4 and Fig. 6-24, respectively.) Note, especially, the difference in table headings, titles, graph paper used, and the units of flow involved. But note, too, that the basic process of analysis is the same for each. The flow duration analysis must include the extremes of flow because the investigator may be as interested in maximum volumes of flow as in minimum reliable supplies.

---

[9] Other time periods may be used, of course, but the day is most commonly used.

Table 6-3 Daily Flow Duration Analysis for 3113-sq-mi Eel River Watershed at Scotia, California 1911–1955

Lower Flow Class Limit (cfs)	Number of Days in Class	Accumulated Number of Days	Percent of Total Days
12	49	15,735	100.0
21	66	15,686	99.7
32	539	15,620	99.3
64	1661	15,081	95.8
110	1735	13,420	85.3
200	1285	11,685	74.3
340	1108	10,400	66.1
600	967	9592	59.1
1000	997	8325	52.9
1800	1342	7328	46.6
3200	1517	5986	38.0
5400	1667	4469	28.4
9400	1385	2802	17.8
17,000	656	1417	9.0
29,000	432	761	4.8
50,000	199	329	2.1
86,000	100	130	0.8
150,000	28	30	0.2
440,000	2	2	0.0

Source: Smith and Hains (1961).

Within a hydrographic region, flow duration curves are similar for watersheds that are geographically similar, that is, have the same size, shape, aspect, slopes, etc. For example, Fig. 6-22 shows flow duration curves for four drainages on the North Coast in California: three of these curves are close together, showing the expected relationship between flow magnitude and drainage area, that is, mean daily discharge in csm decreases with increasing size. The fourth, the Klamath River basin, is the largest, but the reason for its curve being different in orientation from the other three would seem to be the fact that it has a markedly different physiographic orientation and shape: it is more compact and with a dendritic drainage pattern of tributaries that flow generally south before joining and turning northwest, whereas the other three are all long, narrow drainages that flow toward the northwest.

The differentiation of the curves suggests that it may be advantageous to somehow stratify the drainages under consideration by geomorphic characteristics. Dingman (1978a) reports one such successful study for three hydrologically different regions in New Hampshire, and Reich's study (1971) suggests further geomorphic stratification for 60 watersheds in Pennsylvania. Eagleson (1972) included several climatic and watershed characteristics in flood frequency determination from rainfall parameters for three Connecticut basins. How this might be accomplished for both flow duration and flood frequency analysis is developed in the next section.

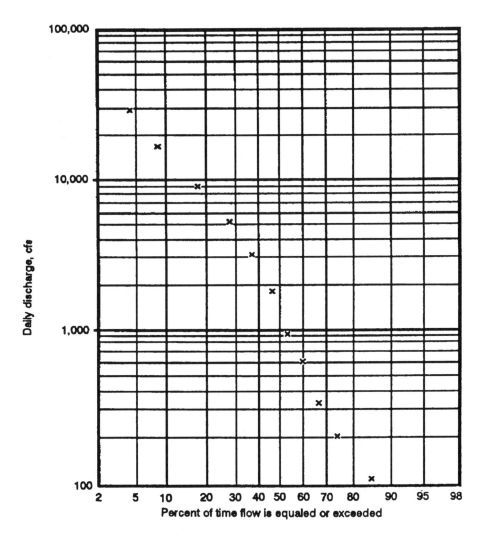

Figure 6-21 Flow duration curve for data shown in Table 6-3

Flow duration curves that are steep[10] represent "flashy" streams, ones that have higher peaks and lower minimum flows. Curves for small mountainous watersheds typically exhibit this type of curve. Larger watersheds with ground water storage that attenuates the abrupt input of frequent storms and/or snowmelt events exhibit curves that are more nearly horizontal, with lower peaks and higher minimum flows. A typical change in the flow duration

---

[10] One must exercise caution in interpreting "curves" or even straight lines on nonarithmetic graph paper: lines that appear to be parallel may not be, in fact. The only way to ascertain the slope of a line, or to compare the slope of two seemingly parallel lines, is to read values at two or more different locations along the curve and to compute the slope at or between the places desired.

curve as a result of change from rural to suburban or urban land use is to make the flow duration curve more steep: the runoff is flashier as a result of both reduced infiltration capacity and soil storage. Regulation, on the other hand, as intended, dampens the extremes of flow duration curves. Flood retarding basins in suburban areas are installed for the purpose of maintaining the natural slope of the flow duration and flood frequency curves. An example of natural variation in flow duration curves appears in Fig. 6-23, which shows curves for four streams in the Mohawk River Basin in New York. The two unregulated streams (188 sq mi and 113 sq mi in area, Fish Creek and the East Branch of Oneida Creek, respectively) exhibit flashier (steeper) flow duration curves; the slopes of the regulated streams (1348 and 150 sq mi, the Mohawk River at Little Falls and at Delta Dam, respectively) are substantially modulated. In each pair, the smaller stream displays larger mean daily discharge values.

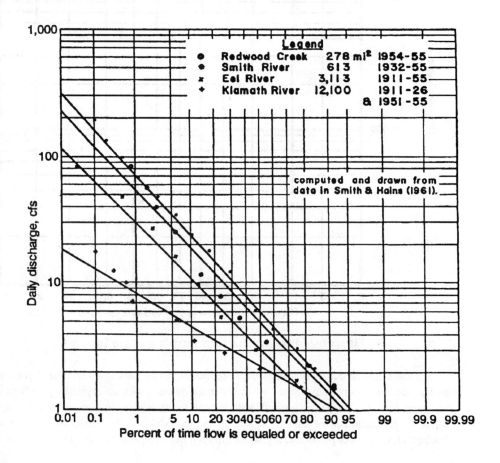

Figure 6-22 Flow duration curves for four California north coastal streams

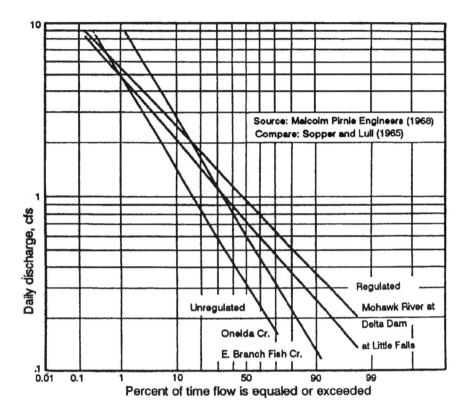

Figure 6-23 Flow duration curves for four Mohawk River streams in New York

Flood Frequency

The **flood frequency curve** is very much like the flow duration curve but, instead of using runoff data in cfs/day, the data are the instantaneous flow values. The flood frequency curve may be derived from streamflow measurements that include peak discharges. Maximum annual peak flows for a series of years constitute a data set that may be analyzed for information about the likelihood and magnitude of occurrence of floods. Such a set of data is particularly useful for analyses related to flood prevention, control, and protection, man's occupation of floodplains, and to economic analysis. However, Linsley, Kohler, and Paulhus (1949) point out that, "such a series ignores the second highest event in any year, which, in some cases, may exceed many of the annual maxima." As a consequence, the annual series is likely to underestimate flood magnitude.

A set of monthly flood peaks, or a "partial duration series," refers to all floods over a specified minimum flow, and is useful for more thorough scientific ("pure") hydrologic information. Its major drawbacks are that the data do not represent independent events and that it is difficult to relate the resultant temporal information to practical management time applications.

For undisturbed streams, flood frequency data tend to plot even closer to a straight line than do the flow duration curves, and thus they lend themselves to more reliable extrapolation. This is especially true for the annual series, which is a series of independent and, therefore, randomly distributed events. As a consequence, the flood frequency curve may be based on as little as 20 years of data, yet with confidence provide an estimate of the 100-year flood.

In any case, it is interesting to note that the annual flood peaks tend to plot as do the data from a normal curve, yet the data clearly represent anything *but* a normal distribution! They are series of (annual) maximums of the maximum annual instantaneous peaks, and represent a mighty small sample of all the instantaneous discharges.

The **100-year flood** is that instantaneous magnitude of flow that, on the average, can be expected to be observed once in 100 years at a specified point on a stream. While there are similarly-designated floods of any magnitude, the 100-year flood is most commonly used for floodplain management, flood insurance regulations, and many engineering designs. If there is a chance of great loss of life and/or property, standards may be set higher, at the 500-year or 1000-year flood magnitude. Many of these terms are often not clearly presented to the public. The **probability** of the 100-year flood is 1 in 100, or 0.01.

A **flood** is really any peak flow, but the term has been restricted to mean a flow that exceeds the capacity of the banks of the stream at a given location. (Thus, when a stream is "in flood" at one locale it does not mean that it is "in flood" all along its full length, although that is frequently the case. Storm and runoff distribution in time and space, land use, stream management, and other disturbances to the stream/watershed system may alter the normal event.) A storm affecting the central portion of a watershed may cause local flooding, even of severe magnitude, but may cause only minor flooding downstream. It is the watershed-covering runoff-producing event (cyclonic precipitation, a large hurricane, or widespread snowmelt) that generates a major flood. Indeed, it is the annual high flow itself that erodes the geologic material through which the stream flows, and that creates and maintains the stream channel size.

On the average, streams reach bank-full stage (height) once a year and overflow those banks once every 2.33 years. The latter figure has been established by repeated flood frequency analyses throughout the Temperate Zone. The distribution of flood peak sizes above bank-full stage is what has been the target of many hydrologists, for the relationship between magnitude and frequency is not always a perfectly straight line and numerous techniques have been developed for more precise flood frequency analysis. For a complete discussion of flood frequency analysis, including the several different analytical variations, and interpretation of the results, see Chow (1964) and Riggs (1968b). There really is no agreement on the definition of "flood" in terms of bank-full discharge, or even on what "bank-full" means (Williams 1978), who reports no less than 11 definitions of that term.

Another way to state the definition of the 100-year flood is that, on the average, once in any consecutive 100 year-period, one can expect that a flood of that magnitude will be observed once. This way of stating the situation promotes understanding by illuminating an important feature of the whole frequency concept: during that 100-year period, there is also a finite chance of observing the 1000-year flood, and the ten-thousand-year flood. There is also a finite chance of observing a 147-year flood, and a 852-year flood, and so on. If one adds all the finite chances (probabilities of occurrence), and then looks at the analysis from the standpoint of how many times during a 100-year period the 100-year flood is likely *to be equaled or exceeded*, the answer is 2.33 times, or once in about 43 (100/2.33) years. That is a lot more frequently than once in 100 years. When one is trying to keep breathing, the height of the water above the nose for a minute is of considerably more importance than the mean daily flow!

The flood magnitude that has a **recurrence interval** (**return period** or frequency of occurrence) of 2.33 years has been specified as the **Mean Annual Flood**. The ratio of any N-year flood to the mean annual flood varies from river basin to river basin, and may be related by standard regression techniques to the drainage area for each basin or for a region. "Particularly in mountainous regions, it may be necessary to add mean basin altitude as a second variable to predict the mean annual flood. Where strong gradients of precipitation occur in a region, it is usually necessary to use mean annual precipitation or mean annual runoff as a second variable along with drainage area" (Dunne and Leopold 1978). Markowitz (1971) recommends not using the term "Recurrence Interval" because, as the reciprocal of probability, it is a dimensionless ratio, "not an interval at all."

Like flow duration curves, flood frequency curves are also a characteristic of a particular stream/watershed system, reflect regional similarities, differ from other stream/watershed systems because of varying geologic and geomorphic characteristics, and may be altered by land use practices and stream disturbances. Flood frequency data are stroked the same way as the flow duration data, only instantaneous peaks are used, either in cfs or csm. A typical flood frequency curve on log-probability graph paper is shown in Fig. 6-24 based on data shown in Table 6-4.

Note that the comments concerning the derivation of the flow duration curve apply here as well. There need not be more than ten classes to obtain a good curve; the size of the classes really does not affect the curve derived; the classes do not need to be the same size, nor continuous. For the flood frequency curve, however, the extreme low point on the curve is somewhat arbitrary and need not be plotted: its exact position is usually not clearly identified, and trying to include it in the plot only confounds interpretation. The interpretation really concerns the other end of the curve anyway. The graph paper used in Fig. 6-24 is known as log-probability paper, and uses nearly the full normal distribution of percentages on the X-axis. Annual flood peaks are considered to be independent events, and thus are more likely to be randomly (normally) distributed than the mean daily flows used in flow duration analyses.[11] Since there is a lower limit on flood peaks, however, there may be a greater degree of skew, thus, the elaborate means of flood frequency analysis presented in engineering texts and handbooks (cf., e.g., Boughton 1980; Matalas and Wallis 1973; McCuen and Rawls 1979; or Todorovic and Woolhiser 1972; Water Resources Council 1978).

Variation of the flood frequency curve[12] can be reduced by stratifying the data. This is because different types of runoff-causing events typically occur at specific times of the year, and because each hydrologic season reflects unique storage conditions. The runoff-causing events and the hydrologic conditions that apply when that type of event typically occurs may

---

[11] The latter are not independent; in fact, they often are serially correlated.

[12] This discussion applies to flow duration curves as well, although the differences between drainages may not be as pronounced as they are with the more volatile flood frequency data.

be confounded. Nevertheless, flood frequency curves for each of the several flood-causing events can be expected to be different. For example, computing the flood frequency curve for a stream/watershed system in the northeast from floods caused by summer thunderstorms will (1) reduce the variation around the curve, and (2) increase the accuracy and confidence with which summer flood peaks may be predicted. Similarly, analysis of floods caused by spring snowmelt events or by hurricanes improves understanding of both the stream/watershed system and the overall flood regime (Black 1989c).

Table 6-4 Annual Flood Frequency Analysis for the 3113-sq-mi Eel River Watershed at Scotia, California 1915–1965

Peak Discharge Size Class (csm)	Number of Peaks in Class	Accumulated Number of Peaks	Percent of Total Peaks
10–19	6	50	99.99
20–29	6	44	88
30–39	8	38	76
40–49	9	30	60
50–59	2	21	42
60–69	4	19	38
70–79	4	15	30
80–89	3	11	16
90–99	2	8	16
100–109	1	6	12
110–119	3	5	10
170–179	1	2	4
240–249	1	1	2

Source: Geological Survey (1967).

A first attempt at both regionalization and grouping by season throughout the northeast was presented by Sopper and Lull (1964). The authors concluded that:

Classification of watersheds by physiographic units and zones of latitude and elevation provides a first and elementary approach to an understanding of stream flow and its variation in the northeast. Significant differences between physiographic units in annual and seasonal flows, flow duration discharges, and storm flows suggest that the physiographic groupings selected had distinctive influence on stream flow; classification of watersheds based upon arbitrary divisions of equal latitudes and elevation provided no better separation. Significant differences within physiographic units suggest a diversity of stream flow derived likely from precipitation, topographic, and geologic influences and land use.

In some regions flood peaks may in fact have only one cause, such as summer thunderstorms in the arid southwestern states. However, in many regions, maximum annual peak discharges may be caused by hurricanes, snowmelt events, or rainstorms. And the rainstorms may be lengthy, rather low-intensity cyclonic storms or intense convectional or orographic storms. Stratification of long-term annual flood peak data can be readily accomplished: Black (1989c) found that simply stratifying the data for two New York watersheds by month, presumably to select distinct groups of hurricanes (September 1 to November 30), snowmelt (December 1 to March 30), and summer thunderstorms (April 1 to August 31), reveals (1) less scatter of points around the flood frequency curve for each cause, (2) different causes for the maximum flood peaks on different size watersheds, and, therefore, (3) confounding of flood peak cause with hydrologic season, since the months selected for each storm type coincide with the Season of Maximum Runoff, Season of Maximum Evapotranspiration, and Season of Soil Moisture Recharge, respectively. Thus, the hydrologist must go beyond the readily-available superficial precipitation intensity-duration-frequency data and study weather and antecedent moisture conditions at the time the storms being used in a flood frequency study occurred.

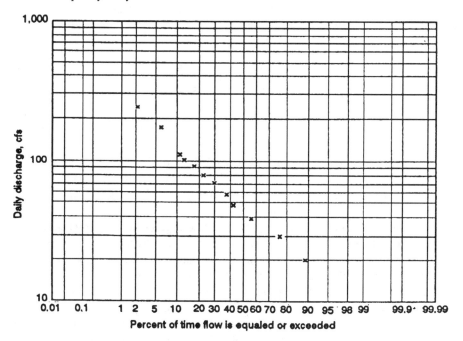

Figure 6-24 Annual flood frequency curve for the 3113-sq-mi Eel River watershed at Scotia, California, 1915–1965

Fig. 6-25 shows a series of flood frequency curves for four subwatersheds of the Susquehanna River in New York. Background data are shown in Table 6-5. There are several observations that may be made from this set of data. First, although not shown owing to excessive clutter, the variation (spread) of plotting points is much less for the stratified curves for each drainage than for the conglomerate curves (not shown). Second, the curves for both snowmelt and rainfall peaks are higher (and approximately equal in slope) for the smaller drainages. Third, the rainfall peaks curves are noticeably steeper than are the snowmelt runoff peaks. Fourth, the rainfall peaks curves from the larger drainages are plotted as one since they overlapped considerably and were markedly different from the small drainage rainfall runoff peak curve. Fifth, the snowmelt curve for the 1370-sq mi drainage reminds the researcher that even usually-uniform and easy-to-interpret hydrologic data occasionally leave something to be desired. And sixth, rainfall peaks are progressively more important at lower frequencies of occurrence with smaller watersheds.

Table 6-5 Background Data for Drainages Used for Flood Frequency Analysis in Fig. 6-21

Stream Name	USGS No.	Area in sq mi	Number of Peaks	
			Snowmelt	Rainfall
Owego Creek	5140	186	21	17
Tioga River	5205	770	21	34
Tioga River	5265	1370	21	38
Chemung River	5310	2530	21	17

Source: Tice (1968).

The most important conclusion from these observations is that to determine the mean annual flood for a specific watershed or for a region based on unstratified data is likely to severely underestimate the 100-year flood, thus exposing life and property to unnecessary risk.

Once the 100-year flood's instantaneous discharge, Q (in cu ft per second) is identified by flood frequency analysis for a given reach of stream, that discharge estimate may be converted into a site-specific figure for practical use. Dividing it by the velocity, V (in ft per second) of flow, which is largely a function of channel (or floodplain land) roughness and slope leaves the cross sectional area, A (in sq ft). Once the width of the floodplain is surveyed, then, the depth of water at such a discharge may be determined, and damages may be calculated anywhere along the stream or river. The relationship between discharge and damages on a particular tract of land is at the heart of the Floodplain Information Series published by the U.S. Corps of Engineers and these, in turn, serve as the bases for floodplain regulations, flood insurance coverage, and community planning.

It is important for the watershed manager — the watershed hydrologist — to understand how this analysis is conducted, as well as how the flood peak is affected by factors on the land that may be manipulated; land use can profoundly affect flood-peak magnitudes (Schneider 1981) and frequency. Urbanization was noted to cause a dramatic increase in flood peak magnitude for small floods with only 30 percent paving of the watershed; the effect declines as recurrence interval increases (Hollis 1975).

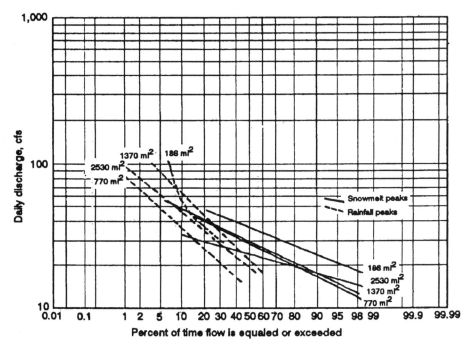

Figure 6-25 Flood frequency curves for data shown in Table 6-5

Ultimately, flood frequency curves and accompanying analyses are used to determine that section of a stream's floodplain that may not be used for permanent structures. Based on some known flood discharge against which we wish to protect and on a survey of the section of the stream in which we are interested, the inherent properties of the floodplain may be ascertained. The **floodway** is that portion of the floodplain which actively conveys the expected discharge. The remainder of the floodplain, referred to as the **flood fringe**, usually provides temporary storage of flood waters and may, under proper conditions, even convey water upstream. The dividing point between the floodway and flood fringe is the Standard Project Flood, often the 100-year flood, unless unusual circumstances dictate a different magnitude. No permanent structures that reduce the area through which the flood discharge should be permitted in the floodway. Open uses of land where no great value will be lost if flooding occurs may be permitted in the flood fringe (Corps of Engineers 1976).

Since the primary causes of floods, excess precipitation and snowmelt events, are often regional in scope, flood frequency curves may be ascertained for regions based on the unit of the cu ft per second per sq mi (csm). Once a discharge value has been estimated for a given stream/watershed system, the figure may be multiplied by the number of square miles in the basin to estimate total peak runoff. Since rainfall intensity tends to fall off from the point estimate (Weather Bureau 1961), larger watersheds may be expected to have lower peak discharges in csm than smaller watersheds. Contributing to this diminution of peak discharge on larger watersheds is the added length of time that it takes for water to reach the mouth of a watershed from where it has fallen simply because the watershed is larger (Black and Cronn 1975). Methods for creating and calibrating regional flood frequency curves are presented by Reich and Hiemstra (1967) and Dunne and Leopold (1978). Through the use of a curve showing the ratio of peak flow to mean annual flood and a curve relating the mean annual flood to drainage area, flood frequency curves may be estimated for ungaged watersheds (Dalrymple 1960).

Finally, it is important to note that the complex integration of hydrologic season, type of runoff-causing event, and any additional unique conditions surrounding a runoff-producing event may result in the storm and flood frequencies being quite different. Thus, Hurricane Agnes in June of 1972 is generally considered to be a 500-year storm,[13] yet produced up to 1000-year flood magnitudes on some Pennsylvania streams (Reich 1973): the soil was fully charged, and the Season of Maximum Evapotranspiration had only just begun. Had the storm occurred during the more normal (fall) hurricane season instead of in June, soil moisture storage would have been low, and the flood magnitudes would, in all likelihood, have been considerably less. The importance of these complex interactions cannot be underestimated: for example, dams had been planned for the Tioga River, a tributary to the Chemung River above Corning, New York. These dams were planned for recreation and flood control, but normal operating procedures would have dictated that the reservoirs behind the dams be *full* at the end of the spring runoff period so as to meet the summer's recreation demand. Indeed, had the dams been constructed, they would likely have contributed to the downstream flooding because, being full, they would not have been able to store the excess runoff and would have been overtopped. In a study comparing floods and rainfall extremes, Reich (1970) concluded that "extreme rains that occur during the spring are more strongly linked to annual floods than yearly maximum rainfalls are."

One of the lessons of the flood frequency and flow duration analyses is the high degree of integrity of hydrologic data: were regression coefficients to be calculated, values in excess of 0.9 would be typical. Yet it seems that an inordinate amount of time is spent trying to refine flood frequency techniques, especially when more thorough consideration of the hydrology of the stream/watershed system is likely to produce greater benefits in the form of more reliable water resource management practices. Note, too, that although the hydrologic data do in fact exhibit high correlation coefficients, hydrologists cannot predict *when* extreme events are likely to occur; only that they *will*, and when they do occur, what the *probability* of their occurrence is — or was.

Drought

Drought is literally continued dry weather, but that is not equivalent to **aridity**, which is merely regularly expected low precipitation. Drought may be a normal characteristic of a particular climate; it may be a common occurrence, or it may be an unusually severe and extended period of low precipitation, often accompanied by high temperature (and low humidity) that exacerbates the perceived degree of dryness. Severe droughts, like floods, are

---

[13] It is difficult to evaluate the frequency of the storm: it was unprecedented in terms of total amount of precipitation, duration, and areal extent. Once a 100-year flood has been observed, the rating curve must be redetermined to comply with certain regulations concerning floodplains and to ascertain the extent of any changes to the channel itself or the floodplain (Corps of Engineers, nd).

recounted in the Bible and have occurred throughout history. Like floods, they are a natural part of the hydrosphere. Thus, their probability may be calculated in much the same manner as flood frequencies are determined. Determination of their magnitude is moot: the ultimate drought is zero precipitation and/or runoff. Droughts are, of course, at the opposite end of the spectrum from floods but, unlike floods, each drought is not a single, clearly identified, or isolated event.

Dunne and Leopold (1978) define drought in general terms as "a lack of water for some purpose." Sadeghipour and Dracup (1985) define drought by its severity, duration, and magnitude. "Magnitude" means average water deficiency, and "severity" means cumulative water deficiency (Dracup, Lee, and Paulson 1980). A regional extreme drought method is developed by Sadeghipour and Dracup (1985) that "is capable of generating a series of drought events which, although they have not occurred historically, are more severe than historic events."

Tree-ring analysis has been used extensively to evaluate drought frequency (L. M. Thompson 1973). Reconstruction of annual precipitation over long periods has been accomplished with these records (Duvick and Blasing 1981). Relationships with sunspot cycles and the moon's 18.6-yr declination cycle are also apparent (Kerr 1984). There are, of course, a large number of factors, each with a unique cycle, that affect our weather and, by definition, our climate. If all the factors that have a negative influence on precipitation peak at the same time, a catastrophic drought (or, at the other extreme, flood) will occur. With a large number of such factors, the resultant frequency-magnitude relationship, as illustrated in the flood frequency and flow duration curves, exhibits characteristics of a series of independent or random events: it is largely these relationships that are used to define, evaluate, and predict floods and drought. With virtually no agreement on the meaning of the term, a universal approach to drought identification, evaluation, or prediction is unlikely.

S. Lee (1985) points out that there are a large number of definitions of drought: three classes may be identified. **Meteorologic drought** is precipitation below some specified levels of amount or frequency, or both. Considering the wealth of precipitation records, it is not surprising that drought is most often cited in terms of meteorological conditions, including number of days without precipitation, amount of precipitation, high temperatures, or low humidity. Other than fueling mankind's fascination with extremes, meteorologic drought of and by itself is not a very helpful tool in water resources management.

Agricultural drought is evaluated in terms of crop growth, demands for water, and the amount of soil moisture available (Hofmann and Rantz 1968). It is best known owing to the creation of the Palmer Drought Index (Palmer 1965, 1968) that is computed regularly by the National Weather Service and is widely used by the agricultural community. This index is complex in computation, and owing to the nature of the data it requires, often lags a month or more behind the conditions that might warrant drought emergency measures.

Hydrologic drought is what municipalities often report in the form of low river (or runoff) data, and low reservoir levels. Hofmann and Rantz (1968) point out that hydrologic drought is basically a deficiency in precipitation or runoff, and that its precise definition varies from country to country. Hydrologic drought might best be considered as a combination of meteorologic and agricultural drought with added consideration of runoff, soil moisture, and long-term deficiencies in precipitation. Perhaps the most important consideration of hydrologic drought is *when* the reduced precipitation occurs. This is important in relation to normal precipitation patterns and storm types, long-term runoff records, and recent river flow trends. The hydrologic seasons tend to integrate several of these characteristics. An hydrologic drought index would recognize that several inches of rainfall in April, for example, will be more effective in filling a reservoir than in July. In July, most of the rainfall would go to recharging soil moisture, not recharging ground water aquifers or appearing directly as runoff which, subject to minimum evapotranspiration losses, could be stored in a reservoir.

Lee (1985) developed a hydrologic drought index that could be updated on a weekly basis, thus is neither as cumbersome as a daily index, nor as potentially obsolete by the time it is published as a monthly index. It is based upon a combination of the water budget and the hydrologic seasons identified thereby, and includes information about precipitation, evapotranspiration, and streamflow. Field moisture is used rather than soil moisture: the concept includes soil moisture, interception, and depression storage. When precipitation does occur, it is distributed between evapotranspiration and streamflow dependent upon existing conditions. As a management tool, the index is designed to indicate not only the level of drought but, given the hydrologic season and existing conditions, the means to answer the question: How much precipitation is necessary to "get out of the drought" situation? and what the probability of that occurrence is. The method requires calibration on each watershed on which it is to be used, although the allocation of excess precipitation to evapotranspiration and runoff is uniformly determined. "It was demonstrated that the weekly computation of the cumulative deviation and drought index provides a timely and relevant evaluation of a hydrologic drought which was not treated adequately in previous drought indices" (Lee 1985).

PRECIPITATION-RUNOFF RELATIONS

Length of the reporting period is an important consideration in hydrologic analyses. Annual analysis of runoff as a percent of precipitation, for example, can be a useful descriptive attribute of a watershed or even a region. Monthly runoff may also be expressed as a percentage of monthly precipitation.

In addition, runoff per unit of precipitation must be examined in light of the broad cycle of the hydrologic seasons (Chapter 4). Changes in storm runoff per inch of precipitation reflect the annual cycle of storage during the year and different responses are characteristic of the different seasons. Data in csm per in. of precipitation for 12 hydrologically similar and isolated storms (amount, duration, intensity, and antecedent moisture conditions) during 1972 in the Mohawk Valley in New York are shown in Fig. 6-26. The annual pattern is what might be expected, reflecting the fact that during the season of maximum evapotranspiration, any precipitation will go first to replenish depleted storage. In contrast, after soil moisture has been recharged in the fall, precipitation goes directly to runoff, resulting in higher runoff per unit of precipitation. However, if the precipitation is in the form of snow and is temporarily stored in the snowpack until melt occurs, the runoff is delayed, as in January. In spring, snowmelt events caused in part by warm rains on the pack can cause large quantities of runoff. If the data are expressed in inches of runoff per inch of precipitation, the values may exceed unity. Further, the month is a rather inexact counting period, especially in view of the fact that precipitation may occur near the end of one month and the runoff may not occur until after the rather artificial end of the reporting period.

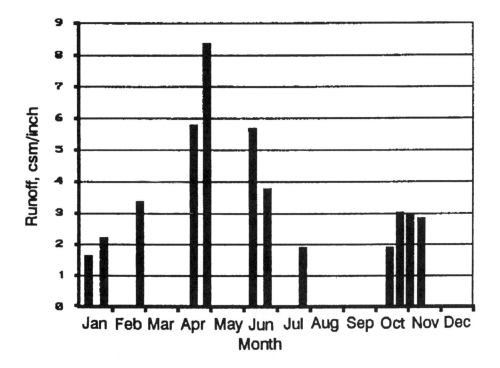

Figure 6-26 Runoff per unit precipitation in the Mohawk River watershed

The expression *P-RO*, as noted in Chapter 4, may be used on an annual (preferably Water Year) basis in order to estimate the total amount of evapotranspiration. Consideration of the expression is useful in understanding the role of storage in the hydrologic cycle in general and on the watershed in particular. Thus, monthly values of *P-RO* do not reflect monthly E_t values owing to the possibility that precipitation may be going to replenish soil moisture storage, into the snowpack, or may even be negative as precipitation water that was stored as snow finally melts and runs off the watershed.

Horton presented a classification of stream rises that was summarized by and cited in Wisler and Brater (1949). It is reproduced in Table 6-6. Interpretation of the table hinges upon relations between rainfall intensity and infiltration capacity, *f*; and total precipitation infiltrated, *P*, and field moisture deficiency, FMD (which, expressed positively, is detention storage). Where runoff occurs, it is expressed quantitatively as $Q_s = P_e$ (surface runoff equals precipitation excess). The classification is based upon observation of hydrograph behavior between two successive points on the recession curve, *m* and *n*, as interrupted by a pulse of quickflow or storm flow, actually consisting of surface and subsurface runoff.

Table 6-6 Classification of stream rises
(overland flow is shown cross-hatched)

Characteristic	Stream Rise Class			
	0	1	2	3
Discharge configuration				
n & m on curve	(fig)		(fig)	
$n < m$		(fig)		(fig)
$n = m$		(fig)		(fig)
$n > m$		(fig)		(fig)
Rainfall intensity	$<f$	$<f$	$>f$	$>f$
Field Moisture Deficiency, FMD	$>P$	$<P$	$>P$	$<P$
Total infiltration	$<$ FMD	$>$ FMD	$<$ FMD	$>$ FMD
Overland flow	None	None	$Q_s = P_e$	$Q_s = P_e$
Ground Water Accretion	None	$P - FMD$	None	$F - FMD$
Change in Discharge, if any	None	Ground water flow only	Surface water flow only	Surface and ground water runoff

Adapted from Wisler and Brater (1949).

The *type 0* rise is identified in contrast to the other types. In fact, there is no change in the depletion curve of the stream during this event, although in the real world, there will be a slight if undetectable blip on the curve representing any channel interception. Sequential storms may eventually lead to an increase in soil storage to the point where one of the other types of rises occurs.

The *type 1* rise is characterized by the fact that rainfall intensity is less than infiltration, but total precipitation exceeds the field moisture deficiency, that is, satisfies retention storage and contributes to detention storage that flows directly to the water table and hence to a stream as base flow. One of three consequences will be apparent dependent upon whether the point at which the depletion curve is rejoined is equal to, below, or above the point at which the deviation from the curve commenced. Note that if point n is higher than m, the rate of depletion will be greater, that is, the curve will, as shown, be steeper. This type of rise is the controversial one: accretion of water to detention storage may produce subsurface flow which behaves similarly to surface runoff. Thus, there may be a small storm flow peak associated with this type of rise.

In the *type 2* rise, rainfall intensity exceeds infiltration, but the total amount infiltrated does not exceed the field moisture deficiency, and surface runoff occurs without accretion to ground water.

The *type 3* rise is a combination, then, of types 1 and 2. The classification is an example of the use of unit hydrograph concepts to further understanding of the complex runoff process.

MODELS AND MODELING

My ancient college dictionary defines **model** as "a representation, generally in miniature, to show the construction or serve as a copy of something" (Barnhart 1949). In essence, the model is simulating reality. Dingman (1984) points out that there are two classes of hydrologic problems that require models: the first is **hydrograph simulation**, or predicting discharge and its variation at some specified cross-section, including prediction of watershed yield change as a result of land use alteration, which involves evaluation of all parts of the watershed affected by the change. The second is **streamflow** or **channel routing**, which is prediction of the discharge pattern at some particular point along a channel when the discharge hydrograph upstream is given.

Pritsker and Pegden (1979) claim four levels of use for simulation models: (1) explanatory or descriptive, (2) analytical, often involving quantification as well as identification of related components, (3) design based on an established descriptive or analytical model, and (4) prediction models that are useful in forecasting, such as the weather. There are many other definitions which include some constraint in the nature of the model or in its purpose. The general definition is fine. And, in general, the simplified *purpose* of the hydrologic model may be either (1) to illustrate (explain or expedite understanding of) a large or complex system in a simplified and readily comprehended manner, or (2) to permit prediction of hydrologic events (and resultant design of management control systems) once the basic relationship(s) between the components of the model are established.

Given the foregoing litany of complaints and complexities of evaluating and representing floods and other hydrologic events, one is tempted to simply build a model of a stream/watershed system and generate rainstorms while monitoring the runoff. However, while it is possible to build a physical model of a watershed, it is not possible to model the water: water is water. Nor is it possible to quickly establish a formula much less more complex representations of the relationships between two or more variables in the hydrologic cycle. A large number of models have, in fact, already been presented in this volume: the graphic presentations in Figs. 1-1 and 1-2; the unit hdyrograph; the recession curve formula; the energy laws and energy budget; the water budget; Dalton's Law; the Bowen Ratio; Penman's E_t equation; the Blaney-Criddle equation; the Thornthwaite Water Budget; Darcy's Law; the Rational Formula; retention and detention storage concepts; and the many diagrams which represent (model) natural phenomena.

A model is constructed of at least two variables and of one or more relationships between those variables. A **variable** is a quantity that may assume a succession of values, such as annual precipitation. Model variables may be independent or dependent. Annual runoff is dependent upon annual precipitation which, for purposes of a simple linear hydrologic model, is independent. In turn, of course, annual precipitation is dependent upon a host of other variables, such as distance from the ocean, latitude, elevation, time of the month, season, and sunspot activity. Some variables may also vary together (for example, the inverse relationship between temperature and relative humidity) so that the model (or the modeler) should be capable of detecting and evaluating *c*ovariance.

Measures of dispersion, such as the standard deviation, and of central tendency, such as the mean, are called **parameters** if they are characteristics of the entire population (these do not vary), and **statistics** if they are characteristics of the sample (subject to variation dependent upon which member set of the population is sampled). Populations are sampled because measuring all the individuals may be impractical, impossible, or unnecessary. The process of determining which individuals in a population will be sampled can be systematic, stratified, or random. The criterion for a random sample is that every individual in the population have an equal chance of being selected for the sample. Randomness is required in order to properly evaluate the statistics of the sample and therefore infer characteristics of the population.

In the absence of randomness, a sample may exhibit **bias** or **aliasing**. A biased sample of annual runoff might be obtained by measuring streamflow during a drought, or sampling soil moisture at 4 PM. A model that predicted evapotranspiration based on such soil moisture measurements would produce incorrect results. Aliasing occurs when a cyclic phenomenon is monitored on a regular basis with a frequency that is different from that of the natural cycle. Aliased data may determine the true mean of the population in the long run, but the natural cycle cannot be discovered by the measurements alone, and maxima and minima will not necessarily be detected either. Model construction and use must be tempered with awareness of these potential pitfalls.

The form of the relationship between variables may be linear, exponential, or logarithmic, direct or inverse, etc. The model must represent the proper relationship between the independent and dependent variables. The degree to which the model represents the truth, that is reality, is measured in terms of **accuracy**; how closely the input variables are measured, or how fine the output measurement units are, is called **precision**.

Verification of a model is accomplished by a combination of **calibration** and **validation**. Calibration is accomplished by utilizing existing data to establish the components that will be used in the model; validation is using the model to obtain new estimations which may be compared with another set of existing data. Often, available data will be divided into two or more samples, with one part used for calibration and the other(s) used to compare with model output as part of the validation process. This may be risky in hydrologic data because the existing data may be biased and not truly representative of that part of the real world being modeled.

Models come in different forms: **physical** models, **stochastic** (or **probabilistic**) models, and **deterministic** models. The latter two types may be incorporated into computer models, also identified generally as **electronic** models. No single model is a panacea, nor is one type: all fill some role and are useful in various portions of water resources management. Models also may be static or dynamic. A **static** model represents reality at some particular point in time (Law and Kelton 1982), as does the earlier presentation of the energy budget at local solar noon; typically, deterministic models are static, while stochastic ones are usually dynamic. A **dynamic** model depicts the condition of the system over a period of time.

Physical Models

Physical models include iconic or "look-alike" models. These have been used extensively in laboratory research on the relationships between precipitation and runoff. Physical models often are used as the basis for establishing the fundamental relationships between model inputs and outputs that are incorporated into more elaborate mathematical and computer models, as was done by Parkes and O'Callaghan (1980) in estimating changes in soil water modeled by lysimeter and climatic data.

A large number of physical models of stream/watershed systems have been developed and reported in the literature. These are discussed in the section on models in Chapter 7.

Deterministic Models

The primary characteristic of a deterministic model is that it has no random component. An example is Dalton's Law. Once the vapor pressures at the evaporating surface and at some finite distance above it are known, the potential evaporation is fixed, or *determined*. There is no chance for it to be something different.

Linear regressions, for example one which provides estimates of net precipitation, throughfall, or interception loss from gross precipitation, are also deterministic models. An early treatise on the use of linear, curvilinear, and multiple regression in hydrologic forecasting established the nature of such analyses (Wilm 1950). They are widely used in hydrology because high correlation or regression coefficients indicative of nearly direct cause-and-effect relationships are common, and because output from one of the major types of hydrologic research, paired watersheds that are gaged for monitoring precipitation and runoff (where one is treated and the other is a control), may be rapidly and reliably analyzed by these methods.

The Thornthwaite Water Budget is a deterministic model as well, albeit a complex one: it actually contains several discrete models (e.g., unadjusted potential evapotranspiration from heat index, or potential evapotranspiration from UNPET and latitude). Such a model serves descriptive, analytical, and predictive purposes. It should be kept in mind, however, that deterministic models, while they appear to be absolute in that they have no error term, can mislead the user into thinking that there is no error. There is. Dingman (1984) uses the term "inhomogeneous" to describe an incomplete, or *empirical*, equation. Also, the regression coefficients in hydrologic relationships are often so high that they seduce the researcher into ignoring the error term(s). This is where many engineering solutions to hydrologic problems, such as the flood frequency curve, ultimately mislead the public, because it appears as if the description, analysis, and prediction are absolutely correct or cast in stone. Far from it.

For example, the relationship

$$RO = a + bP$$

(where RO = runoff and P = precipitation, both for a year, for instance) is a deterministic model. Once we know what P is, RO is determined. If the regression coefficient $r = 0.90$ for this relationship, we may say that $r^2 = 0.81$, or 81 percent of the variation in runoff is accounted for by observing precipitation. We may find that this could be improved by adding another variable, say soil moisture content on June 17th (S), such that, now

$$RO = a + bP + cS$$

and another 14 percent of the variation in runoff is accounted for by monitoring soil moisture content[14] The model might also be a binomial expansion, including exponential terms. It is still a deterministic model. In either case, it is assumed that there is no error, or that the error is so small that any variation is acceptable and the error may be ignored. The model may be used to describe historic relationships between the dependent and independent variable(s), and may also be used for prediction if one is aware of the fact that the actual value of the predicted runoff based on the regression may deviate from the determined value. How much it may vary may be evaluated from proper representation of the error terms of each of the independent variables and expressed by confidence intervals, and assumptions about the variation of precipitation. It is thus possible to convert a deterministic model into a stochastic one.

Other deterministic models cover virtually all components of the hydrologic cycle. Holtan (1965) presented an **infiltration** model that utilized soil moisture deficit *volume* instead of infiltration *rate*, and showed how infiltration capacities could be modeled and predicted. **Subsurface storm flow** was successfully modeled by Beven (1981) using a kinematic wave equation based on Darcy's Law, applicable where soils have high permeability and there is a steep hydraulic gradient owing to ground water mounds or topography. Eagleson (1978a) presents a model of **soil moisture movement** in an unsaturated soil, accounted for by storm infiltration, "exfiltration" (interstorm evapotranspiration processes), percolation, and capillary rise. The complex formulas are useful in water balance computations. Birtles (1978) presents a method for **stream hydrograph separation** based on transpiration, soil moisture deficit, and aquifer exchange rates, assuming rainfall rates and amounts if data are not available. The model was used to "reconstruct" actual streamflow data in England. Singh (1971) presented a method for "analyzing the streamflow variability in terms of flow duration on a regional basis within a state and on an areal basis within one region"; an intermediate step to the goal of determining where data gaps were and thus where gages were needed is the derivation of **regional flow duration curves** and their use at ungaged sites, as well as determination of hydrographic region boundaries. A small watershed **water yield** model based on four parameters derived from a total of 14 geologic, soil, and geomorphic watershed characteristics was calibrated and then successfully tested on seven other watersheds in Kentucky (Jarboe and Haan 1974). **Runoff** from mountain glaciers in Pakistan is modeled from satellite imagery by Ferguson (1985):

> A simple but flexible parameterization of orographic variation in snow accumulation and potential snowmelt enables prediction of annual meltwater

---

[14] If soil moisture content is also related to precipitation, as is most likely the case, the model may have to be made more complex to account for the nonindependence of the "independent" variables. This might be accomplished by an analysis of covariance.

runoff from the extent of winter snow cover, with bounds attached to all for variation in summer heat input to glacier ablation zones.
(The value of this type of model to those who rely on vast ungaged areas of the globe for water supplies is tremendous.) A "simple **rainfall-runoff** model"[15] is presented by Higgins (1981) for tropical areas with neither precipitation nor runoff data. In this presentation (as with Jarboe and Haan's, above), errors are acknowledged, but in records, instruments, observer, and processing, not in the fundamental relationship between precipitation and runoff, whence the source of randomness introduces probability concepts. Instrumental errors, etc., are usually of the noncompensating type and may bias the data: they are certainly *not* random. **Low flow** was estimated successfully in a humid mountainous region by Chang and Boyer (1977), and the classis development of low flow models is presented by Riggs (1968a). A snowmelt-runoff model (SRM) for microcomputers has been developed and made available by Rango and Roberts (1987). Finally, it would appear that with all the foregoing models, one would be able to build one large deterministic model that simulates the **stream/watershed system**. Knapp et al. (1975) present just such a model, but introduction of a choice of upper layer recharge models takes the model out of the deterministic category and gives it a stochastic label.

Time series analysis is a specialized example of a deterministic model that is often used in hydrologic and meteorologic research. The technique makes use of the observation that some given characteristic varies cyclically with a relatively high degree of regularity over time. Thus, the characteristic for time period t_n can best be predicted from t_{n-1} or from t_{n-x}. This procedure admits no random variation at all, relying instead upon the observation that each value depends upon the value of some preceding value(s). Double mass analysis is also time-dependent, and therefore not subject to methods of analysis that are based on a premise of random selection of data.

Stochastic Models

The essential characteristic of the stochastic model is that it contains a random component, and this enables making some probabilistic statement about any prediction from the model. The TV weather forecaster reports a prediction for tomorrow containing a probabilistic statement about the occurrence of snow showers: the prediction is based upon a model that includes consideration of the combined likelihoods of there being sufficient moisture in the air to supply precipitation, a temperature differential that includes the dew point, and the dew point being below 32°F. Worldwide climatic models are stochastic in nature, and have been used for estimation of the impacts of changes in one of the atmospheric factors, for example, carbon dioxide, particulate contents, or the impact of climatic change itself on regional hydrology (Lettenmaier and Burges 1978).

As noted in the section on deterministic models, a linear or multiple regression model that explicitly includes an error term may be stochastic, that is, the predicted value may be

---

[15] It should be noted that none of these models are really simple: all involve complex differential equations, at least.

estimated to be within a certain range of values with a certain degree of probability or simply have a probability estimate attached to it.

As with deterministic models, there is a host of stochastic models that also provide wide coverage of the hydrologic cycle. Scheidegger (1970) described "cyclic growth" and "random configuration" models used to evaluate and predict inputs such as **watershed characteristics**: mathematical modeling of drainage networks is one example. Parmele (1972) identifies errors present in the potential evapotranspiration input and are magnified in the modeling of "actual" **evapotranspiration**. On the same theme, Duckstein and Kisiel (1971) discuss efficiency of **hydrologic data networks** and how it may be improved for the purpose of accurate modeling. Chow and Kareliotis (1970) present a watershed system model that generates **synthetic streamflows** for use in water resource development analysis, as well as an overall analysis of the stochastic nature of hydrologic systems and model-building. Methods of **filling gaps** in hydrologic records are the subject of a paper by Aron and Rachford (1974), and evaluation of **stream water quality** data estimation methods is given by Krishnan et al. (1974). **Snowmelt** forecasting based on current streamflow, temperature, and precipitation observations (Tangborn 1980), **river basin hydrology** (Hill, Huber, Israelsen, and Riley 1972), and **reservoir operations** models (Datta and Houck 1984) are also available. Some of the most complex models are those that evaluate or predict the movement of contaminants through an aquifer. These models are also quite difficult to verify (Kazman 1987).

If one needs to generate synthetic streamflows, ones that will provide a better long-term set of data to use for planning purposes, a stochastic model must be developed and used. This is often a desirable planning practice because the relatively short historic record may contain serially correlated values, be for too short a period, or be for a period which includes abnormal values (a portion of a cycle, for example) thus biasing the model's output. Of course, deterministic and stochastic models may be combined: for example, the precipitation values for the linear regression model above might be derived from a stochastic model that predicts probable precipitation based upon a host of variables (1) that are allowed to vary at random, and (2) whose relationships to runoff and/or to each other are not wholly known.

In contrast to time series analysis, **spectral analysis**, wherein the cyclical nature of independent variables is not known and are assumed to vary about some mean, may attach an error term (e.g., "$\pm e$") to the predicted characteristic, or provide a statement of probability along with its value.

Electronic Models

As soon as computers became generally available, attempts were made to model stream/watershed systems electronically. Both analog and digital models have been created. The former operate as a black box, that is, one applies an electrical signal (current) and, by manipulating dials, adjusts the output signal to some desired characteristic. The input is a "storm"; the output is the hydrograph. The black box is the watershed in all its mystery: we don't really know how all the temporal and spatial factors vary, nor how their interactions affect runoff. Yet it is possible, for example, in the Stanford Watershed Model (Crawford and Linsley 1966) to apply a storm input and to adjust the knobs (variable resistances) of the model so that the output looks like the real hydrograph of some known stream. A thorough presentation, analysis, and practical presentation of a long list of models are given by Viessman et al. (1977). The basic difficulty with the analog computer process is that if we really wish to know what happens to a particular part of the precipitation-runoff process as a result of some activity of civilization, we must know how each relevant factor affects the process: if that information is known, the model is not really necessary.

Digital computer models are especially useful for descriptive, analytical, data conversion, and simulation purposes. The digital computers can readily handle the large masses of data required for hydrologic simulation over regions (Hill, Huber, Israelsen, and Riley 1972). A large amount of computer software for use on personal or desktop computers is now available to the hydrologist. Probably the greatest demand on computer modelling at present (and, no doubt, for a long time to come) is ground water quality models. The natural quality characteristics of water must be understood before man's perturbations can be detected, evaluated, and predicted. In conclusion of a treatise on stream dynamics, Heede (1980) stated that "the science of fluvial hydraulics has not been developed to a level that permits accurate prediction of stream behavior."

It may be that we will *never* be able to completely model the hydrologic cycle, water budget, storm flow for even a small watershed, or a drop of water:

> I can foretell the way of celestial bodies, but can say nothing of the movement of a small drop of water.
> — Galileo Galilei

That profound observation may, indeed, be extended to the forecasting of weather, which is found to be *chaotic*, that is, "order without periodicity" (Gleick 1987). If such be the case, and more and more examples, particularly in the hydrologic field (e.g., Rodriguez-Iturbe et al. 1989; Tyler and Wheatcraft 1990), are pointing in that direction, then all we can do is approximate and clear the court calendars for the litigation that is now as sure as death and taxes.

WATER QUALITY

The sciences of hydrology and limnology matured rather separately at mid-century and, as a consequence, the topic of water quality was introduced into each with different perspectives. In hydrology, the topic was in the "must do" category, with demands on professional engineers to find practical solutions to pollution problems and to design water and waste water treatment facilities. In limnology, the approach was more from the standpoint of pure science, with extensive and more laid back research on the quality aspects of water bodies, although there were occasional urgent demands for information that would resolve some specific pollution problem. But there also was a growth of interest and expertise in between the applied civil engineering approach and the pure science of water body research: nutrient balances, large-scale chemical problems such as the fate(s) of insecticides and herbicides, and early interest in the impacts of acid deposition created a demand for a broader, integrative approaches to environmental chemistry. Essential to that approach is a solid foundation in the chemistry of water in the natural environment. A classic reference is the work by Hem (1970), upon which much of the following is based, with additional references cited, as appropriate.

PROPERTIES OF WATER

Water is a unique substance. It is the only substance that exists naturally in all three states, solid, liquid, and gas, on the Earth. It also has some unique properties that stem from its molecular constitution.

Molecular Structure

The water molecule consists of two hydrogen atoms and one oxygen atom, with the former two arranged on one side of the latter, giving the molecule the appearance of Mickey Mouse (Fig. 6-27). This arrangement means that the two types of atoms are sharing electrons in one region: thus, the molecule is unbalanced, and the lattice that the molecules form is quite stable (Dingman 1984; van Hylckama 1979). The result is that the molecule has a net positive charge on the hydrogen-dominated side and a net negative charge on the opposite oxygen-dominated side. The molecule is therefore *dipolar*, a characteristic that explains some of water's related peculiarities. For example:

Figure 6-27 The water molecule

- The molecule dissociates into OH⁻ and H⁺ (in the liquid state) providing for a high degree of ion exchange, making water an excellent solvent. It is thus useful in domestic, commercial, and industrial processes, as well as being a top notch weathering agent.
- Its consequent high electrical conductivity when it contains dissolved materials is important as a means by which to measure total dissolved solids.
- Strong hydrogen bonds are formed, ramified in medium viscosity and high surface tension, and in its consequent balance of adhesive and cohesive forces that together create its capillarity, so important in soil storage and plant physiology.

The molecular structure of water further profoundly affects water's density, specific heat, viscosity, and related characteristics.

Density

Perhaps water's most unique and important characteristic is that its density does not continuously increase as it solidifies at 0°C. Water is one of four similar compounds, but does not fall into the expected pattern of freezing and boiling points based upon molecular weight (van Hylckama 1979). Water is most dense at 4°C, thus *increasing* in volume while it is still a liquid from that temperature, and until it reaches the freezing point. The fact that water is less dense at the freezing point means that:

- the decrease in temperature that causes surficial soil water to expand upon freezing causes loosening of the soil surface, thus maintaining or increasing surface porosity and, of course, infiltration; and
- ice floats, a phenomenon of paramount importance to the life cycles of aquatic ecosystem fauna that cannot survive being frozen and of importance, no doubt, to the Earth's evolutionary processes.

In its crystalline form, water is also unusual in that the saturation vapor pressure over *ice* is less than it is over *water* (see Fig. 3-6). This feature means that water vapor will move toward ice crystals or bodies of ice crystals such as snowflakes in the atmosphere owing to a vapor pressure gradient. This fundamental precipitation-formation process has been exploited and artificially emulated in the cloud seeding process. The same observation may be made in soils in which frost has formed, or on the surface of snowpacks, where condensation occurs as water vapor is attracted to the snow surface. Upon condensation, it melts and re-freezes as a crust.

Viscosity

Viscosity is the "property of a fluid which resists relative motion and deformation in the fluid and causes internal shear" (Chow 1964). It is temperature-dependent but, within the normal range of temperatures of natural streams, it does not change greatly. Viscosity is of importance only to a fluid in motion. It is particularly important to water quality considerations owing to its effect on the type of flow, *laminar* or *turbulent*. As a consequence, the amount of bed-scouring that occurs and the degree of mixing that takes place is affected by viscosity. Under low velocity, flow in a uniform channel may remain laminar, therefore reducing the mixing of dissolved and suspended solids and preventing levels to build up and/or be distributed widely throughout the stream. However, in most streams, the roughness of the channel and the velocity combine to produce turbulent flow.[16]

Specific Heat

Water's high specific heat is ramified in its high latent heat of vaporization and condensation. Specifically, while it takes 1 calorie to raise the temperature of 1 cu cm of water 1 degree Centigrade at standard temperature and pressure (the definition of a calorie), and approximately 1 calorie per degree of temperature change to warm it to 100°C, it takes *540* calories just to change that cu cm to steam without changing its temperature any further. Similarly, it takes *80* calories to change 1 cu cm of water from the solid to the liquid state. Water vapor and liquid water must *give up* these amounts of energy as the temperature of water decreases through the boiling and freezing points, respectively.

The release of tremendous quantities of energy in the condensation process gives rise to the hot, dry Chinook and Santa Ana winds, for example, as noted in Chapter 3. And, of course, it gives rise to the observation that "evaporation is a cooling process": it takes energy to

---

[16] See Chow (1964) for a particularly lucid presentation of the relationship between the three forces acting on water, gravity, surface tension, and viscosity, as well as hydraulic similitude.

evaporate the liquid, energy that comes from the surface from which the evaporation takes place, thus cooling it.

The combined facts that water can thus store (and yield) large quantities of energy, covers the planet in one or more of its natural forms, and can be readily moved, makes water important in the energy balance of the Earth, as noted in Chapter 2. Water's high specific heat also makes it useful as a climate modifier, as a cooling agent, and as a ready means to transport heat energy in homes and industrial processes.

WATER QUALITY CHARACTERISTICS

Generally, the characteristics of water can be categorized into physical, chemical, and biological categories, but there is some overlap.

Physical Characteristics

Water is wet. It also has temperature, color, taste, and suspended and dissolved solids and gases. Some of the most important physical characteristics such as molecular structure, density, viscosity, and specific heat, are discussed above.

In the natural environment, water **temperature** varies from 0°C to 100°C Ground water is pretty consistently at 12.7°C (55°F), while snowmelt runoff will be considerably colder. Temperatures of surface runoff in summer may reach 21°C and more.

Color of water, which in its pure form is colorless, is usually caused by dissolved solids, salts or phenols such as tannin, frequently found in the natural environment, for example, from hemlock stands. With no dissolved (or suspended) materials, water is also tasteless. Water that has a "good" taste is said to be **potable**. Most municipal and domestic supplies have some distinctive taste (in addition to the often objectionable taste resulting from chlorination): water without some dissolved solids would not be potable, and distilled water could not be used for (regular) drinking because of the extreme volume of water that would enter the bloodstream to equalize salt levels by osmosis (diffusion of fluid through a porous material, driven by differential concentrations of dissolved solids).

Suspended solids are usually referred to as **turbidity** and more often than not are primarily sediments either washed down from watershed slopes or picked up from banks or stream beds. The natural erosion processes of streams ensure that virtually all streams will contain sediment at some time or other during the year or during the stream's life history. Both the diameter of particle that can be carried and the total volume of sediment, or sediment load, are complex functions of stream velocity (Chow 1964) and other factors (Stein 1965). Settling rate varies for particles of different diameters and is dependent upon turbulent flow in order to provide the upward motion that ensures continued suspension. The relationship between velocity and particle size is shown in Table 6-7.

This information is of use in designing riprap outfalls from conduits that resist erosion: since the volume and velocity of water coming from a pipe of known slope, size, and roughness (see Manning's equation) is fixed, the rock used to line the outfall basin need be "one size larger" than that which can be carried by the stream.

While the hydrograph of a stream during a storm period is as shown in Fig. 6-13, the sediment *concentration* remains neither constant nor proportional during the period of time that the flow is changing: the varying velocity and Q mean that it changes, too. The plot of concentration of sediment concentration over time is referred to as a **pollutograph**: a typical sediment pollutograph is shown in Fig. 4-6. Pollutographs differ for the different substances in suspension or dissolved in water. For example, Kennedy (1971) reports an inverse relationship between silica concentration and specific conductance during storm runoff periods in California.

The concentration of **dissolved gases** in water is inversely related to temperature, so that as water increases in temperature, gases are driven off. Usually, increases in water temperature cause a decrease in dissolved oxygen content while simultaneously increasing fish demands for oxygen by increasing metabolic rates, thus putting severe strain on the fish. Occasionally, however, water may be heated without driving off dissolved oxygen, for example, and the water may be supersaturated.

Table 6-7 Relationship Between Particle Diameter and Stream Velocity

Speed of Current (cm/sec)	Diameter of Objects Moved (mm)	Class of Object
10	0.2	Silt
25	1.3	Sand
50	5	Gravel
75	11	Coarse Gravel
100	20	Pebbles
150	45	Small Stones
200	80	Fist-size Stones
300	180	Small Boulders

Source: Krumholz and Neff (1970).

Chemical Characteristics

The dividing line between physical and chemical characteristics of water is not always clear: there are too many interactions. However, dissolved gases typically remain in their stable, molecular form; dissolved solids, however, owing to the physical characteristics discussed above, dissociate in water, thereby altering the chemistry of the water body.

Dissolved solids include the many elements and compounds that are soluble in water. Sources of ions in water include the atmosphere, the terrasphere, and the hydrosphere itself: Gibbs (1970) observed that "the three major mechanisms controlling world surface water chemistry can be defined as atmospheric precipitation, rock dominance, and the evaporation-crystallization process."

Total dissolved solids (TDS) may be monitored by **specific conductance**, measured in micromhos[17] or, more recently, millisiemens per m (mS/m). It is considered to be "useful measure of TDS for routine water quality analysis" (Foster, Grieve, and Christmas 1981). Field monitoring by a relatively inexpensive (from $30 to $400) *conductivity meter* can rapidly and, if desired, continuously, provide useful information on sources and distribution of total dissolved solids. However, Young (1973) notes that "conductivity should be used with great caution since it is an inferential measurement which, by itself, gives no information about the *nature* of the ionic carriers of current" [emphasis added].

In the atmosphere, both gases and particulates are present. Carbon dioxide, ammonia and nitrogen oxides, and oxygen dissolve readily in suspended water droplets (clouds) and in precipitation. The dissolution of CO_2 in atmospheric water produces a slightly acidic pH[18] value of about 5.7 for rainfall. Nitrate concentration in rainfall is typically higher in the vicinity of thunderstorms, it apparently being "fixed" from free nitrogen into nitrates and nitrogen oxides by lightning, although this accounts for only about 10 percent to 20 percent of the nitrate concentration in rainfall (Junge 1958). Oxygen, of course, occurs in water both from dissolution and dissociation, but the latter produces an ionized form.

Particulates are lofted into the atmosphere by anthropomorphic activity, wind, and vulcanism, and also enter the atmosphere from space. They may be washed from the atmosphere by rain and snowfall ("wetfall" or "washout"), the precipitation particles often serving as condensation nuclei for precipitation. The amount of particulate matter in direct (surface) rainfall runoff is a function of the length of time between storms, wind movement, and a large number of related factors (Black 1983). "[D]ryfall can be a major source of chemical loading to forest ecosystems for some ions ... and [a]nnual contributions of dryfall total inputs exhibit substantial year-to-year variability" (Swank and Waide 1988).

The dissolution of elements and compounds that occur naturally in the terraspere is enhanced by the slightly acid precipitation. The gross distribution of several of the more important elements is shown in Table 6-8.

Igneous rocks are poor aquifers, but are often exposed at headwaters of streams. They are thus often steep and, if old, as is the case in the Adirondack mountains of New York State, no longer contain much in the way of soluble materials near or at the surface. Thus, water that comes in contact with soils derived from these rock runs off without many dissolved solids. Such runoff in the northeast has the dual stigma of low pH and low buffering capacity (indicated by alkalinity or carbonate content) and is, therefore, the source of much concern over the acid rain "problem."

Sedimentary rocks are more prevalent near the surface, and often are good water-bearing strata. They consequently provide a rich supply of silica (which is highly soluble in water) from sandstone or calcium or magnesium from limestone. Tilted sandstone beds in the foothills of the Colorado Rockies provide extensive recharge to important Great Plains aquifers, which often have the water's origin "fingerprinted" with typical water quality characteristics.

---

[17] The unit "mho" is the literal reciprocal of "ohm," the unit of resistivity, the reciprocal of conductivity.

[18] The log of the hydrogen ion concentration in moles/liter, ranging from 0 to 14, with 7 being neutral.

The elevation of watersheds affects concentrations of SO_4, showing the interaction of that gas's vertical distribution in the atmosphere and the watershed characteristic. In addition, acidic coniferous vegetation is more typical (than deciduous vegetation, which is more likely alkaline) at higher elevations and, with shallower soils to buffer low pH leachate or percolating waters, the pH of runoff water tends to be lower at higher elevations.

Table 6-8 Summary of Rock Composition

Elements	Igneous (%)	Sedimentary Sandstone (%)	Carbonate (%)
Silica	28.5	35.9	0.0
Aluminum, Iron, Calcium, Potassium, Manganese, Titanium, Magnesium, and Phosphorous	20.8	9.7	9.4
Carbon	0.0	1.4	11.3
all other elements	50.7	53.0	79.3
Totals	100.0	100.0	100.0

Derived from parts per million data in Hem (1970)

The "evaporation-crystalline process" is driven by both simple evaporation (which leaves behind increasing concentrations and, thereby, precipitation of salts) and various chemical reactions which may regulate crystallization and precipitates. There are innumerable reactions among various species in the water. Virtually all types of possible reactions are present at one time or another (Hem 1970). Principal types of reactions are as follows:

Reversible reactions that occur without any chemical change in the water itself include simple ones such as the solution-deposition change involving common salt:

$$NaCl \leftrightarrow Na^+ + Cl^-$$

Reversible reactions can also occur with the water breaking down chemically, as is the case with calcium carbonate, yielding precipitatible calcium, bicarbonate, carbonic acid, and the OH^- radical:

$$CaCO_3 + H_2O \leftrightarrow Ca^{++} + HCO_3^- + OH^-$$

or just carbon dioxide and water:

$$CO_2 + H_2O \leftrightarrow H^+ + HCO_3^-$$

And, finally, reversible reactions in which there is a change of the oxidation state, for example, with iron:

$$Fe(OH)_3 + e^- + 3H^+ = Fe^{++} + 3H_2O$$

Reactions also can be ones in which equilibrium is not readily achieved due to slow rates, one-way supply of solute, energy availability, or decomposition. An example is:

$$C_6H_{10}O_5 + 6O_2 = 6CO_2 + 5H_2O$$

One of the most complex and important reactions is that which reflects the basic inputs of water, carbon dioxide, and calcium, and illustrates the wide range of influence in and around the aquatic environment.

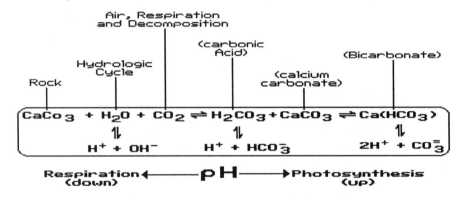

Here the balance continually shifts owing to flux of inputs of nutrients and energy, and to natural diurnal, monthly, and seasonal variations in the influencing factors. This reaction also clearly shows the complexity of the linkage between the physical, chemical, and biological characteristics of water. For example, if the day is cloudy with rain, there will be more input on the left side of the equilibrium, and photosynthesis will be decreased in the absence of solar energy; as a consequence, the aquatic environment will show a decrease in pH.

Biological Characteristics

The general biological components are, of course, the flora and fauna of the aquatic ecosystem. They are dependent upon and reflect the nature of the physical and chemical environment. For example, rapidly-moving water tends to be cooler and, as a consequence, is favorable to a different range of species of fish than are the relative still waters of a lake, pond, or wetland. Similarly, plants thrive in different aquatic environments influenced by water depth and velocity (Patrick 1970).

Forest communities play a major role in water quality in part because they grow where there is sufficient moisture to sustain forest vegetation and because they are a reservoir of nutrients that undergoes predictable, natural change during natural succession, and which react to forest cutting and reforestation, sometimes also in predictable fashion. "Forests have a large capacity to store nutrients because of the large dry weight of the standing crop. Leaves, twigs, and stems become part of the 'free nutrient pool' when they fall, but much of this free pool is absorbed again by the roots of trees and by faunal populations" (Hewlett and Nutter 1969). Thus, if the natural processes are interrupted by harvesting, thinning, fire, clear-cutting and reforestation, or land use conversion, drastic changes in on-site nutrients and stream water quality will occur. These changes are not always predictable owing to the large number of influencing factors, not the least of which may be the *sequence* of factor occurrence. These influences are the subject of much research at present; they tend to be greatest and most dramatic as a result of land use changes at the rural-urban interface, and thus are the focus of many mitigative and preventative best management practices (BMPs), which are discussed in Chapter 7.

Hydrologic conditions such as floods or low flows create habitats that may represent limiting conditions for flora and fauna, both vertebrates and invertebrates. As a consequence, indicator species have been identified and are the subject of much of the limnological literature (cf., Needham and Needham 1962; Macan 1974; and Petts 1984). Management of fish habitat is targeted on the floral composition of the stream (Stalnaker 1979). Reservoir releases may be made from either the surface or the bottom of the impoundment, with resultant opportunity to control downstream water temperatures (Stroud and Martin 1973).

In wetlands, hydroperiod is often a factor controlling the vegetation and, as a consequence, fauna (Mitsch and Gosselink 1986). Water that flows through wetlands will exhibit characteristics of the wetland. For example, downstream from a wetland dissolved carbon dioxide concentration will be higher and dissolved oxygen will be lower than upstream. The result is that the pH is reduced downstream of a wetland. It is also likely that the temperature will be cooler, as the water stands in the shade of the wetland vegetation. Since plants take silica out of solution to build plant walls, that chemical will often be lower downstream from a wetland, too. Often, when conducting a stream water quality survey, the presence of a wetland (or other ecological niche) may be accurately inferred from one or more water quality characteristics.

Some of the basic interactions between physical and biological characteristics include the observation that metals accumulated by vegetation will be present in runoff following plant decomposition; bicarbonate tends to be high in runoff from wetlands; and total precipitation and runoff are inversely related to concentrations. Seasonal and diurnal fluctuations in temperature affect water quality: higher temperatures generally increase solubility of minerals, but decrease solubility of gases. In the case of dissolved oxygen, an increase in temperature severely stresses fish because while the concentration of DO is decreased, the metabolic rate of the fish and their demand for oxygen is increased. Further interactions between land use and water quality characteristics are discussed in Chapter 7.

Monitoring Water Quality

Establishing and maintaining a program of water quality monitoring is a huge job. It entails collecting, processing, analyzing, and storing for retrieval tremendous masses of data that are collected across a wide range of time and space and may involve up to one hundred fifty substances that are measured at each sample point and time. Remote sensing, models, and computers are vital components of the task, along with field work. It is important to commence with the latter.

Analysis of field water quality characteristics has traditionally been accomplished by removing a sample of the water body being tested and executing the necessary tests in a laboratory to which the water sample is transported. However, certain characteristics can be monitored directly in the field with a choice of test equipment: water temperature, Secchi disk transparency, electrical conductivity, pH, and dissolved oxygen are among the water quality characteristics for which there are relatively inexpensive field instruments or meters.

Some of these characteristics *must* be measured in the field because they change during transportation to the laboratory and, in fact, may even be influenced by the act of withdrawing or measuring the sample. Thus, temperature and pH should be measured *in situ*. Samples of water for evaluation of dissolved oxygen (and other gases') content may be withdrawn and fixed prior to transportation to the laboratory. The procedures to be followed by field and laboratory analysts have been prescribed by an inter-agency committee of federal agencies (Geological Survey 1977 ff), based on long-documented procedures by the American Society of Testing Materials' *Standard Methods*. Included are minimum testing and sample control requirements in order to meet the Environmental Protection Agency's effluent standards or the various states' receiving water quality classifications (as mandated by the 1972 Water Pollution Control Amendments). Even the best of intentions have problems, however: a recent study designed to test the reliability of laboratory analysis procedures has shown that identical samples of the same water sent to different laboratories can produce different reports on water quality (Litten 1990).

What is *not* well documented throughout the procedural documents is the *representativeness* of the sample, the issues of natural variation in time and space, relationship to hydrologic characteristics of the water body being sampled, what is in the sample bottle, and the ubiquitous issues of accuracy and precision. Greeson (1978) observed that "the weakest link in the chain of events leading to production of reliable microbial-monitoring data is a poor or inadequate sample." He notes that:

> There is not now, nor is there ever likely to be, a single method of sampling which can be used to describe all microbial aspects of the hydrological environment [the word "microbial" could justifiably be changed to "water quality"].

Greeson suggests a few simple guidelines:

1. define the intended use of the data to be obtained;

2. evaluate the hydrologic conditions under which the sample is to be taken;

3. give consideration to the measurement of significant properties, and

4. use common sense.

The use of common sense includes response to the first guideline which should be answered in relation to the need for concentration means, maxima, minima, or to total loading, and the timing of occurrence of the condition. The second guideline is particularly important in light of concern over the representativeness of the sample to be taken. Considering all the influencing variables and the potential for taking incredibly large numbers of samples, Edgington and Rolfe (1974) caution that "the capacity of the analytical laboratory will be an important factor in determining the number of samples to be taken for analysis."

One of the first considerations in water quality sampling is whether the investigation is undertaken for the one of the principal model construction purposes (describing, analyzing, designing, or predicting) or for detecting violations (Sanders and Adrian 1978). For the former purpose, a National Water Data Network (Langford and Kapinos 1979) was established to maintain a catalogue of information on water data, review water data requirements, plan for efficient utilization of data, and design the data network itself. Three levels were defined for national and regional needs, subregional needs, and local level operations and management needs. For the first level, resource accounting, water use accounting, flood surveillance, and water quality surveillance elements are all in place. Yet, Hill (1986) points out that even with six *years* of nitrate data, records from two rivers are required "for both rivers to ensure that an error of $> \pm 20$ percent would occur in only 5 percent of these observation periods." For the spring runoff season, the number of years is between 6 and 7. One must ask whether that level of error is acceptable, or even of value in comparing nitrate concentrations in the first place.

The next issue is "what is in the sample bottle?" Once in the laboratory, the analyst can readily ascertain exactly what is in the sample bottle, although only a few studies have been accomplished on the variance of the testing process or the equipment (cf, Pitlick 1988). The basic concern, however, is whether the sample is really representative of or integrates all of the conditions in the water body being investigated. The Geological Survey (1977 ff) recommends:

> Sampling is the process of collecting a representative portion of the environment to learn about the whole environment. A representative sample is one that typifies the rest of the environment. To collect representative samples, one must standardize sampling bias related to site selection; sampling frequency; sample collection; sampling devices; and sample handling, preservation, and identification.
>
> Collecting a representative sample and maintaining its integrity until it is analyzed is important because the validity of each measurement begins with the sample. Regardless of scrutiny and quality control applied during laboratory analyses, reported data are no better than the confidence that can be placed in the representativeness of the samples....

An equally important issue concerns inclusion of the hydrologic conditions (rising or falling storm hydrograph, type of runoff event, hydrologic season, etc.). While recording the gage height is suggested for surface chemistry samples (Geological Survey 1977 ff), there is no such suggestion under the guidelines for biological water quality samples. The recommendations for surface waters suggest including "any other observations that may assist in interpreting water-quality data." Frequency of sample, progressive *upstream* sampling (to avoid re-sampling the same water if floating downstream in a boat), and annotation of time since last rain ought to be included.

A few papers have been published reporting on sampling (Edgington and Rolfe 1974), natural variation (Kennedy 1971), and instrument errors (Boyd 1977), but most of these are for certain locations or specific water quality characteristics. Sampling frequency is "one of the most important tasks in the design of a regulatory water quality monitoring program (Casey, Nemetz, and Uyeno 1983). That is true to environmental quality monitoring, too. Kennedy (1971) recommends greater frequency of sampling during the rising limb of the hydrograph, noting that variability is higher at this time as well. There have been some studies specifically on the variability of the *method* of water quality parameter evaluation, on measurements taken from the same sample (aliquots), or the standards (these are documented in the Geological Survey handbook list of references). No studies have been reported, however, on the variation of 25 nitrate samples in the upper foot of a stream that is 3 ft in depth, or what the *actual* value of that water quality parameter really is. And that is only *one* water quality characteristic, and *one* stream size, at *one* point in time, under *one* of a wide variety of hydrologic conditions . . . .

For the most part, natural resources management disciplines have matured coincidentally with the development of a system of monitoring and archiving for later retrieval meaningful field data upon which sound decisions may be based. Data on land surveys, tree heights, diameters, and numbers of individuals or of species in wildlife populations, and so on, were collected early in the history of wildland management activities in the United States, and have been effectively used for educational, scientific, and management goals. Simultaneously, the discipline of statistics matured, and provided researchers with the tools to evaluate field situations and models.

Such has *not* been the case with water. Investigators of water body characteristics have typically assumed that one sample of the surface water in a lake represents the water over the entire surface of the lake, or that a sample in the hypolimnion accurately represents that layer.[19] *We cannot blindly make that assumption.* Even in a (relatively) still lake or pond, radiation and photosynthesis affect pH, dissolved oxygen, and carbon dioxide balances; ions available for hydrogen bonding; interact with the carbonate-bicarbonate balance, and affect buffering capacity. These change from hour to hour, day to day, month to month, and season to season. To further confuse the matter, the water may be — and usually is — in motion, flowing through the lake/watershed system or river section.

Methods of reporting the data are also important. Three-dimensional plots that highlight temporal and spatial trends are useful for proper interpretation. Kennedy (1971) suggests that reporting temporal variation in terms of a ratio of the dissolved constituent to the specific conductance "can be especially helpful because dilution is eliminated as a factor." The concentration of a pollutant in a stream undergoing some temporal change in discharge such as storm runoff is shown in Fig. 4-6. But, as noted also, *pollutographs differ for the different substances in suspension or dissolved in the water.* Thus, one really cannot assume *anything* about the water being sampled on the basis of a water quality measurement made at any arbitrary time during a runoff event. Further, the **load**, or total amount of the pollutant, may be of greater importance than its concentration. It is thus imperative that hydrologic conditions at the time of sampling are noted, when water quality samples are withdrawn.

The result of this inattention to basics is that (1) vital water management decisions are being made daily based upon inadequate data, causing expensive restorative measures, and (2) drastic changes in water policy are being made based on some erroneously monitored parameter of water quality. Liebetrau (1979) points out that a water quality control agency has many objectives in water quality monitoring, and that a "good sampling design is important for achieving the first two objectives:" (1) provide broad ecological information, (2) evaluate water quality changes over time, (3) detect water quality problems, determine specific causes, and evaluate corrective actions, and (4) law enforcement. The magnitude of achieving the ecological and temporal objectives is astronomical considering (1) the many water quality

---

[19] In fact, the samples at lower depths are more likely to be representative of the water at that layer because intuitively it would seem that the water is more uniform and less disturbed by wind, radiation, and diurnal oxygen fluctuations than water near the surface. But note that the word "intuitively" is not scientifically based!

characteristics about which information is desired, (2) the tremendous number of influencing variables, and (3) the wide variety of hydrologic conditions under which the sampling must take place. Researchers have reported on variability of several water quality characteristics: "pollutants contained in precipitation accumulate in the snowpack to be released during a short period in the Spring," often with highly variable percentages of the annual loading being released in a few days (Johannessen and Henriksen 1978); dissolved ocean salts vary inversely with rainfall intensity and may be carried inland up to 10 km (Kennedy, Zellweger, and Avanzino 1979), who also suggest that "very large amounts of sea salts may be transferred from ocean to land during high-wind storms such as hurricanes"; and vertical distribution of several pollutant/nutrients is influenced by thunderstorm activity (Dickerson et al. 1987). An intensive study of water quality sample size determination is presented by Casey, Nemetz, and Uyeno (1983), but it is for the purpose of specifying sample size to evaluate violations.

To further confound the issue, the equipment that is available often is of a different order of precision from that of the standards set by the government. In some cases, we are measuring water quality characteristics crudely in comparison to the requirements of the law; in others, excessive dollars are being spent to gather information that is far more precise than is necessary; in still others, standards have been set that are lower than state-of-the-art equipment's capability to measure.

Finally, water quality data from EPA-approved laboratories often take several weeks before becoming available and, in light of health concerns, that length of time may not be acceptable. Some on-the-spot field kits for determination of some water quality characteristics are available and have been approved by the EPA. One was noted to compromise rapidity and convenience with some small degree of precision, although its results were very closely correlated with those of Standard Methods (Boyd 1977). Field kits generally offer somewhat less precision than laboratory methods, but provide the opportunity to verify anomalous measurements in the field and to replicate samples where necessary. An advantage of the laboratory, in addition to the greater precision, is the objectivity of the analyst. Nevertheless, a competent analyst *can* be objective and use of an unbiased approach to field sampling can be effectively applied with much of the field equipment available at the present time.

Some examples of recent research include the observation by Lettenmaier (1979) that "almost all of the existing water quality networks have been designed on an *ad hoc* basis"; in the absence of expensive continuous recording turbidity meters, natural variation of sediment loads could introduce errors of up to +280 percent in the sediment rating curve (Walling 1977); and sediment samples stratified by hydrologic condition (e.g., greater sampling intensity during periods of peak loads) still underestimate sediment loads by as much 51 percent (Thomas 1985). Lynch, Hanna, and Corbett (1986) report that, as noted in the comments above about pollutographs, "stream pH and alkalinity levels were found to react inversely to stream discharge during storm flow periods, with the lowest levels occurring almost simultaneously with peak flow. In comparison, storm flow acidity was directly related to discharge rate, with the peaks nearly coinciding." And Hirsch, Slack, and Smith (1982) report that time series analyses of water quality data may be complicated even further by non-normal distributions, seasonality, relationship to discharge, missing data, values below detection limits, and serial correlation.

It is important to question water quality data: it may be insufficiently documented with regard to hydrologic conditions at the time of sampling, inadequately replicated, not a representative sample, or misleadingly sampled or analyzed in field or laboratory.

USING WATER QUALITY TO HELP UNDERSTAND HYDROLOGY

Clearly, it is appropriate to consider the three classic parameters of the water resource — quantity, quality, and regimen — concurrently. It is practically impossible to manage one without affecting one or the other to some degree. Beyond that, it is worth noting that water quality may be useful in studying hydrology-related long-term trends in meteorology (Harris, Cartwright, and Torii 1979), the natural and man-accelerated eutrophication process (Hasler and Ingersoll 1968), and the continuing acid deposition problem (Robertson, Dolzine, and Graham 1979). Brown (1986) reports that measured ion concentrations can be used "to reduce the uncertainty in hydrologic model parameters"; McGuiness et al. (1971) describe a relationship between rainfall energy and sediment and how that relationship varies between large and small watersheds, and Pilgrim et al. (1979) explain how storm runoff may be separated into its flow components by use of specific conductance observations.

By the same token, it is clear that certain water quality characteristics will aid in the understanding of the current hydrology of a specific water body being investigated. Dissolved oxygen and carbon dioxide, silica, pH, temperature, bicarbonate, and several other chemical species provide "fingerprinting" of water that has passed through or been in contact with an ecological niche. Simultaneously, it is apparent that manipulation of the water body will affect the water quality, too.

This interaction provides the focus of environmental impact analysis, for no matter where the disturbance to the environment, the effect will, sooner or later, turn up in the water. This is especially true where land use change is taking place, for the ecological niches will be undergoing drastic alteration as lands pass from rural to suburban or urban lands, for example. Understanding the nature of the interaction, then, between water quality, quantity, and regimen provides the foundation of comprehensive environmental impact in the hydrosphere. The concept underlies the material in Chapter 7.

SUMMARY

The hydrosphere *is* the realm of water. It extends its tentacles into the other spheres, both providing them with important characteristics that depend upon the behavior of water, and being influenced by them so that the water that emanates from the watershed is characteristic of the environment with which it has been in contact: its quantity, its quality, and its regimen.

Water has a myriad of characteristics, some of them inherent, owing to the nature of the water molecule, and some of them imparted to the water body by the substances with which the water comes in contact. Physical, chemical, and biological characteristics often interact to create unique water quality conditions. Water quality monitoring is an evolving science and art, as is the basis for water quality research, standard setting, and enforcement.

PROBLEMS

1. Discuss possible reasons for the 186-sq-mi watershed curve to plot erratically (out of sequence with regard to the other watersheds) in Fig. 6-25.

2. Plot a water years' worth of daily mean temperature, precipitation, and runoff data on the same graph page. Describe the relationships between the three variables with particular attention to explanation of the runoff curve.

3. How many inches of runoff would there be in one 31-day month from a 2232-sq mi watershed if the mean daily flow were (a constant) 440 cfs? (You may verify this problem in Fig. 6-3, for the month of July.)

5. Is the pH of the surface water of a wetland (not a bog) likely to be higher in late afternoon or in the morning? Why?

6. Identify three nearby waterbodies and sample (or estimate) several water quality characteristics, including pH, temperature, pH, and dissolved oxygen. How would they vary during a day? the year?

7 Water on the Watershed

*Watershed management is the planned manipulation
of one or more of the factors of a natural drainage
so as to effect a desired change in or
maintain a desired condition of
the water resource*

This chapter introduces the watershed into the hydrologic cycle, and provides an introduction to the principles and practice of watershed management. In contrast to the detailed, close-up presentation of Chapters 1–6, this chapter "backs off." It presents a more distant view and perspective of water on the watershed: what role watershed characteristics play in both the movement and storage, and how their consideration affects the hydrograph.

The chapter is divided into two primary sections: the watershed, and watershed management. The chapter concludes with a brief discussion of nonpoint sources of pollution and best management practices.

The section on the watershed considers basic concepts, including (1) a definition of "watershed," (2) organization of the storage components on the watershed, (3) definition of the basic watershed management unit, the "small" watershed, and characterizing the watershed, the detail of which illuminates the bases for the remainder of the material in the section.

The section on watershed management considers basic principles, perspectives, and practical concepts that may be applied to achieve some specified goal of wildland management, with special attention to vegetative cutting and conversions, snowpack management, municipal watersheds, and wildland water quality control.

The brief summary discussion of nonpoint sources and best management practices with which nonpoint sources are controlled does not substitute for competent management courses in forestry, range management, agriculture, or urban hydrology.

THE WATERSHED

The **watershed** is defined as a unit of land on which all the water that falls (or emanates from springs) collects by gravity and fails to evaporate and runs off via a common outlet (Fig. 7-1).

Chapter 7 Water on the Watershed

The watershed is the basic unit of water supply. Often, the definition is extended to include man-made watersheds such as parking lots, roof tops, and artificial drainageways created for any of a variety of purposes. Size is not a factor in the definition, and watersheds vary from fractions of acres to thousands of square miles. Unless a watershed discharges directly into the ocean, it is a part of one that does, and may be referred to as a subwatershed.

In the foreign literature, the watershed may be referred to as a "drainage basin" or a "catchment." The geologist refers to the outlet of the watershed as the **base level**, defined more specifically as that elevation to which higher topography will be reduced by erosive forces (land-surface-degrading processes) of gravity, ice, water, and wind. Once the base level is identified, all the other physical characteristics of the watershed are determined, and are discussed in subsequent sections.

WATERSHED MORPHOLOGY

The watershed, of course, evolves over time. One of the sciences that aids in understanding this phenomenon and most fundamental to watershed hydrology is **geomorphology**, the study of land forms. This (usually second) course in geology includes the theories, classification, and details of the several aggradation and erosion processes that sculpt our landscape. Major, broad-scale geologic processes are those by which the land surface is lifted and prepared for the processes that wear it down. Locally, aggradation occurs when the stream velocity is diminished such that the water can no longer carry as much or as large-sized particles, a process called **sedimentation**.

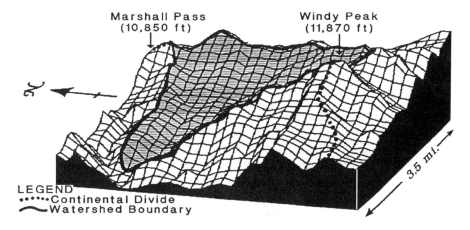

Figure 7-1 Millswitch Creek Watershed, Colorado.
(This drainage is discussed and analyzed in the text and provides the basis for the illustrative watershed analysis calculations. The plot was done from 940 points taken from the Bonanza, Cocheta Creek, Mt. Ouray, and Pahlone Peak Quadrangle maps of the Geological Survey, on an IBM Compatible PC, using the software SURF©; the output was further modified with TOUCH UP©.)

The degradating processes are driven by forces associated with the erosive agents **gravity**, **ice**, **water**, and **wind**. The processes do not stop when the erosion occurs, for they include the transportation and deposition of materials that have been eroded, as noted above. Temperature may also be a geomorphic agent in that it affects ice formation and melt, along with evapotranspiration, which in turn affects soil water content and, therefore, how much water may infiltrate and percolate before saturation forces surface runoff and sheet erosion.

The landforms created by each of these processes, in combination with the original aggradation sequence, exhibit attributes that are characteristic of the rock material and the changes that have occurred over the ages. Type of bedrock, steepness of slope, aspect, soil depth, distribution of rock particles in the soil profile, elevation, elevation range, and stream drainage pattern are all functions of the geologic history of the watershed.

The Millswitch watershed, for example, has been worn down from a previously existing nearly level plain by the action of water. It currently exhibits a typical, mature topography in that there are virtually no flat areas; slopes and stream gradients are rather continuous, indicative of the absence of resistant rock strata that might interrupt the smooth grades; the V-shaped valleys are well-incised, and the ridges are sharp, thus, the watershed boundary is clearly identified, which is not always the case. The relief is at a maximum in the geomorphic history of the watershed; it is at relatively high elevation, which suggests that cooler temperatures keep fluctuating across the freezing point, maintaining a well-aerated soil with a high-infiltration capacity that supports good vegetative growth; incision, and the slopes are somewhat between the maximum that occurred near the earlier stages of stream incisement and the minimum that will be evident when the area is worn down to a near level plain, its ultimate fate.

The physical characteristics of the watershed are important considerations in stream behavior.

STORAGE ON THE WATERSHED

As seen in Fig. 1-2, there are several different forms of storage on the watershed. On and above the ground surface, there is depression storage (puddles, ponds, and water detained on vegetative surfaces), storage in vegetation, and channel storage. Below the ground surface, there is water in both the saturated and unsaturated (aerated) zones: ground water storage below the water table, and retention and detention storage in the aerated zone above it.

Surface and subsurface runoff waters contributing to a hydrograph may have inherently different qualities and **regimen** (rate of delivery, or temporal distribution), many of which may be recognized when the water is in the stream. The proper hydrologic analysis and control of runoff waters is the fundamental goal of practically all hydrologic research and management.

THE "SMALL" WATERSHED

The wildland watershed manager frequently restricts management considerations to a "small" watershed. The implication is that immense engineering works are required for manipulating water resources on a large scale, presumably on or from large watersheds. This has been a source of continuing friction between land managers and civil engineers, in particular, dating back to the 1936 Omnibus Flood Control Act (49 Stat 1540) where responsibilities for upstream and downstream flood control were divided between the Soil Conservation Service and the Corps of Engineers. Competition for government funds, between professional education units, and between individuals perpetuate this feud.

Numerous attempts have been made at trying to delimit the "small" watershed, either by actual size (e.g., 100 sq mi) or function (e.g., response to precipitation inputs), or types of storage (e.g., no ground water storage). Some runoff calculation formulas specify a watershed size limit, as noted in Chapter 6. Chow (1964) quotes the Runoff Committee of the American Geophysical Union (of which he was chair):

> From the hydrologic point of view, a distinct characteristic of the small watershed is that the effect of overland flow rather than the effect of channel flow is a dominating factor affecting peak runoff. Consequently, a small watershed is very sensitive to high-intensity rainfalls of short duration, and to land use. On larger watersheds, the effect of channel flow or the basin storage effect becomes very pronounced so that such sensitivities are greatly suppressed. Therefore, a small watershed may be defined as one that is so small that its sensitivities to high intensity rainfalls of short durations and to land use are not suppressed by the channel storage characteristics.

Chow's definition is based upon a combination of the function and response concepts, specifically, the interaction of rainfall intensity and channel storage. This definition is fine in principle because it is a "floating" one rather than being specifically tied to some arbitrary, finite area. However, the definition is untenable in that it uses **overland flow** (runoff over the surface of the soil before becoming channelized): overland flow is, by and large, not a natural feature of wildland hydrology. Further, it identifies "channel" and "basin" storage separately: they might be better combined as "channel and ground water storage" or, simply "ground water storage," since that is most commonly connected directly to channel and bank storage.

Recognizing that there are broad groupings of factors that affect runoff and storage extending from the large-scale atmospheric and climatic factors, through weather, hydrographic, geomorphic/basin, soils-vegetation/land use, and channel/ground water storage factors, one might best define a small watershed as follows:

> A small watershed is one where channel and ground water storage are not sufficient to attenuate a flood peak primarily influenced by weather and land use.

With this concept of the small watershed in mind, attention may be directed at characterizing the watershed; discussion of three broad principles of watershed management; an overview of and introduction to the practice of watershed management and, finally, the all-important linkage between land use and water quality.

CHARACTERIZING THE WATERSHED

Watersheds are often not immediately discernible from a map or on the ground.[1] The first step in watershed analysis is to identify the watershed outlet (lowest point or base level)

---

[1] On maps of steep, geologically mature terrain, the watershed boundaries are usually quite obvious.

on a map. Once the watershed has been identified, a number of parameters can be calculated that aid in describing and quantifying the characteristics of the watershed. The determination of several watershed parameters provides information that is useful in making decisions about how to manage the watershed in addition to simply describing it.

As implied in the definition of "watershed" earlier in this chapter, the area of the drainage basin is "set" when the base level is identified on a topographic map. Most common of these maps are the "quadrangle sheets" issued by the U.S. Geological Survey. These excellent maps typically cover 7½, 15, or 30 minutes of arc (RF = 24,000, 62,500, and 125,000, respectively), and show streams, wetlands, forest vegetation, and several cultural features in addition to the contours. Cultural features include useful surveying details such as latitude and longitude, map names, and, where appropriate (as in the General Land Office surveying system for the western states), boundaries that are marked on the ground, benchmark elevations, and elevations of peaks and water bodies.

The technique for determining the watershed boundary on a topographic map is to start at the base level and, working uphill, mark the ridge on one side or the other. Errors will almost assuredly result if one attempts to establish the boundary by working downhill. The decision as to whether a particular piece of ground is "in" or "out" of the watershed may be determined by applying the test: "Does the water from this piece of ground flow to the stream above the base level?" It is "in" if it does; "out" if it does not. Three simple rules help in this determination:

1. Water tends to flow perpendicularly across the contour lines.

2. Ridges are indicated by contour "V"s pointing downhill.

3. Drainages are indicated by contour "V"s pointing upstream.

Once the main ridge is "attained," the analyst should start again at the base level, working up the other side of the watershed. In areas of relatively flat topography, field investigation may be needed to ascertain the location of the watershed boundary.

Unfortunately, the **topographic** boundary (divide) of the watershed thus determined may not, in fact, be the true hydrologic boundary: the latter may be determined ultimately by fixing the location of the **phreatic** divide, where subsurface rock configuration may divert water that percolates through the soil away from the stream, effectively reducing the size of the watershed (Fig. 7-2). Conversely, the watershed may also be larger than indicated by the topographic divide because waters

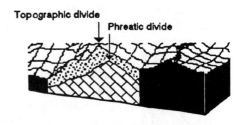

Figure 7-2 Topographic and phreatic divides

are diverted into it by a phreatic divide outside the watershed boundary drawn on the map. On many watersheds, the analyst assumes that the area inside the watershed from which water flows out is compensated by a like-sized area outside that diverts water in. The existence of non-conforming topographic and phreatic divides may have to be verified in the field with seismic equipment.

Chapter 7 Water on the Watershed 253

Upon establishing the watershed boundary (Fig. 7-3), several parameters may be determined, including size; maximum, minimum, and mean elevations; distribution of elevation; aspect; orientation, perimeter length, shape, and drainage network. Perimeter is used in determination of shape and drainage network. Along with other features that must be determined on the ground or from other sources, these physical parameters are useful in evaluating hydrologic characteristics, such as spatial and temporal amounts and patterns of precipitation and runoff. As noted in Chapter 6, it would seem beneficial to model the watershed so as to ascertain the interactions between the several physical characteristics and runoff behavior. The difficulties, results, and opportunities for watershed model research are discussed in the section on size.

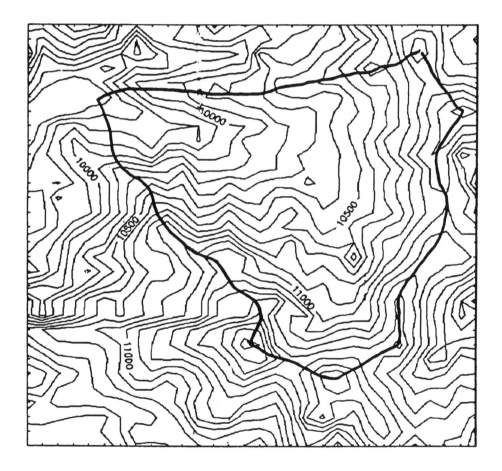

Figure 7-3 Topographic map of the Millswitch Creek Watershed using the software TOPO©

Size

Size — and the watershed boundaries necessary to its determination — is important in order to estimate water resource parameters such as total annual yield and flood potential, to identify ownership of the land being managed, and to evaluate how, when, and where to apply land management measures that control water quality, quantity, or regimen. Most importantly, size is an essential consideration in the initial evaluation of the watershed's hydrologic behavior. Size is the most difficult of the several watershed features to model: once the problems have been overcome, physical models under a rainfall simulator provide some useful information.

Method of Evaluating and Presenting

The area of the watershed may be determined by any of several methods. The most precise method involves the use of a **planimeter** or **computerized digitizer**. Less precise techniques include the **dot grid** (or graph paper). Counting dots (or grid squares) can be tedious, but can be simplified by tracing the outline of the watershed onto the dot grid itself, "checking off" (with an erasable grease pencil) counted squares and converting the total number of dots with the aid of a map scales and equivalents table (Appendix B). Two or more counts (within ±1 percent) should be made to ensure accuracy and to provide reliability in the count. An additional count may be obtained in the process of determining elevation distribution (below) and may be used as a check.

Area is reported in standard units of acres, square miles, or hectares. While it is recognized that a good portion of the watershed is, in all likelihood, on a slope, the area that is reported is the horizontal projection of the watershed boundary.

Effects on Average and Extremes of Flow

Unless influenced by interaction of ground water storage and evapotranspiration, average annual runoff is not affected by size of the drainage. Other things being equal,[2] side by side large and small watersheds may be subjected to the same precipitation and radiation: if the vegetation on each also has equal access to water for evapotranspiration processes, the residual runoff from each will be the same. Because of larger ground water storage reservoirs, however, the minimum flow on the larger watershed will be more sustained than that of the smaller watershed. Streams on very small watersheds, on the other hand, may dry up entirely during dry periods.

The size of the drainage basin has a significant effect on the flood hydrograph. Sherman (1932) proposed that in order to derive a hydrograph for any size drainage basin similar to an original basin for which a runoff hydrograph was known, the hydrograph dimensions vary as the square root of the areas. Brater (1940) reported that as the amount of area contained in the drainage basin increased, the peak rate of runoff occurring during the peak interval expressed as a percentage of the total runoff *decreased*, and that the time base of the hydrograph decreased as the size of the drainage basin decreased. The study basins ranged in size from 4.24 acres to 1876.7 acres. He was careful to point out that the effect of size on the

---

[2] Such as aspect, slope, soil types, etc.

peak and base length of the hydrograph can be masked by differences in vegetative cover, which is notable for such small watersheds.

Other researchers (Hertzler 1939; Laden et al. 1940; Hathaway 1941; Taylor and Schwartz 1952; Wisler and Brater 1959; Minshall 1960; Getty and McHughs 1962; Viessman and Geyer 1962; and Bruce and Clark 1966) present data which show that as the drainage area increases, the rate of runoff per unit area at the peak of the hydrograph decreases. When graphed on log-log paper, the relationship is close to a straight line. Hertzler (1939) and Laden et al. (1940) substantiated Brater's observation that the length of the time base of the hydrograph increased as the size of the drainage basin increased: this relationship has been reported for very small watersheds of only a few acres, such as those investigated by Viessman and Geyer (1962), all the way up to watersheds of a size as large as 25,000 sq mi, as reported by Bruce and Clark (1966).

Wisler and Brater (1959) show that as the size of the watershed being considered gets larger, the average intensity of precipitation per storm and per unit area decreases. This relationship has a profound effect on the peak flow per unit area when the range of watershed sizes is large.

Peak flows of record on larger watersheds, typically from spring snowmelt and/or broad climatic patterns or events, are greater, as measured in absolute flow units (cfs). However, given the assumption that the watersheds in question are uniformly, instantly, and completely covered with a rainfall event, resultant peak flows are *lower* in units per unit area (csm). There are two reasons for this: (1) the intensity, duration, and amount of rainfall decrease with increasing storm area (above) so that the smaller watershed is more likely to be receiving precipitation and delivering runoff from its entire area simultaneously, and (2) it takes longer for water which falls on remote portions of the large watershed to reach the outlet, thus not all of the watershed is simultaneously contributing to the peak runoff, which also delays the time of the peak runoff. Evidence now indicates that the relationship between peak flow and drainage area may *not* be constant. The equation for the relationship between maximum peak, Q_p and drainage area, A, is:

$$Q_p = kA^z$$

where k is the coefficient and z is the exponent. Reporting on watersheds ranging from models 2 sq ft in size to real watersheds over 3000 sq mi, Black (1988) reported that the relationship is not a constant one, and exhibits the classic chaotic characteristics of complex systems (Gleick 1987), namely scaling, fractals, and self-similarity: for example, both k and z were also found to be statistically significant functions of A; and there appears to be a strange attractor throughout the range of relationship between z and drainage area. Black concluded that "the maximum peak/drainage area relationship is a complex, non-linear system, with the exponent, z, *increasing* towards unity, and the coefficient, k, likely *decreasing* towards zero as area increases."

In terms of runoff *per unit area*, the *peak flow is lower and later on larger watersheds*. Small watersheds are said to have "flashy" hydrologic behavior, that is, they exhibit higher high flows and lower low flows. Calculation of the ratio of maximum to minimum flows reveals higher ratios on small watersheds, an interesting but unstandardized measure of "flashiness."

Models and Modeling the Watershed

Conceptual, mathematical, and computer models and their theoretical bases and limitations are discussed in Chapter 6. The other type of model that is used extensively in hydrologic research is the physical, or iconic (look-alike) model.

A prime difficulty with using physical models is that, while it is possible to scale the several dimensions of the watershed for the model, such as length, soil depth, slope, etc., it is not possible to scale water. The issue is identified under the general heading of *similitude*. **Hydraulic similitude** refers to the scale problem as applied to water, primarily in open channels. It is a continual (but not insurmountable) problem for manufacturers and modelers of flumes, penstocks, dams, flood gates, and other water control structures. **Hydrologic similitude** (Chow 1967) refers to the scale problem from the standpoint of the watershed and its behavior.

The problems of hydraulic similitude have been summarized by King et al. (1960). The physical conditions required for similitude (equal behavior) between model and prototype (real-world unit) impose modeling criteria that are shown to be mathematically exclusive in terms of the length (L) dimension. It is thus impossible to simultaneously model length for the watershed and for water's parameters. In the real world, water is acted upon by three primary forces: gravity, viscosity, and surface tension. Under model conditions, those three forces still operate, but their respective dominance under given conditions and, as a consequence, their overall balance, is not necessarily and (most often is not) the same. For models where gravitational forces predominate, Froude's model law describes the criterion for similitude, in terms of $1/L^{0.5}$; for models where viscous forces predominate, Reynold's law is used to determine the relationship between model and prototype in terms of L^2; and for models where surface tension forces predominate, the Weber Number is used to describe the conditions for similitude in terms of $L^{1.5}$. The three L-relationships are not compatible.

Three alternatives are offered to avoid this dilemma: (1) the surface tension, viscosity, and density of the liquid used on the model may be varied (Chery 1968); (2) the problem may be avoided by calling the model a **prototype**, that is, maintaining a one-to-one ratio between model and prototype (in which case there really is no model); or (3) models may be created from materials that, when exposed to water, keep the three influencing forces in their real-world balance. The latter is accomplished by noting that the model does, in fact, duplicate real-world prototype behavior. Quantification is given by Wooding (1966) in that criterion for watershed similitude is that the ratio for subsurface flow on the model to time for subsurface flow on the prototype must have equal or similar ratios for channel or any other type of flow: they behave similarly. Carefully constructed watershed models can exhibit "blackbox" similitude: they behave the way real-world watersheds behave (Black 1975).

According to a series of studies on laboratory models that exhibited hydrologic similitude under a rainfall simulator and related testing on several real watersheds, Black (1975) found that:

1. As drainage increases, maximum peak increases.
2. The time of concentration (period from beginning of rainfall until peak occurs) increases non-linearly as size increases.

3. Decay time, an index to the recession curve (and total runoff time), increases with increasing size.
4. Total time of runoff increases as size increases.
5. Model response to variation in size is similar to that of natural watersheds.
6. The models appear to simply be very small watersheds, which suggests that modeling the watershed may be possible.
7. There appeared to be a mathematical relationship between the models and the real-world watersheds that can be quantified (a variant of this relationship is referred to above as a strange attractor).

These results agree with the earlier studies cited above, and subsequent testing of other watershed model characteristics such as slope, soil depth, drainage pattern, and shape further reinforced field results, as reported in the following sections.

Elevation and Slope

Elevations of specific points on the watershed may be read directly from the topographic map, and interpolated/extrapolated for other points. Slope is simply the **gradient**, or vertical difference between two points whose elevations are known divided by the horizontal distance between them. Elevation is important because precipitation generally increases with increasing elevation due to an orographic effect (see Chapter 2). Slope is important because it is a prime factor in infiltration capacity (see Chapter 4). Combined with elevation, slope can be an important factor in orographic effects, and combined with aspect, slope is also important in insolation considerations that play a role in evapotranspiration and snowmelt.

Method of Evaluating and Presenting

Simple inspection of the map reveals maximum and minimum elevations. The mean elevation may be accurately determined by recording and averaging a large number (about 100) of systematically spaced points on the watershed. A more convenient technique involves evaluating the area between each pair of successive contour lines (about 10 to 20 zones for the watershed). The data thus obtained (Table 7-1) provide an area-elevation or **hypsometric curve**, illustrated in Fig. 7-4. The median (and maximum and minimum) elevations may all be read directly from this curve, which also provides a visual representation of the watershed's profile.[3] Plotted on graph paper, the median elevation is 10,610 ft and, since the watershed slope (the curve) is nearly uniform, the mean and median are close together.

With an elevation-precipitation relationship for the area under study, precipitation can be estimated for each of the watershed's elevation zones. This yields a more reliable estimate of precipitation than would be the case for an estimate based solely on the mean elevation. Application of an orographic-based elevation-precipitation regression should be limited to the combination of aspect, slope, and elevation (and proximity to precipitation sources) where it is known that there is a measurable (and significant) orographic effect.

---

[3] For regional analytical geomorphic research, the axis is plotted as ratios of each zone to the total relief or area.

Table 7-1 Calculation of Hypsometric Curve, Millswitch Creek Watershed, Colorado

Elevation Zone (ft)	Number of Points	Accumulated Percent	Percent of Area Above Elevation
9520 – 9754	11	322	100.0
9755 – 9989	28	311	96.6
9990 – 10,224	37	283	87.9
10,225 – 10,459	53	246	76.4
10,460 – 10,694	50	193	59.9
10,695 – 10,929	50	143	44.4
10,930 – 11,164	43	93	28.9
11,165 – 11,399	32	50	15.5
11,400 – 11,634	14	18	5.6
11,635 – 11,870	4	4	1.2

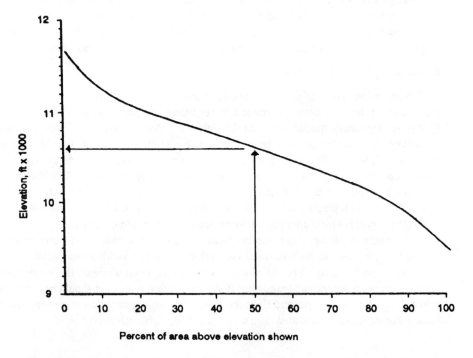

Figure 7-4 Hypsometric Curve for Millswitch Creek Watershed (data from Table 7-1)

For the region of the Front Range of the Rockies, and based on 12 observations stations, a linear regression for annual precipitation, P, in inches on elevation is:

$$P = -31.35 + 6.04E$$

where E is the elevation in thousands of feet (the value of the correlation coefficient, R, is 0.948, d.f. = 10). Inserting the mean elevation of 10.610 thousand feet for E, one estimate of annual precipitation for the watershed is 32.73 in. Another method of determination of mean annual precipitation for the Millswitch watershed shown in Figs. 7-1 and 7-2 is shown in Table 7-2.

Table 7-2 Calculation of Elevation-Weighted Mean Annual Precipitation, Millswitch Creek Watershed, Colorado

Elevation Zone (ft)	Midpoint of Elevation Zone	Estimated Annual Precipitation	Percent of Total Area	Zone Precipitation (in.)
(1)	(2)	(3)	(4)	(5)
(From Table 7-1)	[From (1)]	(by regression)	(From Table 7-1)	[(3)x(4)/100]
9520 – 9754	9637	26.86	3.4	0.91
9755 – 9989	9872	28.28	8.8	2.49
9990 – 10,224	10,107	29.70	11.5	3.42
10,225 – 10,459	10,342	31.11	16.4	5.10
10,460 – 10,694	10,577	32.53	15.5	5.04
10,695 – 10,929	10,812	33.95	15.5	5.26
10,930 – 11,164	11,047	35.37	13.4	4.74
11,165 – 11,399	11,282	36.79	9.9	3.64
11,400 – 11,634	11,517	38.21	4.3	1.64
11,635 – 11,870	11,752	31.35	1.3	.41
Total	-	-	100.0	32.65

Note: The fact that the elevation-weighted estimated annual precipitation is so close to the direct regression estimate of 32.73 in. is indicative of the regular slope of the watershed, as illustrated in the hypsometric curve, Fig. 7-4.

Slope may be calculated by measuring the length of regularly spaced contour lines over the entire watershed. The formula is:

$$S = (DL/A) \times 100$$

where S is the average slope, in percent, D is the contour interval used in the determination (not the contour interval of the map), in feet, L is the length of the contours, in feet, and A is the watershed area, in square feet.

Effects on Average and Extremes of Flow

The impact of elevation is best illustrated by working with actual streamflow records. Results are varied, but tend to follow expectations based on common sense.

The effect of elevation on average flow is likely to be considerable: higher elevations tend to receive higher amounts of precipitation, both per storm and on an annual[4] or other temporal bases. In general, high elevations yield greater amounts of runoff. The effect is confounded, however, because there are other, related factors that affect runoff in varying ways. For example, temperatures are lower at higher elevations, thus reducing potential evapotranspiration. However, relative humidity is also lower and vapor (and atmospheric) pressure gradients are higher, both of which tend to *increase* evapotranspiration.

Also, in the alpine and subalpine zones, more of the precipitation is in the form of snow, which tends to melt and run off over a rather short period of time in the late spring. Running off rapidly means that the water is not present long enough to be available for evapotranspiration. Runoff from high-elevation snowmelt watersheds tends to be flashy. On watersheds where the runoff is not dominated by snowmelt, higher elevations are likely to produce flashy runoff anyway because the higher elevation watersheds tend to be smaller in size and steeper. The smaller size means less potential for buffering of flows by ground water storage, and the greater slope favors lower infiltration capacity.

Elevation is similarly confounded with slope: land tends to be steeper at higher elevations, thus with lower infiltration rates and more rapid runoff. Further confounding the picture are **soil depth**, which tends to be less at higher elevations owing to shorter time for soil to form and steeper slopes from which soil can move by gravity, and **vegetation**, which is generally not as lush at higher elevations (owing to lower temperature and moisture availability).

The overall effect is that average annual runoff is greater from small, high-elevation, steep-sloped, thin-soiled watersheds. Minimum flows can be greater if there is sufficient snowpack storage, and lower if not all the precipitation is in the form of snow and storage is limited by shallow soils. High peaks are common, as is flashy runoff behavior.

Aspect and Orientation

Aspect is the direction of exposure of a particular portion of a slope, expressed in azimuth (0-360°), compass bearings (e.g., N 47°E) or the principal compass points (N, NE, E, SE, etc.). **Orientation** is the general direction of the main stem of the stream on the watershed. A watershed with an east-west orientation is likely to have slopes that are predominantly north and south in aspect.

Aspect is an especially important feature of the watershed in view of insolation. As noted in Chapter 1, a 45-degree south-facing watershed at 45°N presents a surface that is parallel with a horizontal surface at the equator and perpendicular to incoming radiation. In most situations, the rays of the sun have a greater length of travel through the atmosphere which attenuates their intensity. For example, at the summer solstice, with the sun at its maximum northerly declination of 23½°, the 45° south-facing slope at 45°N latitude and the horizontal

---

[4] Note typical regression of annual precipitation on elevation in preceding section.

Chapter 7 Water on the Watershed 261

surface at the equator receive nearly the identical amount of radiation. At certain times, the south-facing slope is certain to be a great deal dryer, have greater evapotranspiration, and therefore support more xerophytic vegetation than other nearby slopes. Conversely, north-facing aspects will tend to be cooler, have vegetation typical of more northern environs, yield greater annual runoff, and exhibit more flashy runoff behavior. The differences are especially evident in the western United States, where a particular vegetative type *may be stressed*, that is, near the limits of its natural range, this contrast will be more noticeable.

Method of Evaluating and Presenting

Determination of orientation may be simply made by inspection. Aspect is not quite so easy, although in certain cases, aspect may also be readily determined by simple map inspection. Often the 100 or so points used for determining average elevation (or the dots of a dot grid or intersections of an overlaid sheet of graph paper) may be used to record aspect data. Cancelling opposing aspects (e.g., NW and SE; E and W, etc.) may yield a preponderance of residual aspects that can then be reported as the average aspect of the watershed. Usually, however, if the aspect is not readily apparent, it is not likely to be a strong runoff-influencing factor.

A Topographic Sampler (Lee 1963) can be used to rapidly obtain elevation, aspect, and slope data from a topographic map at a large number of points (Fig. 7-5). Recorded on a proper form (Black 1988), they may be readily averaged, as noted above. The technique is especially valuable if a large number of watersheds are to be analyzed.

Effects on Average and Extremes of Flow

The overall effect of aspect is that highly insolated facets are likely to have lower average annual runoff than other portions of the watershed. Soils, if well developed, may increase water-holding capacity, resulting in more sustained low flows, and have ample storage for attenuating flood peaks. Runoff will therefore tend to be less flashy as well. The reverse is likely to be true for aspects with lower insolation.

Watershed Shape

The shape of the watershed can have a profound effect on the hydrograph and stream behavior, particularly from small watersheds, and especially in relation to the direction of storm movement.

Time of concentration can be used to aid in studying the effects of watershed shape on the hydrograph and on stream behavior. In this case, the time of concentration (t_c) is depicted (Fig. 7-6) by lines that connect points on the watershed from which it takes equal time for surficial or channel runoff (e.g., storm flow) to reach the outlet (one of several definitions of time of concentration). On each of the watershed shapes represented in the figure, lines of 10-minute time of concentration are shown. These lines are evenly spaced because it is assumed that (1) the watersheds all exhibit the same soils, slopes, drainage patterns,[5] and stream gradients, and (2) each watershed is instantaneously, completely, and uniformly covered by a

---

[5] This is an unlikely occurrence, but the shapes are extremes, too, shown for illustrative purposes.

rainstorm for a period of time less than the channel time of concentration of the entire watershed (or unit storm).

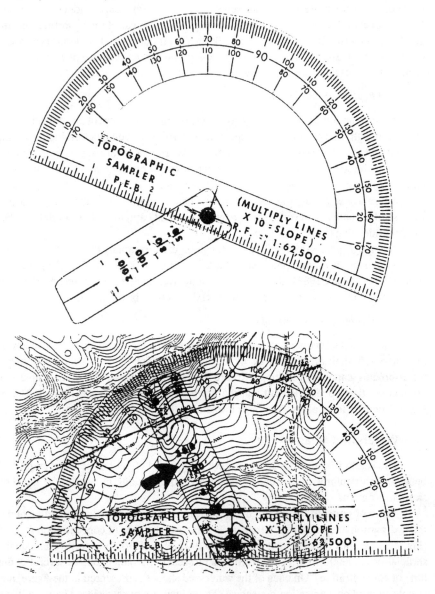

Figure 7-5 The topographic sampler (after Lee 1963). This modified unit may be used for determining slope on 1:62,500 maps with different contour intervals. If the contour interval is 100 ft, then the sample point (arrow) has a slope of 30 percent (3 contour lines intersect the circle), is at an elevation of about 960 ft, and has an aspect of N83°E (north is at the right)

In *a*, streamflow from the entire channel network (which may be extended and expanded according to the variable source area concept) is contributing to the peak flow within 40 minutes, in contrast to the 20 minutes for the circular watershed (*b*); thus, the latter will produce a higher flood peak. Long, narrow watersheds, as shown in *c* and *d*, will also exhibit dramatically different patterns of runoff dependent upon the location of the outlet, as illustrated by the different patterns of lines of equal time of concentration.

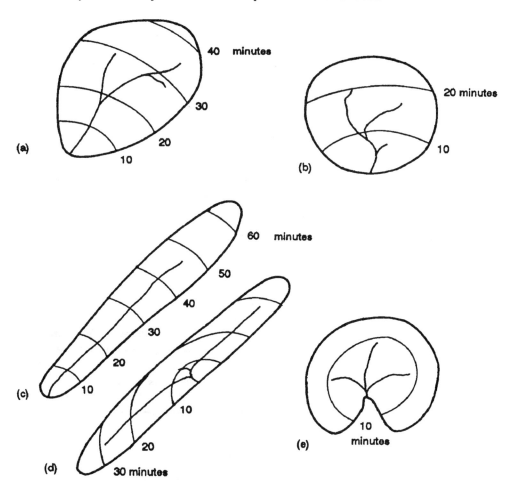

Figure 7-6 Watershed shape and time of concentration

The combination of watershed shape and direction of storm movement is especially important. For example, if the rainstorm moves down the watershed shown in *c* over a 1-hr time period, the peak will be very high because the upper reaches of the watershed will be contributing runoff to the peak at the same time as the storm is over the outlet of the watershed. Conversely, if the storm moves up the watershed, the peak will be greatly attenuated. Hydrographs from a watershed model exposed to a moving storm under a rainfall simulator are shown in Fig. 7-7; the pattern is as predicted by Yen and Chow (1969). The difference on the hydrographs between a fast-moving and stationary storm is reduced (Foroud, Broughton, and Austin 1984).

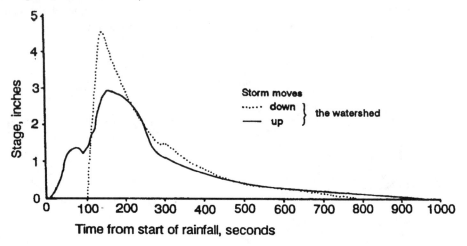

Figure 7-7 Hydrographs from watershed models where storm moved up and down main axis of model

Method of Evaluating and Presenting

The simplest description of watershed shape is accomplished by comparing the drainage to a known shape, such as a circle, pear, banana, rectangle, and so on. While this is often a helpful means of reporting, especially in extreme cases (such as a cigar- or egg-shaped watershed), shape also lends itself to numerical evaluation.

There have been many attempts to evaluate shape numerically and to relate it to runoff parameters such as lag time and flood peak magnitude (Black 1972). The most widely accepted shape variate is Gravelius' **compactness coefficient** (cited in Wisler and Brater 1949):

$$K_c = 0.28 P/A^{0.5}$$

where K_c is the compactness coefficient, P is the perimeter of the drainage basin, and A is the area. This dimensionless parameter compares the perimeter of the watershed with a circle of the same area; thus, if the watershed is a circle, the K_c is 1.00. The perimeter (with a map measure) and area (with dot grid or planimeter) may be readily determined from the topographic map.

It has often been presumed that a circular watershed produces the highest flood peak, but this can be shown not to be the case: note that the (highly unlikely) shape in Fig. 7-6e will produce a peak that is higher than the one from *b* because it takes only 10-15 minutes for the entire watershed channel network to commence contributing runoff. Harbaugh (1966) also reported that a circular watershed model did not have the highest flood peak among the four shapes tested.

The compactness coefficient *does* show the degree of compactness of the watershed, and is useful for that descriptive purpose. But note, too, that the value of K_c for c and d (in Fig. 7-6) is the same for both because the formula does not account for the location of the outlet; the two shapes should have very different runoff behavior.

Ideally, the watershed shape variate should have the following characteristics:

1. Be dimensionless, involving measures which affect runoff behavior, such as watershed length and width.
2. Emphasize the lower one third to one half of the watershed because that is where the flood peak itself is generated.
3. Take account of the location of the outlet of the watershed (e.g., differentiate between c and d in Fig. 7-6).
4. Provide a degree of predictability of relative flood peak magnitude.
5. Be easily measured.

These criteria characterize the parameter, **watershed eccentricity** (Black 1972), which compares the shape of the watershed to an ellipse for which the major axis is twice the minor axis. This is the shape that appears, by laboratory testing and field verification, to produce the maximum flood peak (under the assumptions of instantaneous, uniform, and complete coverage of the watershed by a rainstorm). The formula is:

$$\tau = (|L_C^2 - W_L^2|)^{0.5} / W_L$$

where τ (tau) is the watershed eccentricity, L_C is the length from the outlet to the center of mass of the watershed, and W_L is the width of the watershed at the center of mass and perpendicular to L_C. For a more complete discussion and illustration, see Appendix C.

Watershed eccentricity may be determined by tracing the watershed outline *and the position of the outlet* onto a piece of cardboard. Freely suspending the model at two or more corners from a pin to which is attached a miniature plumb bob will identify the center of gravity which, for a planar representation of the watershed, will be the same as the center of mass.

An added benefit of this determination, if L_C is extended, is a set of perpendicular lines which may be used for the Station Angle Method of determining mean precipitation (Chapter 3).

Effects on Average and Extremes of Flow

Watershed shape has no obvious effect on average annual water yield. The primary effect of watershed shape appears to be its influence on the peak flow during a rainstorm on a small watershed. If storage on the watershed is limited, and there is considerable influence of shape on the magnitude of the peak, then the minimum flow might be affected as well. Such an effect is most likely in the extreme case, for example, where the watershed is long and narrow and exhibits little or no ground water storage.

In an extensive study on models, watershed shape did not have as great an effect on peak flows as other characteristics such as slope or soil depth, and it may be dominated by direction of storm movement, antecedent moisture conditions, precipitation inputs, or other factors (Black 1972). Time of concentration (in this case, time from start of precipitation until the peak flow occurs) was not affected by direction of storm movement, but the lag time (time from start of precipitation until stream starts to rise), and storm peak magnitude was affected dramatically (Fig. 7-7).

Consideration of watershed shape is likely to be important when considering the effect on peak flows, regimen, and movement of pollutants from a portion of a watershed dependent upon its location in the larger watershed of which it is a part. Thus, for example, increased runoff from a small, logged watershed may have a different effect on the peak from a larger downstream watershed (within which the logged area is nested) dependent upon where the logged area is within the larger watershed.

Drainage Network

The drainage network of a watershed is the system that collects the water from the entire area and delivers it to the outlet. It includes the subsurface and surface drainage. In most cases, the entire drainage network is not revealed to the watershed analyst, while the surficial stream drainage pattern is. The pattern of streams is only the surface manifestation of that larger system, and may carry a widely varying percentage of the total runoff. Most of the research into drainage networks has actually been directed at this surface portion: it is readily discernible on the map, can be measured and characterized, and can be described both numerically and verbally.

Method of Evaluating and Presenting

Initial evaluation of drainage networks was on the basis of **stream order** designated by *1, 2, 3*, etc. A stream of order *1* has no tributaries; a stream of order 2 has tributaries of order *1*, and so on. In the European system, a Class I stream is the main stem of the drainage, discharging directly to the ocean or a large water body. Class II streams are **major tributaries** to Class I, and Class III are **minor tributaries** discharging into Class II streams. Wisler and Brater (1959) point out that the original method of designation was using "I" for the smallest tributary, and working downstream assigning the next higher number when two tributaries of like number join. The method is not conducive to comparative use, nor to calculations, as shown. A major difficulty with stream order is that streams of different class may have different flow magnitudes because they have different tributary systems. Conversely, streams of the same class can drain watersheds that are considerably different in size dependent upon which magnitude of stream is designated "Class 1," thus making it difficult to compare or generally inventory the classes. Horton's system of stream order designation commenced at the tributary level (Class I) and the number increased as more and more tributaries were involved (Fig. 7-8); thus, the higher the number assigned to the main stem, the larger the watershed and the greater the number and extent of its tributaries (Linsley, Kohler, and Paulhus 1949). Strahler (1957) modified the system to apply to *segments* of streams between confluences. A great deal of research has been done on stream development theory, network evolution, bifurcation ratios, and relationships between drainage network and geology. While stream order has been shown to be related to other basin characteristics

(Leopold, Wolman, and Miller 1964), no expression of stream order has been consistently or usefully related to *runoff behavior*. Perhaps one reason for this is that as much as 20 percent of the area of a watershed is located outside the drainage area of each numbered tributary in small parcels that are not sufficiently large to support even minor tributary channels.

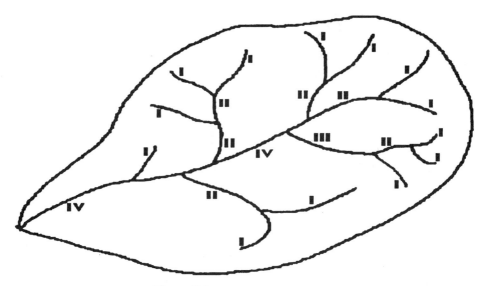

Figure 7-8 Horton's stream numbers

Drainage Density is the number of miles of permanent streams per square mile of watershed area:

$$D_d = L/A$$

where D_d is the drainage density in miles per mile, L is the length streams in miles, and A is the area in square miles. In general, the greater the value of D_d, the more efficiently the watershed is drained. D_d appears to be associated with flashier runoff behavior, greater total surface runoff (lower infiltration), and less ground water storage. It may be estimated from topological variables (Gardiner 1979), as well as from a topographic map, and other means (Richards 1979; Escobar and Rodriguez-Iturbe 1982). However, it has been shown that D_d also varies with annual precipitation (Melton 1957), map scale (Langbein et al. 1947; Young and Stall 1971), watershed shape (Kowall, unpublished), and is likely confounded with rock type as well. An even less-used parameter is **stream density**, which is the *number* of permanent streams or segments per square mile of watershed area. Dingman (1978b) asserts that D_d is of "much more limited value as an independent variable for predicting hydrologic characteristics than has been widely assumed in the literature." Part of the reason for this is that channel characteristics can mask drainage patterns (Black 1972).

Verbal description of the surface **drainage pattern** has not been formalized, but geomorphology texts typically refer to drainage patterns in terms that are derived from describing leaf venation, fruit- or tree-forms, or other well-recognized formations. Thus, the names: dendritic, palmate, pinnate, wye, trellis, radial, and annular are among those most often used. According to laboratory studies on watershed models, drainage pattern appears more important than drainage density in influencing peak flows and lag times (Black 1972). Some examples are shown in Fig. 7-9.

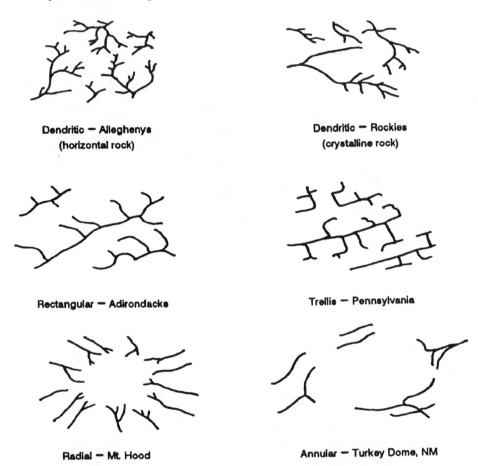

Figure 7-9 Examples of drainage patterns

Streams are classified in geologic texts as being influent, effluent, or intermittent, as illustrated in Fig. 7-10. The **influent** stream (*a*) provides water *to* the ground water storage. The **effluent** stream conveys water *from* ground water storage year round: this is the so-called permanent, or **perennial** stream (*b*). **Ephemeral** streams flow immediately following runoff-causing events, expecially in arid climates; the bed may dry up rapidly, even following torrential runoff, owing to rapid infiltration into the unsaturated zone beneath the streambed or by evaporation (Strahler and Strahler 1973). **Intermittent** streams, which also may flow immediately following a runoff-causing event, provide water to a perched water table, as shown (c), or to deep seepage. Standing on the bank of a stream that is flowing one moment and disappears into its bed the next, it is impossible to determine whether the stream is intermittent or ephemeral by its appearance. A watershed may exhibit any of these classes in different reaches of the stream (van't Woudt, Whittaker, and Nicolle 1979).

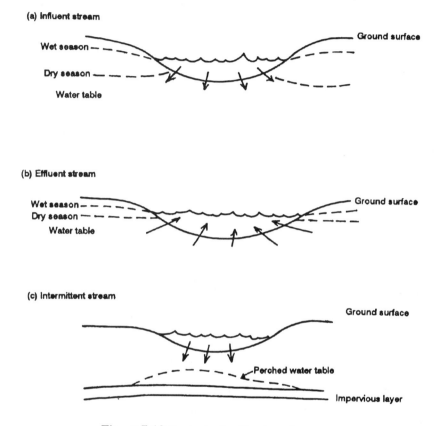

Figure 7-10 Geologic classification of streams

Three channel forms were identified by Brussock, Brown, and Dixon (1985) and found to delimit seven regions in the United States by typical different sequence and consideration of relief, lithology, and runoff. The authors relate the resultant regions to ecosystem models.

Effects on Average and Extremes of Flow

Other things being equal, an efficient drainage pattern, such as a palmate pattern where all stream channels join at the outlet, would produce higher flood peaks than a pattern where all the upstream channels are tributary to one main stem (Black 1972). And, as noted above, D_d appears to be correlated with, more appropriately the result *of*, high annual precipitation, not the other way around: thus, drainage density, stream density, and drainage pattern have no effect on annual runoff, but may be related to it.

SUMMARY

Watershed characteristics have an effect on runoff behavior from small watersheds. An understanding of the impact of those characteristics on stream behavior is essential to successful implementation of land management practices and the principles of watershed management that support them. For example:

Thorough analysis of the Millswitch Creek watershed (Fig. 7-3) reveals that it is pear-shaped, 4.9 sq mi in area, and ranges from 9520 ft to 11,870 ft in elevation. Average elevation is 10,610 ft, and average slope is about 28 percent. The dendritic drainage density is about 1.26 mi/sq mi, compactness coefficient is 1.27, and the watershed eccentricity is 0.48. The drainage has an east-west orientation, and an average aspect that is best described in two large facets, one generally northeasterly, and a more steep westerly one: a small portion of the watershed faces south, but overall, the aspect is best characterized as northerly.

Interpretation of this watershed analysis information requires inclusion of two additional observations about (1) the 32.7 in. of annual precipitation: it is mostly in the form of snowfall; and (2) vegetative cover: the area is largely covered with coniferous forest. It is a moderately high-water-yield watershed, exhibits little or no ground water storage, has shallow to medium-depth soils with good drainage, and steep slopes. Although its annual precipitation is not great, its high elevation and more northerly aspect combine to restrict evapotranspiration losses, leaving more water available for runoff. Runoff in the upper reaches is flashy, a feature that is still in evidence near its confluence with Marshall Creek at 9520 ft elevation (above this point, the Marshall Creek watershed is similar to the Millswitch Creek watershed). The drainage pattern, and low values for both compactness coefficient and watershed eccentricity suggest possible high flood peaks.

A watershed of this type is a viable candidate for intensive management should the owner/administrator desire to increase water yield. Careful reduction of the vegetation cover can release water for added runoff on a continuing basis. How this can be accomplished, and research results that suggest the location, watershed characteristics, and conditions under which such a program would be successful, are described in the section on practice of watershed management below.

WATERSHED MANAGEMENT

Three general principles of watershed management have been identified (Black 1970). They treat: (1) the natural ecology of the watershed as a dynamically balanced system, (2) the factors that affect runoff, and (3) the maldistribution of water in the hydrosphere in relation to watershed management practices.

WATERSHED EQUILIBRIUM

A watershed can be considered as a system in dynamic equilibrium. The definable unit of land on which all water flows toward a common outlet is derived from, consists of, and is characterized by a number of powerful, interacting, and counteracting forces which vary over time and over their natural range with respect to each other.[6]

The watershed derives from the underlying rock and geologic history, the climate, the soil and vegetative conditions and degree of development, and the time over which these forces have acted. (Note the similarity of watershed morphology factors to those of soil formation in Chapter 5.) The underlying rock provides the basic raw material: this limits the effectiveness of erosive agents and time, determines erosion patterns, reflects the past, and controls future geologic history. The erosive elements of climate, gravity, running water, ice, and wind are also linked with the amount of time influencing watershed morphology. However, other climatic events are equally important: means and extremes of insolation, temperature, range of temperature (especially freeze-thaw cycles), humidity, and so forth, all limit vegetative growth and/or soil formation. These, in turn, are closely linked in terms of their development and to the various types of processes and storage throughout the air-soil interface.

The watershed also consists of these elements from which it derives. It thus appears that: (1) there is no specific time when a watershed is formed, and (2) under natural conditions, the various factors of watershed formation interrelate in such a way that there is a near balance of constructive and destructive forces. Therefore, a watershed is continually developing, with a tendency toward a relatively stable equilibrium condition that is a function of all of its elements. Even so, the equilibrium condition may be naturally disturbed by infrequent and unusual climatic events and by regular and irregular activities of animals, including man. These are what cause the equilibrium condition to be constantly changing, thus dynamic.

A change in one of the factors of watershed equilibrium will alter established patterns of runoff. In many cases the alteration in output may be so small that only an expert can detect it; it may even go entirely unnoticed. On the other hand, it is also possible to make an alteration that will severely disturb the balance. George Perkins Marsh (1874) observed:

> Nature, left undisturbed, so fashions her territory as to give it almost unchanging permanence of form, outline, and proportion, except when shattered by geologic convulsions, and in these comparatively rare cases of derangement, she sets herself at once to repair the superficial damage, and to restore, as nearly as practicable, the former aspect of her dominion.

In fact, "derangements" are not all that rare; nor are the damages always superficial: natural fluctuations in the balance of agrading and degrading forces on the watershed occur constantly, and often result in dramatic alterations in the landscape.

As Marsh noted in the title of his book,[7] man has been a major factor in changing the face of the Earth: man is one of the factors of watershed equilibrium, too. While occasionally man may increase watershed stability, his activities more frequently lead to severe degradation through accelerated erosion brought about by drastic change in the watershed's base level or in the soil's protective cover: the former, which is rather infrequent, alters the amount of energy available for erosion, while the more frequent latter disturbance changes infiltration,

---

[6] This section is taken from an article of the same name (Black and Leonard 1968)

[7] "The Earth as Modified by Human Action."

soil storage, and runoff patterns. It is possible, too, for man's disturbance to be indirect, that is, modification of the watershed so that it is more susceptible to one of the natural forces of equilibrium.

When the equilibrium of a watershed is superficially upset, as Marsh observed, the balance may be restored naturally. In fact, the stability of the balance will determine whether natural (or any) restoration will occur. If the watershed is stable, it may be expected that simple removal of the cause of disequilibrium will allow gradual readjustment. But, if the watershed is unstable, disturbance of one or more of the factors of equilibrium will result in accelerated degradation that will not abate even if the original cause is removed; stability may be reached only after a long period of what appears to be irreversible degradation, that is, imbalance between constructive and destructive forces. Note that the unstable state may be either natural or man-caused.

It is possible to define watershed stability in terms of its ability to recover from a disturbance. This permits a dynamic appraisal of watershed damage, recognizing that instability varies with time, watershed morphology, and with the degree of disturbance as well. A **point of no return** can be defined as that degree of disturbance which divides watershed stability and instability: a morphologically stable watershed will recover from a disturbance following the simple action of removing the cause; an unstable watershed will require additional measures to assist it in recovering from a disturbance.

During the early period of watershed evolution, geologic factors predominate and watershed stability may be affected by man to a very limited degree, if at all. The watershed at this stage is highly unstable since all of the factors of watershed equilibrium have not had time in which to act. In contrast, on watersheds where all the factors have had an opportunity to achieve maximum interplay, those factors which are more accessible to man's normal activities are the ones which tend to dominate stability. Consequently, man can play more of a morphological role on well-developed watersheds. At the same time, however, such drainages should be more stable simply because the factors of equilibrium have interacted for a longer period: thus, the point of no return is affected by time. Since watersheds differ with regard to their morphological characteristics, it follows that their points of no return also differ.

The watershed may be characterized by and defined in terms of the very factors from which it derives and of which it consists. Specifically, the watershed is characterized by its geometric properties that are definitively established when identification of the base level is made. The base level is influenced by and influences runoff. *Runoff patterns and stream behavior are thus unique for each watershed, reflecting the interaction of all of the factors which affect it.* Runoff, consequently, is simultaneously a factor in watershed morphology and a useful tool in analysis in addition to being the end product of wildland water resources management whether for water quality, quantity, regimen, or erosion or flood control.

INVERSE INFLUENCE

The natural ranking of runoff-influencing factors according to degree of effect on runoff behavior is inversely related to watershed size and to accessibility by man. The many elements in each of the realms discussed in previous Chapters may be grouped according to their relative scale of influence. Along such a spectrum, the natural categories are *atmospheric-*

climatic factors, *geologic-geomorphic* factors, *soil-vegetation* factors, and *runoff-channel* factors.

Watershed reaction and stream behavior on large watersheds are predominantly influenced by factors over which man has little or no control in contrast to those factors which dominate the smaller watersheds, and over which he may have considerable control.

More generally, over larger watersheds, regional climatic and broad-scale weather patterns dominate stream behavior through influence over annual and seasonal stream yields, length of shortest half-flow and quarter-flow intervals, flood frequencies, minimum flows, etc., In fact, it is a common practice to characterize the hydrology of a region specifically because the prevailing influence thereon is the generalized climatic pattern (cf., Fig. 6-12).

Within the large climatic region, smaller "hydrographic" units are recognized as areas that have some specified degree of homogeneity with regard to hydrologic characteristics as influenced by elevation, front ranges, underlying rock, proximity to lakes, etc. that alter or mask the broad regional pattern. In terms of large engineering works, the hydrographic unit is often the basic planning and/or management unit. The hydrographic unit is intermediate in size between the broad region and the small watershed, which is the basic unit of wildland watershed management.

On still smaller units of the larger drainage basin, soil and vegetation factors such as proportions of retention and detention storage, vegetative density and depression storage, infiltration rates, and evapotranspiration are the major runoff-influencing factors. Runoff responds to differences in soil depth, texture, and structure, to whether forest vegetation is deciduous or evergreen, or is in place of grass cover. These responses are, in turn, superimposed upon and modify the impact of the regional climatic and the subregional geologic/geomorphic patterns found at the hydrographic level of analysis.

Finally, characteristics of the stream, such as alluvial terraces or other floodplain features, represent factors that further refine the already-modified runoff patterns and render each stream unique with its own particular stream behavior and runoff attributes. These characteristics are reflected in the stream's unique storm hydrograph, and in its seasonal and annual runoff patterns, flood frequency and flow duration curves, as well as other runoff characteristics. These include flood frequency curves, flow-duration relationships, modified seasonal distribution of runoff as a percent of precipitation, and the quarter- and half-flow intervals, for example.

Note that it is only the relative influence of the large-scale factor which has been reduced due to the greater influence and masking effect that the local factor superimposes on the regional pattern. The unique stream from a small watershed may be represented by the electrical signal from a complex electronic device: the primary level of output is a function of the level of input, but changing any of the internal elements may result in considerable change in output (see subsection in this Chapter on modeling).

DISPROPORTIONATE PERCENTAGES

Most trivia affect beyond their apparent means. The principle may be observed throughout the natural hydrologic environment.[8] A few examples will suffice to illustrate. For

---

[8] The principle exists and is well documented throughout the biological environment as well.

example, only about one third of 1 percent of the world's fresh water falls as annual precipitation (Table 1-1); at the air-soil interface, the small 1 percent of annual precipitation that flows down tree stems as stemflow makes up a much greater percentage of the water reaching the water table; the amount of water passing through a tree in response to transpiration tension gradients is thousands of times more than the water necessary for growth by being a medium for the diffusion of nutrients; or a large percentage of the annual runoff occurs during a small percentage of the year (expressed in the quarter- and half-flow intervals).

The principle may also be observed in our man-altered environment. Although land use effects are normally quite local in scope, even a small increase in annual runoff due to forest cutting or overgrazing can cause great upset to the equilibrium of a stream, causing erosion and sedimentation downstream, far from the site of the disturbance. Diverting peak flows from one basin to another can cause agrading of the stream from which the diversion is made, and can increase downcutting on the other. Sloppy road construction on a very small percentage of a logged watershed that leaves fill material in a stream can displace the current with consequent bank erosion and sedimentation downstream, again, far from the site of the disturbance. As with the proverbial camel, it is possible to upset the balance of forces with a straw.

SUMMARY

In the case of the watershed, the existence of a balance of agrading and degrading forces (principle of watershed equilibrium), and the variable accessibility of factors that affect that balance (principle of inverse influence) provide the right conditions for implementation of practices that will most likely achieve objectives with minimum effort and cost (principle of disproportionate percentages).

As a consequence, application of the three principles may be used to achieve effective watershed management: understanding the nature of the equilibrium present on a watershed, which factors in that balance are dominating and/or vulnerable to drastic change, and the ramifications of a disturbance make it possible to effect management objectives with a minimum of expense of time and material.

PERSPECTIVES ON WATERSHED MANAGEMENT

This section has the important and appropriate purpose of discussing the question: How does watershed management relate to other methods of water resources management and to large-scale changes in our environment that may mask or make a mockery of man's often skimpy-appearing watershed management efforts?

Methodologies for Modifying the Water Resources Environment

Management of the basic unit of water supply at the interface between the atmosphere and terrasphere is only one of several methodologies for altering one or more components of the water balance for the benefit of mankind. Other methods include desalinization, evaporation reduction, weather modification, diversion, and storage.

Desalinization is a technique that has been undergoing gradual decrease in cost, but is still generally higher than alternative methods of increasing water supplies. It is of primary use where there are no fresh water supplies, where their cost is high, or where energy is cheap. Increasing costs of electricity drive the desalinzation cost up, while improved technology tends to reduce it. However, the practice is only feasible near the salt water supply, since to use desalinized waters elsewhere requires the added cost of transportation to the place of use. And, desalinized water must be diluted with fresh water, anyway, in order to make it potable.

Evaporation reduction is accomplished by establishing a monomolecular film on the water surface so that vaporization is inhibited or prevented. The major research effort was by the Bureau of Reclamation at Lake Hefner, a part of the Oklahoma City water supply. While the technique is effective in achieving about a 10 percent reduction in evaporation, it is difficult to maintain the layer under adverse weather conditions (Bureau of Reclamation 1959). Wind, a primary factor in increasing evaporation loss, tends to drive the monomolecular layer to the leeward side of the large reservoir where it piles up along the shore, doing little to reduce evporation.

Weather modification is a technology that is becoming better understood and more widely used (Hammond 1971 1973a). Major projects that have been undertaken or are currently underway include: hurricane modification and, perhaps, "control" (Simpson and Malkus 1964; Gentry 1970; Hammond 1973b); thunderstorm development modification for lightning suppression in areas of high forest fire occurrence and for hail suppression in areas where damages to certain crops are common (Hammond 1971); increasing precipitation to (1) relieve temporary drought conditions (Woodley 1970; Rudel, Stockwell, and Walsh 1973), or (2) to build up snowpack and, eventually, runoff, as in the San Juan Mountains of southwestern Colorado by the Bureau of Reclamation (Hammond 1971; Sheridan 1981) and Pennsylvania, where "significant increases in monthly and seasonal water yields" have been achieved (Sopper and Hiemstra 1970). Rudel, Stockwell, and Walsh (1973) point out that:

1. fixed costs of increasing precipitation are low and easily reversible,
2. direct costs of operation are readily covered by benefits of increased water availability, and
3. there may be added benefits of more water available for hydropower production and irrigation of forage crops.

Accidental weather modification has been documented for the Kankakee River, downwind of the municipal/industrial complex at the southern tip of Lake Michigan: since 1925 "there has been a close association between the annual steel production and annual precipitation at La Porte," and up to a 32 percent increase in summer runoff from a 1753 sq-mi watershed (Hidore 1971).

Diversion has been practiced world wide since antiquity as a means of providing water where either none existed or the water that was present was already in use and locally unavailable. In the arid western half of the United States this has led to the successful introduction, use, and modification of the Appropriation Doctrine of water rights (Black 1987) that permits withdrawing water from one stream and conveying it to an entirely different watershed for beneficial use. The water need not be returned to the watershed of origin. The effect (and, in fact, the purpose), like that of increasing precipitation, is to increase runoff in the receiving watershed and, of course, result in a decrease in runoff from in the contributing watershed.

Where there has been a direct increase in runoff by diversion, it is to be expected that there will be resultant changes in the streamflow of both watersheds. The increase in runoff may also be indirect, as in the case of increased precipitation and resultant runoff by weather modification. Either case may produce dramatic alterations in erosion and sedimentation rates, and modification of aquatic flora and fauna. Rango (1970) pointed out that "sediment yield will increase substantially in semiarid regions with increasing precipitation until about 27 in. mean annual precipitation; at this point vegetation growth as a result of increased precipitation will begin to reduce sediment yield." If the increased precipitation becomes sufficient to support vegetative growth, the new vegetation may transpire the increased available water. This supports the concept that there is an internal stability to the hydrologic environment, as noted above in the discussion on watershed equilibrium. Further, there is also the possibility of synergistic[9] effects: for example, Hawkins (1970) documented such a reaction where "an increase in annual precipitation, accompanied jointly by a decrease in potential evapotranspiration, was shown to produce more additional streamflow than the additive combination of the two treatments imposed individually."

Storage is the traditional engineering approach to resolution of temporal or spatial water shortages. Reservoirs are the primary means for providing storage, but enhancement of existing natural lake or wetland basins, and accelerated or enhanced recharge of known ground water supplies, are also viable approaches. Reservoir construction is widespread, in large part because it is a politically expedient and economically beneficial technology, and also because the dam and reservoir provide added benefits of recreation, hydropower, navigation, and flood control.

Watershed Management and Large-Scale Changes

Awareness of large-scale changes in the Earth's environment are a product of recent monitoring technology and modeling. They include the so-called "Greenhouse Effect," acid rain, and the impact of the use of certain industrial and commercial chemicals on the protective ozone layer. Both the greenhouse effect and acid rain are normal characteristics of our environment, and life as we currently know it would undoubtedly not exist without these two naturally occurring phenomena.

The **greenhouse effect** or its currently fearsome ultimate effect, **global warming**, is the predicted result of: (1) an increase in the carbon dioxide (and/or water vapor) content of the atmosphere caused by burning fossil fuels, which both uses oxygen and increases carbon dioxide directly, and (2) destruction of wide areas of forest (Woodwell et al. 1983). The consequent increased level of CO_2 restricts outgoing long-wave radiation, restricting back radiation and increasing temperatures, thereby increasing evaporation. However, burning fossil fuels also produces particulates which, along with the greater amount of cloud cover caused by the increased evaporation, *reduce* the incoming short wave radiation, thus providing

---

[9] Synergism is where the reaction from two or more actions is greater than the sum of the expected reactions from either.

a self-buffering element to the entire system. While recent measurements of CO_2 have shown what appears to be an alarming increase, the graphed data have often been presented artificially enhanced by Y-axis truncation and actually show an increase of only about 2 percent to 3 percent (Sundquist 1985) compared with naturally occurring, long-term documented changes of 20 percent (Kerr 1989b). Against such large long range fluctuations, the recent data pale, yet these small increases are of considerable concern and even a 2 percent or 3 percent change may have dire consequences. Nierenberg (1990) contends: "I am certain that most working climatologists believe that there has been no significant increase in temperature in the last 100 yrs." Sundquist (1985) continues: "Over time scales of thousands of years and longer, atmospheric CO_2 is significantly influenced, if not controlled, by the cycle of weathering and sedimentation"; "the earth surface carbon cycle can be viewed as at secular equilibrium with respect to the sedimentary cycle over tens to hundreds of millions of years." Nevertheless, that may be more than a sufficiently important change to us: "learning how carbon cycles through the environment, with and without human intervention, is crucial to predicting the greenhouse effect" (Kerr 1983a), and "in our calm assessment, we may be overlooking things that should alarm us" (Kerr 1983b).

For example, interaction between opposing forces, such as El Nióo and cooling clouds generated by volcanic activity, or each acting according to obscure long-term temperature cycles, appear to be within predictive models ranges; elimination of these effects from the record reveals a cleaner picture of global warming (Kerr 1989a). Bakun (1990) reports that coastal upwelling intensity has increased as greenhouse gases have built up, increasing fogginess of coastal lands where such upwelling occurs. "One might not be surprised if the buildup of carbon dioxide provokes the system into another mode of operation" (Kerr 1988). Ramanthan (1988) points out that "the predicted changes, during the next few decades, could far exceed natural climate variations in historical times." Schneider (1989) points out that model "forecast" of the current level of CO_2 from the start of the Industrial Revolution does not reflect what has actually occurred; again, clouds are an important factor (Philander 1989). Ultimately, water supply is linked to climate (weather and cloud cover), which interacts with the CO_2 levels (Wallis 1977): "Water shortages may be exacerbated by climatic change, but current and forseeable climatologic forecast ability is not likely to be accurate or specific enough in either time or space to be useful to the water resource planner."

The output predictions of currently available global climate models (GCM) generally agree until cloud cover is introduced into the models: then the predictions diverge (Kerr 1989b; Philander 1989; Ramanthan 1988). In addition to the uncertainty introduced by considering cloud cover (which is essentially the interaction of the oceans and atmosphere), hydrologic futures predicted by GCMs (1) are sensitive to time and space distribution of the snowpack, (2) lump hydrologic processes, and (3) consume incredibly long periods of time on even the fastest computers (Gleick 1989).

At present, it appears certain that: (1) different areas of the planet may be warmer and others colder, which suggests directing our concern toward global *climate change*, rather than global *warming*, (2) the natural variation exceeds the observed or likely extremes of global climate change, and (3) steps should be taken to control emissions of those gases that are likely to set in motion some long-term adjustments to which we are not likely to enjoy accommodating. Considerable effort is already being expended to identify and launch research into feasible strategies for coping with global change (McCabe and Ayers 1989): in the Delaware River basin's "humid temperate climate, where precipitation is evenly distributed over the year, decreases in snow accumulation in the northern part of the basin and increases in evapotranspiration throughout the basin could change the timing of runoff and significantly reduce total annual water availability unless precipitation were to increase concurrently."

The **acid rain** issue is a pervasive one, and is highly controversial (Kerr 1982). It has been noted in a variety of circumstances all around the planet and extensively in the United States from smelter activity in the west (Lewis and Grant 1980; Turk and Campbell 1987), east (Winger et al. 1987), and Florida (Brezonik, Edgerton, and Hendry 1980). Precipitation is naturally acidic (Frohliger and Kane 1975), but the excess acidification derives from the formation and deposition of nitric and sulfuric acids in the presence of atmospheric water vapor or droplets. Emissions from automobiles and industries are the primary anthropogenic sources of the nitrogen dioxide and sulfides (Carter 1979), although volcanic activity is a major and highly variable natural source. The term has evolved into "acid deposition" (Kerr 1981) and, of primary interest to hydrologists, limnologists, and watershed managers, into "acid runoff."

The ecological balance of soft water lakes is particularly vulnerable to acid input owing to the low buffering capacity of the low-alkalinity waters (Booty, DePinto, and Scheffe 1988). Lakes may be naturally low in buffering capacity owing either to acid soil or substrate or to accumulation and decomposition of organic material (Patrick, Binetti, and Halterman 1981). (Water bodies that are acid as a result of contact with acid-forming rock and/or soil are likely to be higher in hardness, alkalinity, and buffering capacity.) The most recent research indicates that the toxic effect of acid rain/deposition/runoff is initially an effect on the soil (Krug and Fink 1983): rain (or snowmelt runoff) that is more acid than that which predominated when the soil was formed, leaches aluminum compounds that formerly were left behind — along with iron compounds — in the podzolization process; it is the aluminum that is particularly toxic to the aquatic flora and fauna (Cronan and Schofield 1979; Lawrence, Driscoll, and Fuller 1988). Results, however, are highly variable (Schindler 1988). Direct lake and watershed liming are potential mitigative measures (Warvinge and Sverdrup 1988), but the magnitude of even a herculean effort to lime a watershed pales beside the geographical extent of the drainage area (Rodenhauser 1989).

The **ozone** issue is more recent, even less well documented, less understood, and of less direct interest hydrologically. Nevertheless, both the role of the hydrologic cycle in ozone distribution and the long-term consequences of ozone depletion should be of concern to the hydrologist. A layer of ozone in the stratosphere inhibits ultraviolet radiation, cause of skin cancer and possible mutations. Interactions of chlorofluorocarbons and like substances with sunlight and weather exist but are incompletely evaluated to date (Kerr 1987; Maugh 1987). Highly complex interactions link the oceans, atmosphere, earth's land surfaces, and incoming solar radiation (Kerr 1988). As with the global climate change issue, the magnitude of the natural fluctuations of CO_2 levels, ozone quantity and distribution, and acid rain appear to far exceed the culturally induced changes by man's activities to date: *but we do not know*. Furthermore, neither the stability of the planet's energy balance nor, as a consequence, the limits of the issue are fully known. The role of urban areas, their proximity to vulnerable land or water masses and to strong stable air currents and the oceans (a prime but relatively slow sink for CO_2) and interaction(s) between the ozone and acid rain issues (Miller et al. 1978) remain to be resolved (Abelson 1987).

Summary

The practice of watershed management can readily be accomplished without further reference to these more global problems. However, not to be aware of them and the perspective in which they cast the entire watershed management field is irresponsible. Overall, the temporal and spatial scales of the "greenhouse effect" appear to be many orders of magnitude greater than the effects of most watershed management activities. Knowing the relative importance is crucial in presentation of management practices to the public and to basic understanding of the role and credibility of the hydrologist. The ultimate impact on the environment of extensive enlargement of the "ozone hole" would be to render the earth unliveable for mankind, in which case there would be no need to manage the water resource in any way at all. The importance of the "acid rain/deposition/runoff" issue to the hydrologist thus involves (1) relations between soil and subsurface runoff, (2) impacts of rehabilitation and preventive measures such as land and water treatment with lime or appropriate land use or vegetative cover type conversion, (3) the role of riparian vegetation, and (4) vegetation management to influence water yields, especially in high impact areas. The bottom line is that there may be significant effects on climate and water supplies.

PRACTICE OF WATERSHED MANAGEMENT

This section summarizes results of research and reported administrative experience with a wide variety of watershed management practices throughout the nation. It is intended to provide a comprehensive review of the topic but not a thorough analysis, e.g., for a full course in watershed management. Rather, it provides direction for further study as well as an up-to-date summary and introduction to the field.

There are three general objectives of watershed management activities. The first is **rehabilitation** of abandoned, abused, or even naturally altered lands that produce excess sediment, unwanted soluble materials in runoff, or excess or ill-timed runoff itself. The second objective is **protection** for normal and especially sensitive areas from activities that might lead to the need for rehabilitative measures. Historically, it makes sense to list these first two objectives in this order: the need for rehabilitation of abused lands was indeed noted prior to recognition of the opportunity for protection against or prevention of the abuses. The third objective is **enhancement** of the water resource characteristics. This is accomplished by manipulation of one or more of the watershed features that influence hydrology or water quality.

The three objectives only classify the goals of management; the applicable practices are unrestricted and, in fact, difficult to classify owing to the fact that a given practice may meet all or none of the broad goals at different times and places. The practices that protect, enhance, or rehabilitate the watershed are referred to generally as **best management practices** (BMPs), although they were not so designated when first identified in the 1940s; the term did not occur officially until the so-called "Clean Water Act" of the 1970s (see below). Note, however, that the BMPs cannot be standardized or dictated: they represent a wide range of technical opportunities that can be used through regular adherence to the process of their consideration, not detailed implementation.

Rehabilitation

The need for rehabilitation of abused lands was probably first noted in antiquity, but the relation to the water resource was not specifically made until this century. Widespread denudation of forested slopes for timber products, and widely scattered smelter activity denuded vegetated slopes with attendant increase in on-site erosion, off-site sedimentation, and downstream floods.

Most famous of the denuded areas is the Copper Basin, a 60,000-acre area of the watershed surrounding Ducktown in southeastern Tennessee. Smelter fumes are largely sulphur dioxide, which combines with water to form sulphuric acid, the denuding agent typical of natural or enhanced acid rain. The area was particularly sensitive to denudation owing to the high annual rainfall, on the order of 60 in. to 80 in. and the havoc this amount of relatively high-intensity precipitation wrought on the unprotected, highly erodible soils. This area was denuded in the two decades following 1850 and again after mining and smelting resumed after 1891 (Quinn 1988). Even though reforestation efforts commenced in 1936 (Kittredge 1948), the area can still be readily seen from a jet flying at 30,000 ft. The direct impact on streamflow from the Ducktown area is not known, but was one of the prompts behind locating the first Forest Service's major hydrologic research facility, the Coweeta Hydrologic Laboratory, a short distance to the east in western North Carolina.

One of the major research activities in the early years at Coweeta was to devise and evaluate means of rehabilitation through revegetation following typical southern Appalachian land use practices, including mountain farming, unrestricted logging, and woodland grazing. Thus, three watersheds at Coweeta were dedicated in 1940 to development of (1) a mountain farm, (2) a typical timber sale, and (3) grazing on a fenced watershed. These studies permitted documentation of the actual effects of these types of land uses for the first time, and generally continued for a period of about 10 yrs from 1940 to 1950, following which the areas were rehabilitated because (1) people were by then convinced that that type of land use was, indeed, abusive, and (2) of the demonstration value of the rehabilitation methods.

Methods on Forest Lands

Typical methods for rehabilitating abused forest lands follow well-recognized principles, such as (1) remove or restrict the cause of the disturbance, (2) establish and maintain vegetative cover on bare soils, (3) separate water from roads or skid trails, and (4) close unprotected roads and skid trails to further regular use.

The disturbance on forest lands, of course, is usually logging, which is of limited duration, thus the cause is automatically removed upon completion of the logging activity. Eastern forest lands are usually highly productive of vegetation and require little more than an opportunity to re-establish vegetative growth. This is apparent in the moisture index: forest lands are net annual producers of water ($P > E_t$) and the large amount of available water promotes good growth that quickly revegetates an area and promotes a habitat in which there is biotic activity that obliterates the effects of disturbances. Often, as a result of logging, openings in the forest stand and along roads and skid trails provide the very ingredients — space, nutrients, light, and water — necessary to reforestation. Thus, revegetation is not difficult on areas where there has been selective logging, or shelterwood cutting, or partial cuts or thinnings.[10] Most of the effort can be directed at the small percentage of the total area that is in roads and skid trails because this is the source of the greatest proportion of the sediment.

While western forest lands are generally not quite as lush as are those in the eastern United States, the basic principles for rehabilitation obtain there as well. Often, however, additional aid must be rendered in order to combat the lower growth rates by providing means of retaining infrequent moisture, capturing snowmelt, or paying more attention to aspect. Brush and species that are undesirable from the standpoint of commercial timber management may be quite satisfactory from the standpoint of watershed protection, but may also alter runoff water quality. Drastically different measures may be required to rehabilitate adjacent areas with different aspects, and steep slopes in mountainous municipal water supply watersheds need rapid revegetation measures following denudation by fire. This is particularly important in the southern California mountain ranges where denudation by fires, fed by the Santa Ana wind, and exacerbated by monsoonal rains that follow the fire season, require heroic efforts to preserve runoff values.

These observations do not necessarily apply to the Coast Redwood forest of California, where unusual conditions obtain. The brittle heartwood of the huge, older redwoods requires thorough protection from shattering when felled. That protection is normally provided either by creating a soft bed of young trees and brush or a series of earth ridges to receive the falling tree. The latter is the more common practice today, and is likely to preclude sprouting of the Coast Redwood, its primary means of reproduction. Unusually large equipment in use in the Redwood Region also demands some particularly unique practices in order to rehabilitate — and protect — soil and water relations.

Hand- or machine-scarification of abandoned roads and skid trails prepares the exposed, compacted soil for fertilizer and seeding, often followed with a mulching. The same practices may be used on landings. Roads need to be closed to vehicles because undirected and uncontrolled continued use can lead to deep rut development that produces excessive erosion and sedimentation. Large boulders, logs, ditches, or ridges may be successful in deterring the typical off-highway vehicles, but the newer all-terrain vehicles (ATV) strain the land manager's patience and ability to completely close such areas to all vehicular use. Similarly, persistent wheel ruts caused by irresponsible operation of ATVs and other 4-wheel drive vehicles on mountain meadows invite continued similar misuse, along with accelerated erosion and impaired runoff water quality.

Methods on Rangelands

It is particularly important to remove the cause of the disturbance that necessitated any rehabilitation efforts on natural rangelands. In most cases, the natural rangelands do not

---

[10] These harvesting systems are, by and large, exactly what they sound like, and are used according to demands of the silvical characteristics of the forest and the objectives of the landowner or manager.

receive enough annual precipitation to sustain tree growth ($P<E_t$) and, as a consequence, are not very productive of vegetation. Rangelands are thus likely to exhibit the effects of vegetation abuse for long periods of time. For example, wagon tracks from the westward expansion period in the mid-nineteenth century are still visible on the High Plains.

Also, in contrast to logging activity, the primary disturbance on rangelands, grazing, has no natural end and may continue for many years. This usually means that the rehabilitation program will take longer as well. On the other hand, different grazing intensity levels — all *less* than what would be considered overgrazing — "had no effect on streamflow characteristics" (Higgins, Tiedemann, Quigley, and Marx 1989). J. R. Thompson (1968) reported that seasonal changes in surface soil characteristics had a greater effect on infiltration than did grazing.

Distribution of animals in time and space is the primary means of controlling grazing for either rehabilitation or protection from abuse. The most common means of controlling grazing use are fencing, rotation grazing, and salt and water availability. A specific goal is to keep the grazing animals from congregating in sensitive zones, especially in or near wetlands, and on steep, erodible slopes near streams. To accomplish this, it is preferable to divert the water out of the stream and provide animal access to water at some distance from the source than to permit the animals direct access to the stream.

Methods on Agricultural Lands

In most cases, agricultural lands that are in need of rehabilitation have been abused or abandoned, or both. Commonly, as economic conditions favorable to farming activity decline, the poorest of the farms go out of production first. These are often the marginal farms, ones that exhibit low productivity, in part because they are steep slopes, nutrient-poor soils, and too hot or too dry or too cold or too wet. It is often difficult to rapidly rehabilitate these lands: the fact that they were low in agricultural production suggests that they will also be low in ability to produce weed crops that might readily take over the site and protect the watershed values of the land. Also, long years of farming may have completely removed weed seed sources, or roots and runners by which new plants might re-invade the site.

Specific techniques may involve fencing, addition of seed, organic matter, fertilizer, and water, and implementation of standard cropping systems (contour plowing, etc.) that permit perennial species to take over the site. Mulching may be an important practice that protects the soil from raindrop and wind erosion in addition to holding moisture and preventing dessication of the upper layers of the soil. Overgrazed pasture lands may be treated as are the rangelands (above). Where gullies have formed, headcut control, seeding, fertilizing, and tree planting may be appropriate. Sheet erosion may be controlled by establishing a cover crop on the land, a process that may be hastened by mulching, either with dead organic material or by green manure mulching, that is, leaving the bulk of the crop standing to protect the soil, as with corn stalks that otherwise might be harvested for silage.

Since agricultural use of land produces more frequent and immediate economic benefit, an extensive list of additional best management practices become available to the regulator, including incentives, taxation, permits, and imposition of legal penalties (Frere et al. 1977): these are also available for control of impacts of other land uses, but to a more limited degree. Incentives might include tax write-offs, cost-sharing benefits, or direct credit toward other environmental programs. Taxation could be imposed upon materials, e.g., fertilizer or pesticides, that stress the assimilative capacity of the environment. The same materials might be used only with permits from the appropriate regulatory agency, and legal penalties for compliance failure could be imposed. Frere et al. (1977) also suggest certification by a regulatory agency that particular guidelines or practices are being followed. This is the current approach to the control over new use of highly erodible lands and wetlands in agriculture.

Methods in the Stream Channel

Removing foreign material such as branches, trunks, and other logging debris is not recommended: preventative land use measures — including complete protection by closure to active uses — are preferred. In steep, mountainous terrain where **torrents** (steep channels that exhibit considerable turbulence and shooting flows) are common, debris is often ubiquitous, persistent, and plays an important role in the dissipation of the water's energy. The debris should, therefore, be left in place (Heede 1972; Heede 1985; Hibbert and Davis 1986; and Snyder, Haupt, and Belt 1975). "Minimizing hillslope failure" is a preferable means to managing stream debris than debris removal, according to Swanson and Lienkaemper (1978). Restoration of disturbed stream channels is occasionally warranted but, with good management, should not be necessary. Drainage may be diverted directly from a relatively impervious area (e.g., a road) through a natural filter area of undisturbed forest floor, for example, before releasing the water into the stream.

Small dams may be built in the channel with outlets to 1 percent to 3 percent sloped ditches that allow the water to be led back onto and into the soil for growing crops (Fig. 7-11). The small dams will also trap sediment and stabilize eroding channels.

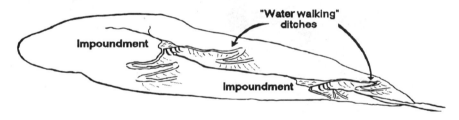

Figure 7-11 Water walking structure system

Upstream, beyond the live streams, in gullies and intermittent channels, techniques of erosion control include gully plugs, channel stabilization with rip-rap, revetments, and brush (Fig. 7-12). Increasing the infiltration capacity of the soils that provide runoff to gullies reduces the amount of flow the gully must convey and enables the small check dams and gully plugs to work. Methods should not reduce the cross-sectional area of the gully because a flood of the magnitude that "sized" the gully would overtop it and cause erosion of the banks. A first step in gully control is removing the excess flow that caused the gully to form in the first place.

Active gullies can be distinguished from those that are beginning to stabilize by the presence of steep, unvegetated banks, pedestalling (a column of soil protected by a pebble or small rock), and erosion pavements (a surface covered with material, e.g., pebbles, that cannot be moved by the surface runoff that removed the soil material).

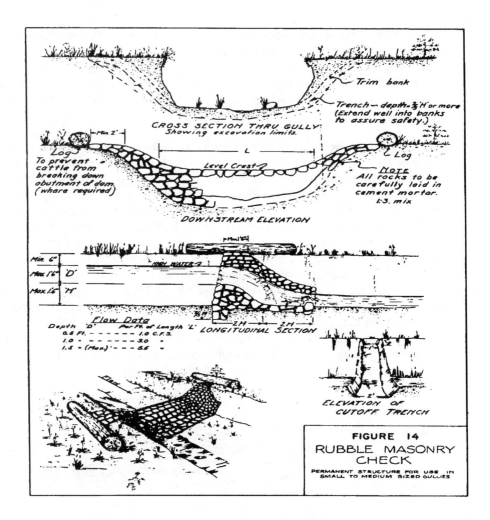

Figure 7-12 Check dam for gully control. These may be made out of concrete, brush, wire, masonry, or loose rock (Forest Service Manual).

Gullies that exhibit upstream extension by drainage net "fingers" and a relatively smooth longitudinal bottom profile are **continuous** gullies. They may be most effectively controlled by constructing a gully plug at the deepest point in the gully, since gully deepening occurs here; if there are tributaries to the continuous gully, however, all must be treated (Heede 1960).

Discontinuous gullies, on the other hand, may exhibit many head cuts, including one at the upstream limit of the gully that extends its length, each of which exhibits upstream migration and promotes gully growth, and a stepped longitudinal profile (Fig. 7-13). Discontinuous gullies are best controlled by constructing gully plugs at the site of each head cut, even if the level line from the downstream check is above the base of the next one upstream (Heede 1960). In all cases, gully control must focus on headcutting and downcutting:

head cuts should be sloped and seeded (if not incorporated into a gully plug), and water should be diverted from entering the gully to whatever extent is possible. Downcutting can be arrested by runoff control and structures. Active growth points may be readily identified by raw beds and banks, often vertical and without vegetation (Heede 1976). Two or more discontinuous gullies may fuse into a single continuous gully (Heede 1967). Heede (1980) has expanded the classification of gullies by including combinations of continuous and discontinuous gullies that are and are not fused into networks. Gully control and water management structures called "trincheras" — low earth dams — were in use in the southwestern United States and Mexico as early as 900 yrs ago (Dennis and Griffin 1971).

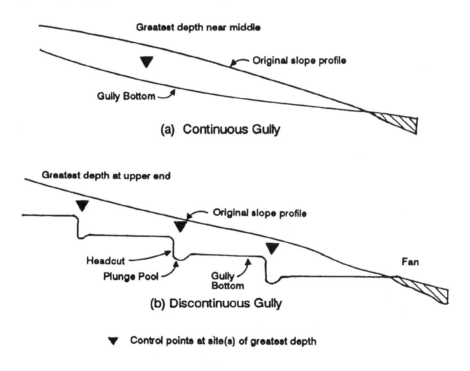

Figure 7-13 Longitudinal profiles of gullies

Protection

Recognition of the need for protection of wildlands from abusive land practices use grew out of the need for rehabilitation of abused lands. Exploiting that observation, the general goal of protection of wildlands from abusive land uses should be to preclude any need for rehabilitation. To accomplish that in the field requires understanding of the consequences of any particular practice. This is the foundation of successful proposed mitigative measures, including BMPs, in the field of hydrology identified during environmental impact analysis.

For example, restoration and protection of riparian areas merge with enhancement of watershed conditions: DeBano and Schmidt (1989) classify upstream and downstream riparian zones as being different in both their contribution to runoff behavior and the application of specific management practices, observing that "these areas must be considered in a watershed context." The upper riparian zones found in "higher elevation uplands" are part of the variable source area that contributes runoff to the drainage network; downstream zones flowing through "hot desert environments" often are net consumers of water. "The most obvious practices benefitting riparian areas are upstream treatments aimed at improving watershed condition, increasing duration of streamflow while moderating flood peaks, and stabilizing channels to reduce erosion" (DeBano and Schmidt 1989). Riparian areas are, of course, the most sensitive zones of most watersheds, and require particular attention; they also produce the greatest returns in terms of healthy watershed conditions for each management expenditure. As a consequence, certain types of watersheds and watershed zones require priority consideration regardless of land use.

Methods on Forest Lands

Since most of the sediment and altered hydrology from a logged area is going to emanate from the small percentage of the area that is in roads, skid trails, and landings, protection should focus on these areas, especially those that are naturally in proximity to water. Three basic principles are:

1. Keep the roads and skid trails out of the streams

2. Keep the water off the roads and skid trails

3. Plan in advance

Some of the BMPs that implement these principles include:

distribute logging shows widely, a practice designed to assure that no one large area will be vulnerable to exigencies of weather, disease, or insects and, as a consequence, no wide spread alteration to the hydrologic environment will occur;

not felling into or across streams, and *not skidding across streams*, designed to keep direct contact with the stream to a minimum;

leave a buffer strip along the stream, a practice that maintains natural stream temperatures, natural stream beds, and water quality;

cross the stream at right angles, a practice which minimizes the contact between water and exposed soil, and tends to preclude having the stream turn onto the road at times of high water;

use a culvert or bridge at stream crossings, to preclude vehicular entry into the water, so as to minimize erosion and petroleum products dripping into the stream;

log uphill with tree-length skidders or an overhead cable system, a practice that promotes dispersal at the landing or deck of water that collects in the skid trails, in contrast to downhill skidding which causes all skid trails to terminate their downslope, runoff-laden length at the same spot (the landing) which, owing to lack of vegetative cover, is thus especially vulnerable to erosion;

locate landings and roads well away from streams, a practice that advanced planning will facilitate, in order to maximize the length of the infiltration zone after water is turned off the road or skid trail;

turn water off the road frequently, at good filter locations and not onto fill, by water
bars, dips, or by gently rolling the grade.

Many of these BMPs are identified more particularly for specified vegetative and soil types, and may even be detailed and required in timber sale contracts. While formulas exist for the spacing of water bars, for example, there is no substitute for having (1) an alert and common-sense hydrologist and a well-educated heavy equipment operator on the site, and (2) good communication between the two, when the road, landing, and skid trail locations are laid out.

It has been frequently pointed out that most of the increase in runoff that occurs as a result of logging activity emanates from the large area from which timber is harvested and that most of the water quality damage is caused by the relatively small percentage of the total logged area that is occupied by roads and skid trails. Further, "a heavily used road segment in the field area contributes 130 times as much sediment as an abandoned road. A paved road segment, along which cut slopes and ditches are the only source of sediment, yields less than 1 percent as much sediment as a heavily used road with a gravel surface" (Reid and Dunne 1984). However, King and Tennyson (1984) showed that small changes in flow duration curves were caused by logging roads alone. Any change on the watershed may be detected in the stream, even if not statistically significant!

Gary (1982) reported that commercially developed campgrounds in forest environments "did not influence or impair" water quality when properly equipped and well-managed facilities were employed. Hornbeck and Reinhart (1964) recommended "timely completion of the [timber harvesting] operation" and observed that "in most respects, practices recommended for watershed protection also contribute to the overall efficiency of the logging operation." In a discussion on buffer strips, G. W. Brown (1970) pointed out that "strip configuration must vary with every stream." Application of the specific criteria for almost any BMP must be flexible, demanding that watershed protection — and rehabilitation and enhancement — are more art than science or simple numerical engineering.

Methods on Rangelands

By and large, any practice that is good for range management is also good for water resource management. This is due, in part, to the fact that the more desirable forage plants are also the "best watershed protectors" (Bailey and Connaughton 1936), and, in part, to the fact that maintaining a moderately sized herd on pasture will ensure both better forage growth and better water yields than either very heavy or very light grazing.[11] Also, infiltration may be higher on grazed lands (J. R. Thompson 1968), and rotating of water supplies can improve forage growth (Martin and Ward 1970; Sharp et al. 1964). As noted above, techniques for controlling grazing animal distribution on the range include fencing, rotation grazing, and salt and water availability. Control should be exercised in consideration of season of growth and reproduction for the particular grasses involved. Highly erodible areas, especially those near streams, should be fenced and grazing animals excluded. Rangeland infiltration may be improved by range pitting, terracing, or trenching, and with judicious use of a sheeps-foot roller. Increased water in the soil will, of course, aid vegetative growth that will both provide forage for the animals and protection for the soil (Sartz and Tolsted 1974).

---

[11] "Water for the West," a Forest Service, USDA film, ca 1953.

With the exception of the Ponderosa Pine and Longleaf Pine forest types, domestic grazing animals should be kept out of the woods (Johnson 1952). In the case of the southwestern Ponderosa Pine, the trees are sufficiently spaced to permit good grass growth between the stems and, as a consequence, the animals will tend to stay with the grass and not under the trees where they might severely compact the soil, thereby restricting tree growth.

Longleaf pine goes through a "grass stage," where the tree seedling may remain for several years before being able to break away from the surrounding grasses and enter the sapling stage. A light fire that kills the grass, or grazing, removes or reduces the space-competing grass and releases the trees, which may suffer some usually minor, direct trampling damage.

One of the lessons learned at the Coweeta Hydrologic Laboratory's grazing watershed was that the water, timber, and animal values would be better served by clear-cutting, fencing, and controlled grazing than by maintaining cattle in the woods where compaction decreased height and diameter growth, and directly damaged reproduction (Johnson 1952).

Fire

Control of fire has a been a long-time concern of wildland managers. Early efforts were directed at putting out fires, then aimed at fire prevention. More recently, practical ecologists have subscribed to the idea that fire is a natural part of both the forest and range environments and, in order to preclude buildup of catastrophic levels of natural fuels, fires should be allowed to burn when they occur naturally, as they frequently do by lightning. For certain species, forest stands appear to be better fuel sources after a certain age; thus it is important to have thorough knowledge of the role of fire, if any, in the ecology of the vegetative type under management.

Fire may also be used to control vegetation and vegetative development. Removal of vegetation usually increases water yield, but the use of fire for this purpose is risky and usually has only temporary effects. In sum, the bad effects of fire, especially on water quality, far exceed the beneficial effects of any increased quantity that may occur. Although fire may occur without water quality damage, especially if there is (other) soil disturbance, the use of fire as a management tool for sensitive water supplies is a risky operation.

The extent of fire damage will vary with:

1. *Severity of the burn* — hotter fires destroy more organic material.

2. *Extent of the burn* — a wider area that is burned may exhibit unique weather and even climatic patterns that are sufficiently drastic to adversely impact future vegetative growth and soil protection.

3. *Season of the burn* — different species are sensitive to extreme temperatures at different stages of growth.

4. *Soil type* — humus type especially responds to being burned and may be drastically altered by fire, further affecting vegetative reproduction.

5. *Topography* — the interplay between solar radiation, slope, aspect, and vegetative cover can play a major and controlling role.

6. *Climate and weather conditions* — these affect the severity and extent of the burn.

Prescribed fire, in the absence of other concurrent disturbance such as timber harvest operations, had "limited effect on soils, nutrient cycling, [or] hydrologic systems" in a coastal plain pine forest (Richter, Ralston, and Harms 1982). If such is the case on the vulnerable, sandy soils of that type, similar results are to be expected, and have indeed been reported elsewhere (McColl and Grigal 1975; Tiedemann, Helvey, and Anderson 1978; Feller and Kimmins 1984). In general, increases in nutrient concentration in runoff water were noted, but neither profoundly impacted receiving hydrologic systems nor persisted for more than a few years. Particularly sensitive municipal watersheds, on the other hand, may need full protection from fire and, if burned, need immediate and often expensive rehabilitation. Immediate soil protection in the form of mulching, contouring, etc., may be necessary in addition to re-seeding with annual or perennial grasses and/or with trees.

Insects and Disease

Defoliation by insects and diseases is a common natural or man-caused occurrence, especially in relatively pure stands of vegetation. Increased water yields in the White and Yampa River basins of from 30 mm to 53 mm per yr from defoliated forest stands is documented (Bethlahmy 1974) in the Colorado Rockies where an Engelmann Spruce/Subalpine Fir forest was damaged by wind and then infested by bark beetles: the effects on runoff persisted for 25 yrs (Bethlahmy 1975). More frequently, increases in water yield are low and more likely to impact seasonal patterns and water quality than annual yields (Lee 1980). However, Potts (1984) reported on the 15 percent increase in annual runoff, a two- to three-week advance in the annual peak flow, a 10 percent increase in low flows, and only slight increase in the magnitude of the peak flow from a 51.5 mi^2 watershed in southwestern Montana that suffered a 35 percent reduction in timber volume by insects; and similar results were reported by logging following a Pine Beetle infestation in British Columbia (Cheng 1989).

Protection against insect and disease outbreaks may be encouraged by diversification of planting, treatment, and land use activities. Preventative spraying on low-value forest lands is not economically feasible from the forest stand value standpoint, but often is done using watershed values as an added justification.

Enhancement

The enhancement goal of watershed management is based upon the recognition that the natural soil-vegetation system is not necessarily the optimum producer of water. Dependent upon the goals of management of the particular piece of land, water yield, regimen, or water quality may be altered. Most of the current enhancement practices and programs have been in the name of water quantity and regimen, whereas most of the protection and rehabilitation practices and programs have been aimed at controlling or restoring water quality.

As early as 1946, the Forest Service (1946) suggested that:

> In some forests, the landowner might obtain maximum returns from timber by cutting out the hardwoods, such as oak, hickory, birch, or maple, and encouraging the faster-growing or more salable softwoods, like pine or spruce. On the other hand, the protection of downstream farms and communities from floods might be

better attained by maintaining a dense stand of mixed hardwoods, which produce the best humus and improve the porosity and storage capacity of the soil.

The emphasis was clearly on protection, yet the Forest Service already had established sophisticated research installations for evaluation of the effects of vegetation conversion and land use changes on water yield.

Vegetative Cutting and Conversion

Already described in Chapter 3 are the original cutting studies at the Coweeta Hydrologic Laboratory in North Carolina. These early studies, of which the one on Watershed 17 was a part, were designed to find out how much water would be released — presumably not evapotranspired — by simple removal of the existing hardwood forest. The experimental cuts were set up as a 100 percent clearcut with annual cutting of regrowth; 100 percent clearcut with natural regrowth; 50 percent[12] of the overstory cut in strips perpendicular to the main axis of the stream (generally east and west across the watershed); and 20 percent of the vegetation, the understory Mountain Laurel (Azalea) and Rhododendron cut. In each case, the vegetation was left where it was cut and not harvested in order to (1) retain the before-cut intercepting surfaces for the first year, and (2) preclude disrupting the natural hydrologic processes at the air-soil interface. A tremendous body of literature on the impact on streamflow of forest cutting has built up over the 50 yrs since the first Coweeta studies (Sopper and Lull 1967; Sopper and Corbett 1975; Leaf 1979; and Gaskin, Douglass, and Swank, 1983). Most of this literature is based upon studies that were undertaken under the best of watershed research conditions, including paired watersheds, sophisticated streamflow and meteorological instrumentation, and continuous soil moisture monitoring. Some studies have also been reported based upon "field data," that is, where long-term (usually USGS) streamflow records are clearly and solely applicable to a particular change in land use on the watershed over time.

Increases in water yield from experimental watersheds of up to 60 percent in rainfall country (Hibbert 1969) and about 24 percent in snow zones (Troendle 1983) have been reported. About one third of the runoff increase is attributed to redistribution and altered melt patterns of the snowpack and two thirds to the reduction in evapotranspiration (Troendle and Leaf 1981).

Regardless of the source of runoff, these increases are from the areas on which forest cutting has taken place; if these areas are considered (as they should be) as a small portion of the larger drainage of which they are a part that is under management, then the percentage increase in flow from each cut area must be reduced by the percentage of the total area in which active harvesting takes place. The reason for this is that the part of the watershed that produces the increases depends upon certain characteristics of those adjacent portions of the watershed that are currently "idle." Those lands are growing a new crop of timber, may shade the newly cut area, and may in fact be producing time-reduced yield increases from an earlier cutting. Thus, by and large, increases in streamflow from managed forested watersheds are not expected to exceed 2 or 3 percent, rarely reaching as high as 6 percent (Kattelmann et al.

---

[12] The percentage applied to the basal area of the forest stand, that is, the sum of the cross-sectional area of all stems 0.5-in. dbh (diameter at breast-height, or 4.5 ft above the ground). Determination of basal area was made by 100 percent cruise of each watershed.

1983). While such increases in streamflow do not of themselves warrant expensive management systems, the fact that timber production provides revenues and can be accomplished without water quality impairment creates an economically and ecologically attractive package. As noted below, there are also some ownership problems.

In most reported field studies, streamflow was found to increase when vegetation was cut, and the increase decreased over time as the vegetation or its replacement grew back. An exception was noted for conditions where the regrowth favored snow deposition, pack retention, and delayed melt following farm abandonment in New York (Black 1968). In another study of long-term streamflow records in the southeast, increases in forest lands of from 10 percent to 28 percent within populated stream basins reduced streamflow by 4 percent to 21 percent (Trimble and Weirich 1987).

Vegetative conversions were the subject of some of the research at Coweeta: north- and south-facing watersheds were converted from hardwood to pine, with considerable reductions in both monthly and annual streamflow (Douglass and Swank 1975). An early study at Coweeta involved the simple removal of vegetation from the riparian zone on watershed 6 (which was later completely converted to grass): the primary effect was to eliminate the diurnal fluctuation in streamflow that appeared to be associated with the evapotranspiration draft of streamside vegetation. The study also suggested that the water available for runoff from Coweeta watersheds was uniformly distributed over the watersheds, since the increase in runoff was proportionate to the percent of the watershed area cut (and to the well-distributed basal area), a result generally found in the other Coweeta watersheds as well.

The substantial reduction in streamflow accompanying the hardwood-to-pine conversion presumably was caused by a combination of sustained high interception loss, reduced evaporation from the shaded forest floor, and year-round transpiration. In contrast, and as expected with the reduced interception and decreased rooting depth in the deep soils of Coweeta, conversion of the 22-acre, north-facing watershed number 6 to grass produced a maximum of 5.8 in. of increased runoff per year in the fourth year following conversion (Douglass and Swank 1975).

These and other research conversion studies have not been verified by reversing the vegetative conversion. Nor can they, for the native vegetation presumably grows on the site owing to the water (and nutrients, light, and space) available: if one tries to plant hardwoods on a natural grass watershed (as was done in the shelterbelt plantings on the Great Plains), the results are not necessarily going to be the mirror image of those obtained by planting grass on a natural hardwood watershed. In addition, many of the forest treatment and vegetative conversion experiments cannot be positively ascertained owing to large experimental error (Hibbert 1967), lack of replication, and lack of control.

For example, research on a pair of 200+ sq mi watersheds in Arizona, where one had 38 percent of its pinyon pine-juniper cover eradicated, showed no statistically significant differences in runoff relations for the water year, summer, or winter periods; however, the precipitation-runoff relationship was shown to be different for the modified watershed, a result that could be attributed to a documented change in total annual precipitation (Collins, and Myrick 1966). Nevertheless, conversion of various vegetative types to grass is operational in the southwest. DeBano (1977) reported that "chaparral and ponderosa pine provide the largest potential for increasing water yield," and research on the Three Bar watersheds in

Arizona showed increased runoff, reduced fire danger, and better habitat for wildlife as well as increased forage when chaparral was converted to grass (Hibbert, Davis, and Scholl 1974). Unfortunately, runoff water also showed moderate to low contamination with the herbicide that was used to effect the conversion. Davis (1987) reported that large increases in nitrate produced by chaparral conversion were somewhat balanced by decreases in bicarbonate, suggesting some other complex reactions are taking place in the aquatic system. Hibbert pointed out that with the normal low annual precipitation, the increase in runoff is not great but that the efficiency of the practice increases as annual precipitation increases: "summer rainfall normally does not recharge the soil mantle and consequently contributes little to streamflow. Evapotranspiration decreases after conversion, particularly during the early growing season. The soil mantle remains wetter, and when recharge occurs, outflow is heavier and lasts longer than it does under brush" (Hibbert 1971). While increases of up to 15 percent have been reported for Big Sagebrush rangelands, concurrent increases in sediment do not appear to justify conversion for increased streamflow: the interaction of the brushy plant and wind, over-snow runoff following diurnal melting to which the vegetative type is sensitive, and observations with snow fences suggest that direct fencing and patterns of residual brush may be more productive management tools in this vegetative type (Sturges 1975).

One summary of the potential water yield increases that might be obtained from the broad regional forest types in the conterminus United States (Fig. 7-14) is shown in Table 7-3.

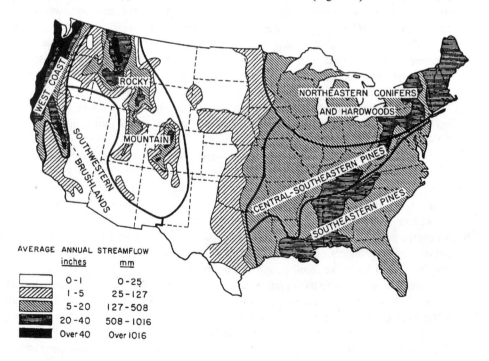

Figure 7-14 Forest types in the United States (after Anderson, Hoover, and Reinhart 1976)

Owing to less demand for major water yield increases in the eastern states and the attendent difficulty with maintenance of water quality, potential increases are not shown for the eastern states (Anderson, Hoover, and Reinhart 1976), although as noted in the discussion of the Coweeta data, the amounts may be substantial. Perhaps more important, the primary means of harvesting in the east would likely be partial cuts, not the clearcut areas that dramatically enhance the amount and distribution of snow captured and the amount of water available for runoff.

Table 7-3 Hydrologic Characteristics of Major Forest Types in the United States

Region and Forest Type	Total area (thousand acres)	Mean Annual... Precipitation (in.)	Mean Annual... Streamflow (in.)	Mean Annual... Evapotranspiration (in.)	Potential Increase in Yield (in.)
Western United States					
West Coast					
Mixed conifers	8,470	44	22	22	4.5
True fir	6,150	60	36	24	6.0
Douglasfir, hemlock, redwood	25,570	75	45	30	15.0
Rocky Mountain					
Lodgepole pine	14,470	33	14	19	3.0
Englemann spruce, fir	7,400	33	18	15	3.0
White pine, larch, fir	6,900	42	20	22	4.5
Aspen	4,000	33	10	23	3.0
Ponderosa pine	34,200	21	4	17	0.5
Douglasfir	9,000	28	7	21	1.0
Southwest brushland					
Southern California chaparral	7,500	25	5	20	1.0
California woodland grass	9,000	25	7	18	1.0
Arizona chaparral	5,500	19	1.5	17.5	0.5
Eastern United States					
Northeastern Conifers and Hardwoods					
New England, New York, Pennsylvania					
White-red-jack pines	5,054	42	22	20	-
Maple-birch-beech	18,665	39	22	17	-
Oak-hickory	-	-	-	-	-
Michigan, Wisconsin, Minnesota					
White-red-jack pines	4,435	29	10	19	-
Maple-birch-beech	9,630	30	11	19	-
Oak-hickory	6,170	30	10	20	-
Aspen	17,882	28	10	18	-
Central and Southeastern Hardwoods					
Northwestern portion					
Maple-birch-beech	3,416	44	19	25	-
Oak-hickory	61,051	40	16	24	-
Oak-gum-cypress	10,919	49	17	32	-
Southeastern portion					
Oak-hickory	25,776	49	20	29	-
Oak-gum-cypress	25,884	50	13	37	-
Southeastern Pines					
Southeastern area					
Loblolly-shortleaf pine	52,008	49	16	33	-
Longleaf-slash pine	25,967	52	14	38	-

Source: (Anderson, Hoover, and Reinhart 1976).

In the western states, the extensive area of Douglasfir, Hemlock, and Redwood show great promise for increasing runoff from managed forests (especially the Douglasfir type), while the especially large area of Ponderosa pine illustrates a very low potential for increased water yield, reflected in the characteristically wide spacing of trees. In the east, on the other hand, the types with the highest evapotranspiration either grow on highly erodible soils that would be vulnerable to the exposure brought by clear-cutting (the southeastern area), are too low in elevation to contribute significantly to runoff (the southeastern portion of the Central and Southeastern Hardwoods), or are not silviculturally appropriate for clear-cutting (the northwestern portion).

In addition to considerable research into how to improve the natural water yield, regimen, and/or quality of water, numerous management operations have been established in order to effect desired changes. The most successful and longest operations occur where (1) there is a high demand for the product or, put more succinctly, where there is sufficient money to pay for improving the water resource in spite of the fact that the improvements often do not directly repay the costs, or (2) there is a great potential for improving the water resource, either owing to poor quality, regimen, or low yield. The two best examples are (1) where municipalities manage and may even own nearby forested watersheds, and (2) in the subalpine zone where the opportunity exists for dramatic increases in runoff from snowpack normally protected by commercial forest cover. In some cases, both conditions apply.

In a 1983 compilation of a series of articles on the practical potential for yield augmentation, Ponce (1983) observed that "the potential for augmenting water yield by manipulating vegetation within a basin has intrigued hydrologists for decades." These papers are summarized as follows:

TITLE	ABSTRACT
The Potential for Water Yield Augmentation from Forest Management in the Eastern United States	"Generally high rainfall and extensive forests in the East combine to produce excellent potential for managing forests for increased water yield. . . . [C]umulative water yield increases can be predicted within about 14 percent of the actual value. However, because of the diverse land ownership patterns and the economic objectives of owners, realizing the potential will be difficult at best" (Douglass 1983).
The Potential for Water Yield Augmentation from Forest Management in the Rocky Mountain Region	"With the exception of the Sierra-Cascade mountain ranges, the Rocky Mountain chain is the only portion of the western United States that consistently yields more than 3 cm of flow annually. Ten to fifteen percent of the land mass in the region produces the majority of the total flow. . . . The optimal harvest design appears to consist of small openings, irregularly shaped, and about 3 to 8 tree heights in width parallel with the wind" (Troendle 1983).
The Potential for Water Yield Augmentation from Forest Management on Western Rangelands	"Increasing water for onsite and offsite uses can be a viable objective for management of certain western rangelands. One approach utilizes water harvesting techniques to increase surface runoff by preventing or slowing infiltration . . . . An attractive alternative, where applicable, is to replace vegetation that uses much water with plants that use less so that more water percolates through the soil to streams and ground water. . . . probably less than 1 percent of the western rangelands can be managed for this purpose. However, where annual precipitation exceeds about 450 mm (18 in.) and deep-rooted shrubs can be replaced by shallow-rooted grasses, there is potential to increase streamflows and to improve forage for livestock. Little or no increase can be expected by eradication of low-density brush and pinyon-juniper woodlands" (Hibbert 1983).

The Potential for Augmenting Water Through Forest Practices in Western Washington and Western Oregon

"Highest demand for water coincides with lowest streamflow levels between July 1 and September 30 when less than 5 percent of annual yield occurs. . . . Most of the increase has occurred during the fall-winter rainy season. Estimated sustained increases in water yield from most large watersheds subject to sustained yield forest management are at best only 3-6 percent of unaugmented flows. . . . [W]ater yield augmentation is limited by its size and by its occurrence relative to the time of water demand" (Harr 1983).

The Potential for Increasing Streamflow from Sierra Nevada Watersheds

"The Sierra Nevada produces over 50 percent of California's water. . . . Vegetation and snowpack management can increase runoff from small watersheds by reducing losses due to evapotranspiration, snow interception by canopy, and snow evaporation. Small clearcuts or group selection cuts creating openings less than half a hectare, with the narrow dimension from south to north, appear to be ideal for both increasing and delaying water delivery in the red fir-lodgepole pine and mixed-conifer types of the Sierra west slope. Such openings can have up to 40 percent more snow-water-equivalent than does uncut forest. However, the water yield increase drops to ½ percent to 2 percent of current yield for an entire management unit . . . . As a rough forecast, water production from National Forest land in the Sierra Nevada can probably be increased by about 1 percent (0.6 cm) under intensive forest management. . . . [D]elaying streamflow is perhaps the greatest contribution" (Kattelmann, Berg, and Rector 1983).

Overall, the potential for increased runoff from forest lands is an increasingly attractive management option. Maximum yields can be obtained by clear-cutting and paving, as has often been done within the orographic precipitation area of tropical islands. Short of that extreme, care must be taken to ensure hydrologic integrity: aquatic ecology of wetlands and the stream system, water quality, surface and streambank erosion, the flow duration curve, the flood peak size and distribution, and other wildland values. In practice, however, and with the exception of watersheds in the snow zones of the Rocky Mountains and Sierra Nevada, the small percent increase in runoff is rarely considered worth the risk of reduced water quality.

Complete consideration of the decision to harvest forest vegetation for increased water yield must include analysis of the costs, both positive and negative (Krutilla, Bowes, and Sherman 1983): they include alternative costs of water, delays in reproduction induced by special harvesting requirements, improved access that often accompanies more careful planning and logging operations, and both on-site and off-site benefits. Finally, increasing pressure for high-quality water from wildland watersheds will undoubtedly result in solutions to current legal and economic problems and more situations where wildlands can be successfully managed for both timber and water (Ponce and Meiman 1983).

Snowpack Management

Enhancement of water yield from snowpacks may be effectively accomplished by manipulation of natural vegetation and by artificial means in the subalpine and alpine zones, respectively. The gradual disclosure of information on the extent and factors involved was the result of a rather orderly program of research, mostly out of the Agricultural Research Service, the Corps of Engineers, the Forest Service, the Geological Survey, and the Soil Conservation Service.

Modification of the runoff from snowpack in the subalpine zone was the consequence of the first forest cutting studies in the United States. The studies were performed at Wagon

Wheel Gap in the Rio Grande drainage in Colorado around the turn of the century (Bates and Henry 1928) and were originally designed to determine the impact of forest cutting on precipitation and, presumably, climate. The impact of cutting the predominantly aspen forest was on runoff, not on precipitation, as expected. A change in the runoff monitoring control system at the same time as the vegetative cut caused poor calibration of the experimental watersheds and precluded application of the specific results.

The basic principle that cutting forest vegetation increased runoff was established, however, and, in order to evaluate it further, and upon establishment of the Fraser Experimental Forest, various patterns of clear cuts were made in Lodgepole Pine stands on the St. Louis Creek watershed (tributary to the Fraser and Colorado Rivers) just west of the Continental Divide on the Arapaho National Forest. Differing degrees of cut, timber stand improvement work, and uncut plots comprised an extensive study that showed the effects of forest cutting on water available for runoff from the snowpack (Wilm and Dunford 1948). Following this intensive study, the entire Fool Creek watershed (adjacent to the earlier cutting studies) was logged in a series of cut strips of varying width, designed to determine the impact of forest cutting on *streamflow* (not just the amount of water *available* for runoff). Subsequently, other research installations throughout the United States launched studies into how streamflow might be increased by planned timber harvesting. In consideration of the high amount of annual snowfall, careful forest management in the snowfed subalpine zone was and is considered likely to be the most fruitful area in which to practice water yield enhancement by manipulation of forest stands.

"There is now ample evidence that both small patch cuts and thinning increase snow accumulation in the treated area. These effects extend well beyond the approximately fifty yrs of record at the Fraser Experimental Forest. The processes responsible are not yet precisely defined, but reduced snow interception losses appear to be a major factor" (Meiman 1987). Snowpack conditions in the southern Rocky Mountains of Arizona and New Mexico exhibit greater annual variability than further north ye have shown increases in runoff of from as little as 7 to as much as 111 percent depending upon the moisture content of the soil mantle (Ffolliott, Gottfried, and Baker 1989).

Conversely, reforestation was shown to decrease runoff from a small forested watershed in central New York State: between 1932 and 1958, from 35 percent to 58 percent of three areas was reforested with pine and spruce and compared with a control. All three showed reductions in runoff during the dormant and growing season and for the year. Average streamflow reductions from Shackham Brook were as high as 66 percent for November, 16 percent for April. "The reductions in peak discharges during the dormant period are attributed largely to increased interception and sublimation of snowfall, and a gradual desynchronization of snowmelt runoff from the wooded and open areas of partly reforested watersheds" (Schneider and Ayer 1961). With the widespread shifts in agriculture that occurred during the last two centuries, such as that in the northeast, long-lasting changes in runoff and stream behavior can and have been documented.

Working in the Sierra Nevada, H. W. Anderson (1956) showed a "high degree of association" between forest cover variables and snowpack accumulation and melt to the water equivalent on April 1 and subsequent melting degree-days (above $35^{\circ}F$). Subsequent suggestions on how management might be done to increase runoff from such stands included the recommendation that forest openings be 1 to 2 tree heights across and L-shaped on east and west aspects (Anderson, Rice, and West 1958). Simultaneous reporting of results from the 1.11-sq mi Fool Creek installation indicated that the increase in runoff was in the "spring freshet period and was associated with an earlier and more rapid rise of the stream and high annual peak rates of discharge" (Goodell 1958). The prodigious work *Factors Affecting Snowmelt and Streamflow* was released the following year (Garstka, Love, Goodell, and Bertle 1958). Conclusions from that extensive instrumentation of the entire 32.8-sq mi St. Louis Creek watershed (which includes the Fool Creek drainage) included principles of snowmelt runoff behavior, detailed methodologies for snowpack melt/accumulation analysis, runoff prediction, and instrumentation methodologies.

Meanwhile, in the Sierra Nevada, Anderson (1960) provided a detailed report on which factors influenced the increased runoff from snowpack manipulation by timber harvesting. These included slope, curvature of the slope, aspect, elevation, exposure, and degree of forest cover (Fig. 7-15). A variety of different areal patterns of cut were also evaluated: savings (reduction of evaporative losses) were greatest in the strip cuts, followed by clearcut blocks, and least in selection cuts. Brush removal was also shown to be effective in reducing evapotranspiration losses as evidenced by increased runoff. On the Onion Creek Experimental Forest at the Cental Sierra Snow Lab, the 2-yr average reduction in soil moisture loss was 3.6 in. under strip cutting, 3.0 in. under block cutting, and 0.8 in. under commercial selection cuts; corresponding increases in the approximately 46-in. annual runoff were 8.6 in. (18.7 percent), 6.3 in. (13.7 percent), and 3.4 in. (7.4 percent), respectively (Anderson and Gleason 1960). Simultaneously, for Adirondack forest stands, it was reported that "the snowpack in hardwood sapling and sawtimber stands contained two more inches of water [about 25 percent] than the snowpack in balsam fir and red spruce stands. In hardwood stands, snow melted about twice as fast as in conifer stands, and snow disappeared about 9 days earlier" (Lull and Rushmore 1960), suggesting that timber harvesting and associated reforestation with less tolerant and more open stands might show increased runoff results in the east as well. Eschner and Satterlund (1963) reported on the influence of several different forest types, brush, and stand condition in the Allegheny Plateau on snow deposition and melt, showing that the greatest accumulation was in brushy hardwoods and least under dense Norway Spruce. Satterlund and Eschner (1965) also reported that streamflow regimen differences between agricultural and forested watersheds were detectable using the half-flow interval.

Martinelli (1965) reported on measuring runoff from the highly productive alpine snow zone, suggesting that some enhancement practices such as the judicious placement of snow fences might be feasible there. He reported the results of such practices (Martinelli 1964) in both the alpine and subalpine zones, pointing out that "relatively few storms account for most of the snow that accumulates under both natural and fenced conditions. Between 50 percent and 65 percent of the annual accumulation took place in the five biggest storm periods, and between 25 percent and 40 percent during the two biggest storm periods each year for the past 4 years." Hart (1966) reported preliminary results from the Hubbard Brook Experimental Forest in New Hampshire that were consistent with those already cited, speculating that perhaps 40 percent of the increased streamflow might become available during the summer months when streamflow was most needed. Clearly, the basic principles of how forest cutting influenced the amount of water available for runoff were becoming widely established. Hoover (1969) pointed out that:

> Sometimes wood-producing practices will benefit water yield, and the increased flow would require no charge to water users. Where maximum water yield is important, practices desirable for water flow may conflict with maximum wood production. In these circumstances, the general public or those directly receiving the water benefit will have to pay the costs .... Knowledge is now available to bracket the upper and lower limits of water yield increases possible from forest land management. Much research is needed to fully understand the processes involved, but that need not delay application of what is known.

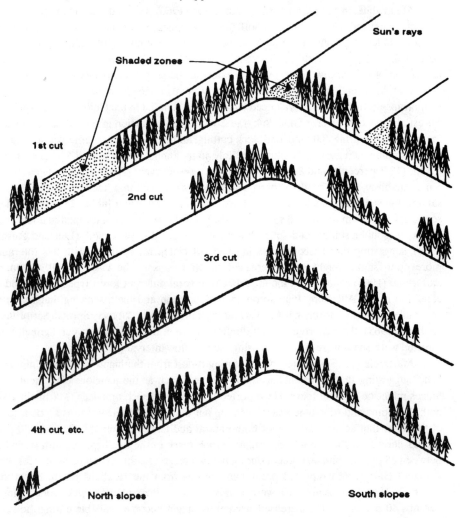

Figure 7-15 Diagrammatic representation of wall-and-step forest in area rotation (from Anderson 1969)

That is exactly what happened: high-altitude, subalpine zones that were slated for timber harvesting operations were managed for both timber and increased water yields. Some of the continuing research was carried on at the experimental forests already identified, and some in fully operational forest management and other areas. Factors affecting accumulation and snowmelt include the factors that affect evapotranspiration (Chapter (4), but the relative importance of several of them change. A few are added, too such as albedo of the snow surface, advection of warm, moist air over a snowpack (for condensation), advection of warm, dry air over a snowpack (for sublimation or melt), sequence of meteorological events (e.g., wind to blow intercepted snow off tree crowns after a storm), density of pack, which varies from time to time and place to place, exhibiting different degrees of uniformity in the open and under forest cover, and pattern of melt (D. E. Miller 1977).

For example, contour trenches were found to affect snow accumulation on southwest slopes, and the slight increase in water yield was estimated to be of greater importance to revegetation than to water yield (Doty 1970).

Detailed analysis of snow accumulation and meteorological, topographic, and forest variables in the Central Sierra Nevada were reported by H. W. Anderson (1967): he emphasized the inherent complexity and analytical sophistication of the situation thus:

> The forest processes acting during snow accumulation have always been better described than evaluated. The goal of separating the effect of the forest on interception, shading, shelter from winds, and back radiation has often been frustrated by the fact that the same trees and even the same part of a single tree can be acting on several different processes almost simultaneously. By using reduced rank models, evaluated by principal component analyis, we have some prospects of separating these natural processes in analyses.
>
> Some years ago [Anderson 1956] a wall-and-step pattern of forest cutting was suggested to make the most effective use of forest shade versus back radiation in snow storage. Later [Anderson 1963] the pattern was elaborated upon to include trapping of cold air. The present analyses tend to confirm earlier findings, but also to add further quantitative results for predicting the effects of shade, back radiation, interception, and shelter on different slopes. Such analysis can lead to more effective designs of forests to maximize snow accumulation.
>
> The results sugest that if we uniformly remove trees within a dense forest so as to reduce the overhead canopy by 60 percent, the resulting reduction in shading will be only 10 percent with a net increase in snow accumulation of 40 mm (1.5 in.). On the other hand, if we cut 60 percent of a continuous forest into openings, leaving 40 percent "margins" [the areas around the openings, two-thirds the extent of the opening], we change the average snow accumulation in dense forest of 310 mm (12.1 in.) to 40 percent margin with 400 mm and 60 percent opening with 470 mm. This forest-opening combination will have 440 mm (17.4 in.) of snow water equivalent, that is, 40 percent more snow than the dense forest, or 30 percent more than a forest of average density. The most effective forest cutting will be on north slopes, being two times as effective as on south slopes. The major disadvantage on south slopes arose from back radiation, which may be minimized by a wall-and-step cutting pattern and/or cutting of narrow strips which will shade the back wall of trees.

Other concurrent and subsequent studies revealed (1) how snow intercepted by a forest canopy was redistributed (D. E. Miller 1966); (2) the effect of forest cutting on streamflow in the northeast (Hornbeck et al. 1970), who noted that "elimination of transpiration by forest clearing causes substantial increases in water yield .... It is important to note that ... the greatest absolute increases came in wet months, and the greatest relative increases came in dry months," and (3) relationships between opening characteristics, wind movement, and snow accumulation in Idaho (Haupt 1972; Gary 1975).

In 1975, the Forest Service issued a series of reports that summarized the state of knowledge with regard to water yield enhancement in the alpine zone (Martinelli 1975), and the subalpine zone (Leaf 1975a; Leaf 1975b; and Leaf and Alexander 1975). In the alpine, snowfencing is presented as the most effective means of manipulating the snowpack (Martinelli 1967), although terrain modification, intentional avalanching, and glacier building are presented as additional, attractive and perhaps feasible practices, verified hydrologically and economically by Tabler and Sturges (1986). Increased yields from the subalpine can be substantial, and depend primarily on the *pattern* of the forest opening. Increased water yields are compatible with sound forest management practices and are ecologically acceptable as well. Computer simulation of snowmelt in a subalpine forest also became available (Leaf and Brink 1973), and a variety of studies reported effects of extending the application of snow management practices to other forest types such as Western White Pine (Packer 1971; Satterlund and Haupt 1972; and Haupt 1979a and 1979b); Sagebrush (Sturges 1977); Lodgepole Pine (Gary 1980); the northeast (Eschner 1980); Engelmann Spruce, Subalpine Fir, and Lodgepole Pine in Alberta (Golding and Swanson 1986); the Lake States (Verry 1986); Douglasfir (Berris and Harr 1987); on other watersheds in the subalpine zone (Troendle 1987; Troendle and King 1987), and even southward into the high elevation Ponderosa Pine forests of the Salt River watershed in Arizona (Ffolliott, Gottfried, and Baker 1989).

A 1980 paper on the Fool Creek studies brought those results up to date, including the observation that the pattern of recovery was not as predictable as the watershed manager might desire (Troendle and Leaf 1981) and, more recently, attention has turned to the natural chemical characteristics of snowmelt (Siegel 1981) and how they are impacted by timber harvesting. Considerable research has been directed at focused technical topics regarding snow: measurement, data collection, analytical techniques, and models (for accumulation, melt, and runoff). A major difference between eastern and western snowpack conditions is that, in general, soil moisture is fully recharged in the east by the time the snowpack develops, thus melt runoff can go directly to runoff (or ground water recharge) whereas, in the west, the melt runoff water goes first to recharge the soil.

Remote sensing of snowpack development, areal extent, and melt patterns have been the subject of research for more than two decades (Bowley and Barnes 1979). Interest has been focused on using various wavelengths, especially microwaves, which permit data acquisition even when clouds obscure the snowpack (Rango 1979). Evaluate the amount of snow water equivalent can be from airborne or satellite imagery alone and in combinations (Yates et al. 1986; Carroll and Carroll 1989). The amount of water in the near-surface layers of soil can now be accomplished by satellite imagery (Ulaby, Bradley, and Dobson 1979), even under vegetative cover (Martinec 1982) and in the snowpack itself (Chang et al. 1977).

Since annual snowmelt runoff is related (dependent upon geographic area) to water equivalent on specific dates, e.g., April 1, ground-determination of water equivalent on the date may be combined with areal extent of snowpack as determined by satellite imagery to provide a prediction of snowmelt runoff (Rasmussen and Ffolliott 1979; Merry et al. 1979). Models have been successfully developed and/or used for different parts of the United States, including Colorado (Shafer, Leaf, and Marron 1979), Maine (Merry et al. 1979), and Arizona (Rasmussen and Ffolliott 1979). Translation of results from a small watershed to a substantially larger one have been successfully reported by Martinec (1982). All of this technology is of great value because acquiring snowpack data is very labor-intensive, or requires expensive on-site instrumentation, or both; the use of satellite-derived data permits rapid worldwide evaluation of water available for spring floods and summer runoff for irrigation, along with the basis for extensive predictive model development. Information on snowpack accumulation and melt is also important in the interpretation of the global climate models (Gleick 1989), discussed above.

In summary, delays of snowmelt flood peaks up to two weeks can be realized by carefully planned forest management activities. This can reduce the peak and increase the potential for storing water that otherwise would bypass reservoirs. Alternatively, peaks may be speeded up from subwatersheds, reducing flood damages on large watersheds that include them. Most of the logging activity that is necessary to effect these desirable changes may be conducted without impairment of water quality, and are particularly dependent upon the pattern of the forest opening both in relation to existing neighboring stands and previous cuts that are now reforested. The level of soil moisture at the time the pack forms and melts are critical determinants in the snowmelt runoff process. The potential for prediction of and planning for worldwide water supplies is the focus of satellite research and holds great promise.

Municipal Watersheds

Numerous municipalities in the western states rely upon runoff from forested watersheds for their primary source of water. Notable among the larger of these are Salt Lake City, Seattle, San Francisco, and all the areas served by the California Water Project, Denver and Colorado Springs, and Portland, Oregon.

The oldest of the operating municipal watersheds that is simultaneously producing timber products is the Cedar River watershed of the City of Seattle (Thompson 1960). The 143 sq-mi watershed has been providing water for Seattle since 1901. Regular patrols ensure compliance with rules that prohibit recreational use, personnel distribution, and railroad train toilet closure while passing over the watershed. Private logged-over lands were purchased with the long-term goal of total watershed ownership, and logging has continued by high lead systems and clear-cutting in strips or staggered areas that meet silvicultural requirements of the mostly Douglasfir timber stands. Timber sales have largely paid for the land acquisition program and provide revenue to the city.

In the eastern states, there are over 1900 watersheds tapped as water sources. More than half are under 10 sq mi, which is generally considered too small for management of the concurrent timber resource: Dissmeyer et al. (1975) consider the practical size limit to be about 100 sq mi. In the southeast, 62 percent of the watersheds are smaller than that limit; 92 percent in the northeast. The largest area held for protection of the water resource are the

Adirondack and Catskill preserves in New York. The Adirondack Park, the larger of the two, actually does not yield water supply for the great masses of population, and could benefit from careful management of the forest to ensure its continued health through vegetative cover diversity (Eschner 1965) and the simultaneous production of good water. The Catskill Park is heavily relied upon for water supplies for New York City and local interests. Neither of these vast areas are currently managed other than for protection.

"In the northeast, approximately 29 percent or 2,000,000 acres of the total watershed areas was owned or controlled by 750 municipalities, private water companies, and state and federal agencies" (Dissmeyer et al. 1975). The holdings average only about 4.2 sq mi, and 87 percent are under government administration. Most of the acres are, indeed, forested, with 26 percent of the watersheds 100 percent forested. Another 26 percent are "90 to 99 percent forested." The land is used for a variety of purposes that range from limited recreational use to mineral exploration and development. Minimal data are collected on most of these lands and, as a consequence, administrative research that might lead to intensive management for improved water yields is limited as well. One large private company operates water supply companies in 20 states, serving over five million residents. Many of these are supplied by forested watersheds of 50 to 2500 acres that are managed by the company for timber products as well.

Stream Improvement

Stream improvement has been a euphemism for many different programs, many of which have provided solid economic benefits. These programs include channelization, some phreatophyte eradication, streambank erosion control, stream access programs, wetland drainage, fishery protection and restoration, and beautification programs (Gillette 1972). Much of the channelization work, however, is short-lived, providing primary economic benefit to the organization that supplies employment in the continuing need to maintain the "improved" channel.

Channelization has been an integral part of larger programs, including the actions under Soil Conservation Service's Small Watershed Protection and Flood Prevention Act; Forest Service's logging debris clearing, aesthetic enhancement, and fishery rehabilitation programs; Corps of Engineers' flood control and navigation improvement and maintenance programs, and numerous other organizations' attempts to clear and maintain floodways so as to reduce natural or post-development flood stages in the flood fringe and the floodway itself.

Channel clearing consists of two main phases: first, removal of materials such as logs, sediment, and debris from the channel. These may be materials that are washed in from disturbed areas, or may be a natural part of the stream-watershed system. In either case, their removal permits the water to flow more freely, reducing backwater effects and reducing flood fringe storage. The resulting increased velocity may also increase streambed and bank erosion, although careful examination of an impeded stream channel will undoubtedly reveal locations where the stream is diverted around an obstruction with resultant high flow rates directed at erodible banks with attendant accelerated erosion. Second, actual removal of streambed material in order to widen or deepen the active channel is often practiced in order to hasten the departure of flood peaks and lower flood flows onsite and upstream. Ultimately, long-term biological damage to the aquatic ecosystem outweighs the short-term economic benefits of the channelization that causes it (Gillette 1972).

Any alteration of the natural balance of stream slope, streambed roughness, and velocity and volume of flood water is likely to fail. The streambed and its natural characteristics developed in response to the existent slope, materials, and the volume of water present. Between the upstream and downstream elevations, a channel reflects the amount of energy that the water converts from potential to kinetic form. Artificial straightening of the channel does not change the amount of energy that must be dissipated, and it takes only a short period of time with normal flow for the channel to restore its pre-deepened depth, its pre-straightened meanders, or its pre-cleared obstacle course.

The fundamental hydrologic issue at the heart of the long-running battle between those who promote flood prevention through upstream land treatment measures and those who prefer downstream flood control measures is often overlooked: flood control does nothing to reduce the volume of water in the flood, only contain it; flood prevention is aimed at reducing that flow volume and, because we often can exercise little control over the volume of precipitation or snowmelt, fail in that purpose as well. Flood prevention is certainly more ecologically focused, based upon the understanding that maintaining or increasing infiltration capacity can dramatically shift excess runoff water through the flood-peak-attenuating soil horizons rather than across its surface.

When a channel is straightened and encased in concrete or other man-made material, it can be stabilized if proper energy dissipating structures are included, and/or if the flow is regulated at the upper end of the channelized section, or if the stabilization is expected only for a limited range of flows. The basic principle still applies, however. More successful are attempts at using natural materials to protect a streambank by understanding its morphology, for example, as shown in Fig. 7-16. If the sediment-laden water is made to flow through stabilized brush mats and allowed places to overflow, eroding banks can be protected from relatively high, but not all, flows.

Watershed Modification

Modification of a watershed can be accomplished within limits by changing its land slope, stream gradient, size and, as already noted (in several preceding sections), by altering its vegetative cover, which can also effect a change in albedo and, as a consequence, evapotranspiration and runoff amounts and patterns.

Land slope is a commonly changed watershed characteristic. The most obvious change occurs under intensive agriculture where the water requirements of rice culture demand level fields. Extensive terracing, contour plowing, and strip cropping all can produce reduced slopes, allowing water to remain on the surface for longer periods of time, thereby increasing evapotranspiration and reducing runoff. Grazing on steep slopes produces altered land slopes in the form of terraced hillsides created by the cattle which cannot move up and down the slopes as well as they can on or nearly on the contour. The decreased infiltration capacity that is the result of the compaction of the soil is offset, in part, by the reduced slope that slows the runoff.

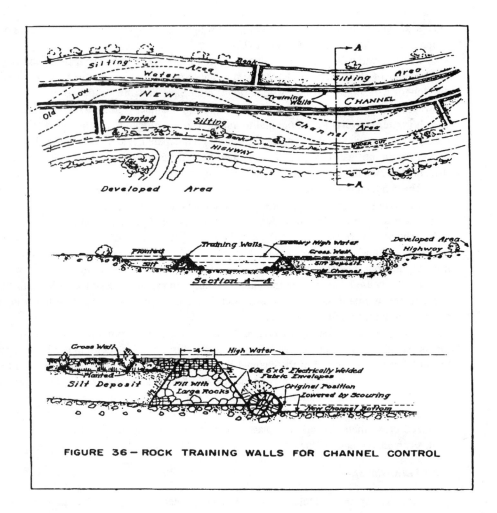

Figure 7-16 Example of stream bank stabilization system (from Forest Service Manual)

Stream gradients are altered naturally by beaver as well as by the more obvious dam-building activity of man. Sometimes that change is subtle, as when a road limits the flow in the floodway; sometimes it is not so subtle, as when a dam completely obstructs the downstream movement of water except over a spillway that is at the same elevation as the point where the stream enters the impoundment, perhaps miles upstream. On smaller channels, gully plugs and various pool-and-riffle development or maintenance projects do the same thing: erosion and sedimentation are affected to greater or lesser extents, with attendent impact on the aquatic ecosystem, up- and down-stream characteristics, and adjacent land features as well.

Changing the size of a watershed may influence the relationship between the magnitude of runoff events that shaped it and its morphological condition, that is, whether it is aggrading or degrading. Size change may occur naturally, as when a stream is "captured," or by man's

Chapter 7 Water on the Watershed 305

activities, such as diversion, deliberate boundary change, or inadvertent change such as what may occur when water is not turned off a stretch of highway until the road crosses a watershed boundary.

Summary

Most of the change that occurs as a result of any of the three watershed objectives is in fact an alteration of the natural rate of change. Nevertheless, it is the change that we consider to be adverse impact. It is important to recall that the water will run downhill whether man is there to help it or not. The conditions under which it moves, however, are subject to considerable control, and almost inevitably affect its quality. On wildlands, that water quality impact is often subtle, showing up delayed in both time and space, often far from the site of the disturbance.

NONPOINT SOURCES OF POLLUTION

"End of pipe" discharges from manufacturing or waste treatment plants are commonly referred to as "point sources" and have come under greater and greater control since the 1948 Water Pollution Control Act (62 Stat 1155). The first water pollution control legislation was actually the 1899 "Refuse Act": it was directed at control over dumping in and obstruction of navigable waters and actually was an amendment to the Rivers and Harbors Act (30 Stat 1152) of that year. Pollution of both surface and ground water bodies was noted throughout the 1960s, and written about specifically in Rachel Carson's *Silent Spring* in 1962. Insiduous, diffused sources of pollution from the widespread use of pesticides were the prime concern, a concern that led to the ban on the use of DDT in the United States by the end of the decade.

Nonpoint sources of pollution are typically variable, unpredictable, and dispersed, as opposed to point sources, which are steady, predictable, and concentrated.[13]

THE LEGAL BASIS

The term "nonpoint sources" was not used officially until 1972. Section 208 of the 1972 Water Pollution Control Amendments is titled "Areawide Waste Treatment Management" and calls for encouraging and facilitating plans to control "agriculturally and silviculturally related nonpoint sources of pollution." To these two widespread land uses was added diffused sources of pollutants from mining, construction, saltwater intrusion, waste disposition and disposal, and pollution from urban-industrial development areas.

The law called for state identification of problem areas, approval by the Environmental Protection Agency of the states' individual management control plans, creation of local or state Section 208 planning organizations, and implementation funding by the federal government. When funding was cut in the early 1980s, the 208 planning organizations

---

[13] From notes taken at a lecture on agricultural nonpoint source pollution by Douglas A. Haith, Professor of Agricultural Engineering, Cornell University, March 9, 1990.

languished, only to revive slightly and receive limited planning funding again in the 1987 Amendments to the Clean Water Act (as the entire body of water quality law had become known).

Initial estimates of the contribution of nonpoint sources were somewhat overestimated for silvicultural activities, and severely underestimated for agricultural activities and urban runoff. Many of the forest management activities have been the subject of research and timber sale contracts for years (Myers 1986). On the nearly two thirds of the nonfederal lands in the United States that is cropland or grazed, nearly 500 million pounds of active pesticide ingredients and 20 million tons of commercial fertilizer are used each year (Myers 1986). The National Water Commission (1973) estimated that the 10-yr program aimed at controlling the nonpoint urban stormwater runoff alone could easily cost $248 billion in 1973 dollars (*almost $25 billion per year*); Congress had (in the 1972 Act) authorized only $18 billion for the first six yr. Typical constituents and extent of the urban stormwater runoff were evaluated in the National Urban Runoff Program (Heaney and Huber 1984): the assessment "demonstrated the critical need for additional short-term and long-term sampling programs."

With the simultaneous phasing out of federal subsidies to the construction of treatment facilities for point sources, expert, government, and public awareness gradually turned to the need for control over nonpoint sources, believed by many to be a much more serious problem. One of the most vexatious aspects of the nonpoint source pollution control problem is that implementation must be of a rather nebulous *process*, not a series of standards against which water quality can be appraised, although the final measure of the success of a nonpoint control program can be and usually is in terms of some water quality characteristic(s).

THE PROCESS OF NONPOINT SOURCE POLLUTION CONTROL

The process consists of the consideration and, where appropriate, the implementation of that management practice that will best control (preclude, mitigate, ameliorate) pollution sources and episodes. Such a process is difficult to create, implement, enforce, and evaluate. Yet all four of these phases must be effected if nonpoint sources of pollution are to be controlled.

Haith (see footnote 13) identified seven strategies for nonpoint source pollution control, grouped in three generalized categories:

A. Waste Management

    1. Reduce waste stream at source

    2. Limit access of waste stream to water

B. Water Management

    3. Reduce runoff (surface or subsurface) or percolation (to ground water)

    4. Divert runoff from entering site

    5. Collection (and treatment) of leachate-contaminated runoff or percolation water

C. Sediment Management

    6. Prevent soil erosion

    7. Remove sediment from runoff

A workable program of nonpoint source pollution control consists of four stages: *recognition of the problem, identification of the source, selection of the strategy,* and *implementation of the BMP*. Candidate BMPs to accomplish the overall pollution reduction objective may be selected from a long list of practices, many of which have been in existence for many years (cf., for example, Department of Environmental Conservation 1990). Many of the candidate BMPs may be appropriately listed under more than one strategy and, furthermore, a particular BMP may not always be appropriate given the particular environmental conditions on hand.

Pisano (1978) suggested that by and large, since nonpoint pollutants are in transit overland or through the soil for considerable distances prior to entering a water body, "the most effective controls [BMPs] are land treatment techniques and conservation practices." He further pointed out that compliance must be on the land rather than in the water body, which is the case for point sources. In addition to verifying that nonpoint sources were important pollution sources in the three states studied (Virginia, New York, and Wisconsin), Peirce (1980) concluded that: (1) the "technical controls are generally available to solve the problems," (2) existing control programs were "not necessarily technologically sound or cost effective," and (3) existing control programs would work only under rather limited conditions, including implementation by the respective states of certain specified actions.

The concept of water-based land management (WBLM) is the natural outgrowth of the recognition of nonpoint sources, best management practices, and control concepts. WBLM was formulated in consideration of agricultural use (Osteen, Seitz, and Stall 1981), and is best administered at the local level of soil and water conservation district level. Agriculture provides the intensive economic activity necessary to sustain an interest in maintenance of production and often a high degree of altruism on the part of the agricultural landowner/operator. However, WBLM is applicable to any of the land uses from which diffused pollutants are discharged. "Ideally," point out Osteen et al. (1981), "WBLM permits interdependent choices of water use goals, water quality goals, and resource management practices by comparing the cost and benefits of alternative plans." Since landowners may not want to or be able to politically affect soil conservation district level decisions concerning water quality programs, and since the districts themselves are not necessarily set up to manage water quality, WBLM requires some special handling, perhaps even changes in the states' district laws, to which there might be considerable political opposition. Osteen et al. (1981) further point out that "the relationship between land management and water quality is difficult to model," which makes control even more costly. Nevertheless, WBLM is a long-term goal on all lands that produce nonpoint sources of pollution. In fact, the current "sodbuster" and "swampbuster" provisions of the 1985 Farm Bill[14], are a first step in a long-run trend that will bring about more and more land use regulation to control water quality (Black 1987).

For normal operations on forest, range, agricultural, and developing urban lands, the best management practices have already been identified under the "Protection" section of this chapter. They tend to be preventative rather than curative (Pisano 1978), and promisary rather than restrictive, that is, they tend to improve the level of management of the land for the primary purpose as well as to improve or maintain the quality of the water that emanates from the land under management. Also, by and large, the basic principles underlying each set

---

[14] The Food Security Act, PL 99-198 1985.

of these primary four land use practices were identified long before the formal use of the term "nonpoint sources of pollution." Some principles may be summarized from a variety of sources (including these references and annotated bibliographies: Department of Environmental Conservation 1989; Field 1985; Forest Service 1980; Government Accounting Office 1988; McClimans et al. 1978; National Council of the Paper Industry for Air and Stream Improvement 1988; Novotny 1988; Water Quality Group 1988; Vigon 1985) as follows:

BEST MANAGEMENT PRACTICES PRINCIPLES

Forest lands: Focus attention on small portion of operated area that is in roads, skid trails, landings, and hot decks; keep water off these areas and keep them out of streams. Where good forest practices are carried out, water quality degradation is not likely to occur.

Rangelands: Use salt, water, and fencing to distribute animal use in time and space; maintain infiltration capacity of grazed lands; remove cause of overgrazing prior to rehabilitation. Sedimentation and animal wastes are the most likely pollutants, and require special engineering consideration in the case of cattle feed lots.

Agricultural lands: Rotate crops and pasture use, utilizing standard soil conservation practices such as terracing, contour plowing, grassed waterways, strip cropping, conservation tillage, and green manuring; recycle animal wastes on land when nutrients will be used by crops; move animal concentrations away from live streams. Water quality degradation is likely to be in the form of sedimentation, and untenable levels of plant nutrients, pesticides, herbicides, and animal waste products.

Urban lands: Separate storm and sanitary sewer systems; provide intercepting vegetated terrestrial or aquatic buffer areas for discharge of surface and subsurface runoff from developed areas (e.g., parking lots, streets, malls) before discharge to live streams or ground water bodies; include flood retarding structures in new development and revitalization plans; consider environmental water as a vital resource to the urban development so that its values will be inherently considered in planning. Urban runoff may include excess concentrations of road deicing salt, petroleum products leached from paving materials, herbicides, pesticides, fertilizers, and pet animal wastes. Dependent upon dry fallout between storms, which is dependent upon wind direction, fetch, and potential sources of airborne particles and gases, pollution from dry fallout may be substantial.

All lands: Involve the public in best management practice process planning, implementation, enforcement, and evaluation. Land application of waste disposal sludges, surface and subsurface mining acitivity, landfills, inadvertent spills, and virtually all land uses have effects that eventually show up in the surface and ground waters.

BEST MANAGEMENT PRACTICES ON WILDLANDS

This book has focused on the wildland watershed, yet the discussion on best management practices has included consideration of lands that are under more intensive management than would be expected for wild or undeveloped lands. It is essential to recall that the developed lands were once wild. Without proper consideration of and respect for the characteristics of the natural water resource, they may once again be wildland watersheds.

Examination of the hydrology of any land surface, of any watershed, of any stream, must commence with consideration of how the runoff behaved under natural conditions, what factors affected it, and how it has been changed by the sequence and extent of land use that now exists. Only with such evaluation will we be able to begin to control water quantity, quality, and regimen to meet our long term goals.

PROBLEMS

1. Identify a local watershed and consider its area, elevation, orientation, aspect, shape, and slope characteristics; delineate the boundary on a map and measure its characteristics for comparison with your estimates. Consider the stream behavior in light of the watershed characteristics.
2. Tabulate storm hydrograph characteristics for two nearby or nested watersheds where the the area of one is about twice the other.
3. Consider the impact of various types of watershed land uses on the quarter-flow interval where snow runoff makes up about half the annual runoff. What assumptions do you need to make?
4. For a local, urban watershed with which you are familiar, consider (1) how the different types of storage on the watershed have changed since it was in an undeveloped state, and (2) how the changes have affected runoff behavior.
5. Consider the impact of urbanization on the just the lower half of a watershed on the storm hydrograph. What assumptions do you need to make?
6. What are the likely effects on wetland (or reservoir or stream) water quality characteristics if there is a doubling of the cloud cover as shown in Fig. 2-4?

Epilogue

> *Scientists, if not careful, can hypnotize themselves*
> *into a state of happy ignorance*
> *based upon fundamental principles*
> *written in the textbooks*
> — *W. U. Garstka*

The watershed is, above all, a natural system, a unit of the landscape that functions in occasionally predictable ways. It functions in certain predictable ways because it both derives from and consists of certain characteristics that react in known ways to external influences. It is also a complex, often unpredictable system owing to both its fundamental probabilistic nature and the existence of numerous feedback loops.

Energy and precipitation are the inputs to the local hydrologic environment. Important input characteristics are embraced by the concepts of precipitation intensity-duration-frequency relationships. These relationships are capable of being analyzed in ways that allow us to predict with no small amount of uncertainty the probable consequences in the natural environment, but not necessarily the *timing* of their appearance. Thus, both risk and uncertainty are integral parts of the hydrologic environment.

The summary outcome is, of course, stream behavior. What occurs in the stream cannot be considered independently of what occurs on the watershed. *That* is the fundamental precept of watershed hydrology. Civilization's water management practices have, for too long, ignored that simple fact. We can no more emplace a dam in a stream-watershed system without major consequences both up- and downstream than we can emplace a cork in the middle of our small intestine without dire bodywide consequences. And we can no longer claim ignorance about intestinal behavior and the factors that affect it, either.

The broad-scale hydrologic events take place on a seemingly immutable geologic foundation, where knowledge and information are out of sight — inverse functions of depth below the surface. The scenery is the highly visible geology-modifying soil and vegetation that we observe, understand, and learn to live with if not actively manage. There is risk and uncertainty, too, in how we ignore, tolerate, control, or husband the hydrologic environment.

Like an intricate play, with all its elements of producer, script, stage, actors, scenery, props, lighting, directing, and audience, one can neither completely control nor precisely operate together, of that we can be sure. And there is no way one can affect part of the production and not have a profound influence on any or all other parts of it. "Most of our failures", says Eugene P. Odum (1969), "can be traced to shortsighted action that considered only the benefits to a part of the landscape rather than to the whole ...." Eisel (1972) observed that "an important aspect [of an hypothetical systems management investigation] ... was integration of the upstream watershed into the downstream basin", especially where traditional objectives yield to more recent ones such as "preserving the quality of natural environments for outdoor recreation". Not just outdoor recreation. Living in an urban environment, the American public has deprived itself of contact with the natural environment. With newfound social acceptability and support instead of the contemptuous attitude toward the environmental do-gooders of the mid-century, the public is now becoming and desperately needs to be informed about water issues. Indeed, professional water managers *must* inform the public in consideration of the practical partnership necessary to do an effective job, if not by law. Thus, the public is on the brink of truly becoming involved in water management in a meaningful, productive way. The hydrologic scientist has a responsibility to provide the substantive basis for decision making.

The population of the United States has, within a scant one hundred years, been transformed from one that was primarily rural and somewhat less than 76 million in number in 1900 to one that as of this writing surpasses a quarter of a *billion*, and is practically all urban or suburban. It doesn't *know* the natural environment. It not only hasn't known where its water comes from, it isn't likely to believe anyone's description of some of the preposterous and unlikely behavior in the hydrologic cycle. Yet, that same population, becoming more and more affluent, mobile, and environmentally aware, will be dictating what is done with our aquatic environment in the years ahead just as surely as that population is now, in part, dictating wildlife management, timber management, and range management policy on grounds that have little to do with the scientific basis of the natural systems that underlie those environments.

Note the word "dictating", not "participating in". The law currently provides for participation by the citizenry in virtually all aspects of natural resources management. That participation won't be *responsible* unless citizens can feel that they are, indeed, taking part in the decision-making process. And they will be doing so alongside the professional managers of the hydrologic environment. These are the experts who have spent years taking courses and gaining experience in soils, geology, climatology, hydrology, limnology, other sciences, modelling, and the fundamental concepts of land use and the social sciences that underlie natural resources management. Those experts will not be heeded unless they do, in fact, *listen* to the public and ensure that the public *believes* that they are playing an active role in water management decisions.

In a description of the hydrologic cycle, E. G. Dunford (1966) used the phrase "When precipitation enters the forest environment"; that expresses an important concept, one inherent in comprehending Eisel's and Odum's quotes (and those of many others) above. Residual forest lands tend to be thin-soiled, too steep to plow, relatively infertile, and not well suited for agriculture. They are at high elevations where orographic precipitation provides a surplus of water for runoff. Thus, writes Wilm (1955), "good forest conditions must be a dominant feature of watershed improvement". We have known that fact since before the

Johnstown and Pittsburgh floods, since the Wagon Wheel Gap studies in Colorado provided the scientific evidence for including in the 1911 Weeks Forest Purchase Act "in order to protect the flow of navigable streams ..." the federal government can get into the forest management business. We have, as a consequence, seen a branch of water resources education grow up in our schools of forestry or natural resources.

With the growth of the agencies at mid-century came fragmentation of water resources at the federal level. There was a simultaneous drive toward "comprehensive planning" based upon scientific information, political expediency, and even logic. Petty jealousies, turf-protection, and dramatic natural and man-caused events, however, have precluded attaining that comprehensiveness that is, in the post-Watergate and post-Vietnam era, obsolete owing to massive distrust in government. Ironically, other forces have further diminished the need for comprehensive project planning: the best sites have been developed, the public's concern over environmental quality has grown, interest rates (that adversely impact the benefit-cost ratio) have soared, and two administrations have imposed a regime of deregulation as well as decreased federal activity.

The integrative planning and management that can come about through communication and consensus must prevail. It can and logically should take place in the natural and flexible environmental unit called the "watershed", based upon a solid foundation of **watershed hydrology**. Doing so will permit effective management to be as local as possible, to the benefit of bio-chemi-physical solutions as well as to the sociopolitical goal of achieving solutions at the lowest possible administrative level.

Watershed management must begin with the responsible hydrologic scientist (1) seeking the essence of his or her complex scientific world and distilling it for public consumption and use, and (2) taking an active part in the available decision-making processes. If this book has made any progress in achieving those two objectives, I will be quite pleased.

References

Abdul, A. S., and R. W. Gillham, 1984. "Laboratory Studies of the Effects of the Capillary Fringe on Streamflow Generation," *Water Resources Research* 20(6):691.

Abelson, P. H., 1987. "Ozone and Acid Rain," *Science* 235:141.

Achuthan, K., 1974. "Man on the Hydrologic Cycle," *Water Resources Bulletin* 10(4):756.

Ackerman, W. C., 1969. "Hydrology Becomes Water Science," *Transactions* of the American Geophysical Union 50(3):76.

Ackerman, W. C., G. F. White, and E. B. Worthington, Eds., 1973. *Man-Made Lakes: Their Problems and Environmental Effects*. American Geophysical Union, Washington, DC.

Adams, W. P., 1976. "Areal Differentiation of Snow Cover in East Central Ontario," *Water Resources Research* 12(6):1226.

Adrian, D. D., and J. B. Franzini, 1966. "Impedance to Infiltration by Pressure Build-up ahead of the Wetting Front," *Journal of Geophysical Research* 71(24):5857.

Aguado, E., 1985. "Radiation Balance of Melting Snow Covers at an Open Site in the Central Sierra Nevada, California," *Water Resources Research* 21(11):1649.

Alley, W. M., 1984. "On the Treatment of Evapotranspiration, Soil Moisture Accounting, and Aquifer Recharge in Monthly Water Balance Models," *Water Resources Research* 20(8):1137.

-------, 1985. "Water Balance Models in One-Month-Ahead Streamflow Forecasting," *Water Resources Research* 21(4):597.

American Society of Civil Engineers, 1949. *Hydrology Handbook*. 33 West 39th Street, New York, NY.

Anderson, E. A., 1968. "Development and Testing of Snow Pack Energy Balance Equations," *Water Resources Research* 4(1):19.

Anderson, H. W., 1956. "Forest-Cover Effects on Snowpack Accumulation and Melt, Central Sierra Snow Laboratory," *Transactions*, American Geophysical Union 37(3):307.

-------, 1960. "Prospects for Affecting the Quantity and Timing of Water Yield Through Snowpack Management in California," *Proceedings*, Western Snow Conference, p. 44.

-------, 1963. *Managing California's Snow Zone for Water*, Research Paper PSW-6, Pacific Southwest Forest and Range Experiment Station, Forest Service, USDA, Berkeley, CA 28 pp.

-------, 1967. "Snow Accumulation as Related to Meteorological, Topographic, and Forest Variables in Central Sierra Nevada, California," International Association of Scientific Hydrology Publication No. 78:215.

-------, 1969. "Snowpack Management," In *Snow*, Water Resources Institute, Seminar WR 011.69, Oregon State University, Corvallis, OR pp. 27-40.

Anderson, H. W., and C. H. Gleason, 1960. "Effects of Logging and Brush Removal on Snow Water Runoff," Extract of Publication No. 51 of the International Association of Scientific Hydrology, pp. 478-89.

Anderson, H. W., M. D. Hoover, and K. G. Reinhart, 1976. *Forests and Water: Effects of Forest Management on Floods, Sedimentation, and Water Supply*, General Technical Report PSW-18/1976, Pacific Southwest Forest and Range Experiment Station, Forest Service, USDA, Washington, DC, 115 pp.

Anderson, H. W., R. M. Rice, and A. J. West, 1958. "Snow in Forest Openings and Forest Stands," *Proceedings*, Society of American Foresters, Salt Lake City, UT, pp. 46-50.

Anderson, M. G., and T. P. Burt, 1978. "Toward More Detailed Field Monitoring of Variable Source Areas," *Water Resources Research* 14(6):1123.

Anderson, P. H., 1980. "Forested Wetlands in Eastern Connecticut: Their Transition Zones and Delineation," *Water Resources Bulletin* 16(2):248.

Anderson, P. H., M. W. Lefor, and W. C. Kennard, 1980. "Forested Wetlands in Eastern Connecticut: Their Transition Zones and Delineation," *Water Resources Bulletin* 16(2):248.

Aron, G., and T. M. Rachford, 1974. "Procedures for Filling Gaps in Hydrologic Event Series," *Water Resources Bulletin* 10(4):719.

Aubertin, G. M., Ed., 1977. *Proceedings "208" Symposium*, Southern Illinois University, 10/19-20/77, Carbondale, IL.

Aubertin, G. M., and J. H. Patric, 1974. "Water Quality After Clearcutting a Small Watershed in West Virginia," *Journal of Environmental Quality* 3(3):243.

Austin, T. A., 1986. "Utilization of Models in Water Resources," *Water Resources Bulletin* 22(1):49.

Ayer, G. R., 1968. "Reforestation with Conifers: its Effect on Streamflow in Central New York," *Water Resources Bulletin* 4(2):13.

Bailey, R. W., and C. A. Connaughton, 1936. *In Watershed Protection* from "The Western Range, a Great but Neglected Natural Resource," Senate Document 199, Separate No. 10, Forest Service, USDA, Washington, DC, pp. 303-39.

Baker, M. B., 1984. "Changes in Streamflow in an Herbicide-Treated Pinyon-Juniper Watershed in Arizona," *Water Resources Research* 20(11):1639.

-------, 1986. "Effects of Ponderosa Pine Treatments on Water Yield in Arizona," *Water Resources Research* 22(1):67.

Baker, M. B., and J. J. Rogers, 1983. "Evaluations of Water Balance Models on a Mixed Conifer Watershed," *Water Resources Research* 19(2):486.

Bakun, A., 1990. "Global Climate Change and Intensification of Ocean Upwelling," *Science* 247:198.

Barge, B. L., R. G. Humphries, S. J. Mah, and W. K. Kuhnke, 1979. "Rainfall Measurements by Weather Radar: Applications to Hydrology," *Water Resources Research* 15(6):1380.

Barnhart, C. L., Ed., 1949. *American College Dictionary*. Random House, Inc., New York, NY.

Barry, R., M. Prévost, J. Stein, and A. P. Plamondon, 1990. "Application of a Snow Cover Energy and Mass Balance Model in a Balsom Fir Forest," *Water Resources Research* 26(5):1079.

Bastin, G., B. Lorent, C. Duque, and M. Gevers, 1984. "Optimal Estimation of Average Areal Rainfall and Optimal Selection of Rain Gauge Locations," *Water Resources Research* 20(4):463.

Bates, M., 1960. *The Forest and the Sea*. Mentor Books, NY.

Bates, C. G., and A. J. H. Henry, 1928. "Forest and Stream Flow Experiment at Wagon Wheel Gap, Colorado," *U.S. Monthly Weather Review*, Supplement.

Baumgartner, D. M., editor. 1981. *Interior West Watershed Management*, Washington State University, Pullman, WA.

Beaumont, R. T., 1957. "Cooperative Snow Surveys and Water Supply Forecasting," *Journal of Soil and Water Conservation* 12(3):115.

Ben-Asher, J., 1981. "Estimating Evapotranspiration From the Sonoita Creek Watershed Near Patagonia, Arizona," *Water Resources Research* 17(4):901.

Benton, A. R., W. P. James, and J. W. Rouse, 1978. "Evapotranspiration from Water Hyacinth," *Water Resources Bulletin* 14(4):919.

Bernard, M., 1932. "Formulas for Rainfall Intensities of Long Duration," *Transactions* of the American Society of Civil Engineers 96:592.

Berndt, H. W., 1964. "Inducing Snow Accumulation on Mountain Grassland Watersheds," *Journal of Soil and Water Conservation* 19(5):196.

Berris, S. N., and R. D. Harr, 1987. "Comparative Snow Accumulation and Melt During Rainfall in Forested and Clear-Cut Plots in the Western Cascades of Oregon," *Water Resources Research* 23(1):135.

Beschta, R. L., and R. L. Taylor, 1988. "Stream Temperature Increases and Land Use in a Forested Oregon Watershed," *Water Resources Bulletin* 24(1):19.

Bethlahmy, N., 1972. *Hydrograph Analysis: a Computerized Separation Technique*, Research Paper INT-122, Intermountain Forest and Range Experiment Station, Forest Service, USDA, Ogden, UT.

-------, 1973. "Estimating the Land Slope of Mountain Watersheds," *Journal of Soil and Water Conservation* 28(5):229.

-------, 1974. "More Streamflow After Bark Beetle Episode," *Journal of Hydrology* 23(1974):185.

-------, 1975. "A Colorado Episode: Beetle Epidemic, Chost Forests, More Streamflow," *Northwest Science* 49(2):95.

-------, 1976. "The Two-Axis Method: A New Method to Calculate Average Precipitation Over a Basin," *Hydrological Sciences Bulletin* XXI:379.

Betson, R. P., 1964. "What is Watershed Runoff?" *Journal of Geophysical Research* 69(8):1541.

Betson, R. P., and J. B. Marius, 1969. "Source Areas of Storm Runoff," *Water Resources Research* 5(3):574.

Betters, D. R., 1975. "A Timber-Water Simulation Model for Lodgepole Pine Watersheds in the Colorado Rockies," *Water Resources Research* 11(6):903.

Beven, K., 1981. "Kinematic Subsurface Stormflow," *Water Resources Research* 17(5):1419.

Beven, K., and P. Germann, 1982. "Macropores and Water Flow in Soils," *Water Resources Research* 18(5):1311.

Birtles, A. B., 1978. "Identification and Separation of Major Base Flow Components From a Stream Hydrograph," *Water Resources Research* 14(5):791.

Bissell, V. C., and E. L. Peck, 1973. "Monitoring Snow Water Equivalent by Using Natural Soil Radioactivity," *Water Resources Research* 9(4):885.

Biswas, T. D., D. R. Nielsen, and J. W. Biggar, 1966. "Redistribution of Soil Water after Infiltration," *Water Resources Research* 2(3):513.

Black, P. E., 1957. *Interception in a Hardwood Stand*, MS Thesis, The University of Michigan, Ann Arbor, MI.

-------, 1963. "Timber and Water Resource Management," *Forest Science* 9(2):137.

-------, 1964. "Climate, Soils, and Hydrology of the Redwood Region," Consultant Report to the National Park Service, Department of the Interior, Washington, DC.

-------, 1967. *The Coast Redwoods, Water, and Watersheds*. Report to the National Park Service, Department of The Interior, Washington, DC.

-------, 1968. "Streamflow Increases Following Farm Abandonment on Eastern New York Watershed," *Water Resources Research* 4(6):1171.

-------, 1970. "The Watershed in Principle," *Water Resources Bulletin* 6(2):153.

-------, 1970a. "Runoff From Watershed Models," *Water Resources Research* 6(2):465.

-------, 1972. "Hydrograph Responses to Geomorphic Model Watershed Characteristics and Precipitation Variables," *Journal of Hydrology* 17:309.

-------, 1974. *Mohawk River Flood Plain, Oneida County, New York*. Oneida County Planning Department, Utica, NY.

-------, 1975. "Hydrograph Responses to Watershed Model Size and Similitude Relations," *Journal of Hydrology* 26:255.

-------, 1980a. "Reasons for Wetlands Preservation," Letter to the Editor, *Journal of Soil and Water Conservation* 35(3):108.

-------, 1980b. "Water Quality Patterns During a Storm on a Mall Parking Lot," *Water Resources Bulletin* 16(4):615.

-------, 1982 "Hardwood Forests as a Public Water Source," Invited Paper, Proceedings of the Annual Meeting, Society of American Foresters, Cincinnati, OH, Sept. 20. 15 pp.

-------, 1983. "Conceptual Modeling of Parking Lot Runoff," Paper presented at the Nineteenth Annual Meeting of the American Water Resources Association, San Antonio, TX. Lithographed.

-------, 1987. *Conservation of Water and Related Land Resources*, Rowman and Littlefied, Totowa, NJ 336 pp.

-------, 1988. "Strange Attractors in the Closet?" Paper presented at Annual Fall Meeting of the American Geophysical Union, San Francisco, CA, December 9.

-------, 1989a. *Computer-Based Hydrology Laboratory*. Second Edition. Faculty of Forestry Misc. Publication No. 16, SUNY College of Environmental Science and Forestry, Syracuse, NY.

-------, 1989b. "The Thornthwaite Water Budget for APL Computers." Poster/Work Station Presentation, IAHS 3rd Scientific Assembly, Baltimore, MD.

-------, 1989c. "Stratified Flood Frequency Analysis," In *Headwaters Hydrology*, Proceedings of the Symposium, Woessner and Potts, editors, p. 563.

Black, P. E., and P. M. Clark, 1959. *Timber, Water, and Stamp Creek*. With P. M. Clark, Southern Region and Southeastern Forest Experiment Station, Asheville, NC.

Black, P. E., and R. E. Leonard, 1968. "The Principle of Watershed Equilibrium," *Water Resources Bulletin* 4(2):49-50.

Black, P. E., and J. W. Cronn, 1975. "Hydrograph Responses to Watershed Model Size and Similitude Relations," *Journal of Hydrology* 26:255.

Black, T. A., 1979. "Evapotranspiration from Douglas Fir Stands Exposed to Soil Water Deficits," *Water Resources Research* 15(1):164.

Blackburn, W. H., J. C. Wood, and M. G. DeHaven, 1986. "Storm Flow and Sediment Losses From Site-Prepared Forestland in East Texas," *Water Resources Research* 22(5):776.

Blaney, D. G., S. L. Ponce, and G. F. Warrington, 1984. *Statistical Methods Commonly Used in Soil Data Analysis*. WSDG Report Series TP-00011, USDA, Forest Service, Fort Collins, CO.

Blaney, H. F., and W. D. Criddle, 1950. *Determining Water Requirements in Irrigated Areas from Climatological and Irrigation Data*. SCS TP-96, USDA, Washington, DC.

Blumenstock, D. I., and C. W. Thornthwaite, 1942. "Climate and the World Pattern." In *Climate and Man, the 1942 Yearbook of Agriculture*, Department of Agriculture, Government Printing Office, Washington, DC, pp. 98-127.

Bodhaine, G. L., 1968. *Measurements of Peak Discharge at Culverts by Indirect Methods*. Book 3, Chapter A3 of *Techniques of Water-Resources Investigations of the United States Geological Survey*, Government Printing Office, Washington, DC.

Booty, W. G., J. V. DePinto, and R. D. Scheffe, 1988. "Drainage Basin Control of Acid Loadings to Two Adirondack Lakes," *Water Resources Research* 24(7):1024.

Boughton, W. C., 1980. "A Frequency Distribution for Annual Floods," *Water Resources Research* 16(2):347.

Bowley, C. J., and J. C. Barnes, 1979. "Satellite Snow Mapping Techniques with Emphasis on the Use of LANDSAT," In *Satellite Hydrology* edited by Deutsch et al. 1981, pp. 158-64.

Boyd, C. E., 1977. "Evaluation of a Water Analysis Kit," *Journal of Environmental Quality* 6(4):381.

Brady, N. C., 1984. The *Nature and Properties of Soils*. Macmillan Publishing Co., New York, NY.

Bras, R. L., and I. Rodriguez-Iturbe, 1976. "Network Design for the Estimation of Areal Mean of Rainfall Events," *Water Resources Research* 12(6):1185.

Brater, E. F., 1940. "The Unit Hydrograph Principle Applied to Small Watersheds," *Transactions of the American Society of Civil Engineers*, 105:1154.

Brezonik, P. L., E. S. Edgerton, and C. D. Hendry, 1980. "Acid Precipitation and Sulfate Deposition in Florida," *Science* 208:1027.

Broadfoot, W. M., 1954. "Procedures and Equipment for Determining Soil Bulk Density," In *Some Field, Laboratory, and Office Procedures for Soil-Moisture Measurement*, USDA, Forest Service, Occasional Paper 135, Southern Forest Experiment Station, pp. 2-11.

Brown, G. W., 1970. "Effects of Forest Management on the Thermal Properties of Streamflow," Paper presented at ASCE Symposium on Interdisciplinary Aspects of Watershed Management, Bozeman, MT.

-------, 1985. *Forestry and Water Quality*, Second Edition, Oregon State University Book Stores, Inc., Corvallis, OR.

Brown, G. W., and J. T. Krygier, 1970. "Effects of Clear-Cutting on Stream Temperature," *Water Resources Research* 6(4):1133.

-------, 1971. "Clear-Cut Logging and Sediment Production in the Oregon Coast Range," *Water Resources Research* 7(5):1189.

Brown, J. H., and A. C. Barker, 1970. "An Analysis of Throughfall and Stemflow in Mixed Oak Stands," *Water Resources Research* 6(1):316.

Brown, M. J., 1986. "Use of Stream Chemistry to Estimate Hydrologic Parameters," *Water Resources Research* 22(5):805.

Brussock, P. P., A. V. Brown, and J. D. Dixon, 1985. "Channel Form and Stream Ecosystem Models," *Water Resources Bulletin* 21(5):859.

Brutsaert, W., 1963. "On Pore Size Distribution and Relative Permeabilities of Porous Mediums," *Journal of Geophysical Research* 68(8):2233.

Buchanan, T. J., and W. P. Somers, 1969. *Techniques of Water-Resources Investigations of the United States Geological Survey, Chapter A8, Discharge at Gaging Stations*, Geological Survey, USDI, Washington, DC.

Buckman, H. O., and N. C. Brady, 1960. *The Nature and Property of Soils*. Sixth Edition. The MacMillan Company, New York, NY.

Bureau of Reclamation, 1959. *Water-Loss Investigations: Lake Hefner 1958 Evaporation Reduction Investigations*, USDI, Denver, CO.

Burton, T. M, 1981. "The Effects of Riverine Marshes on Water Quality," in Richardson (1981), p. 139.

Camillo, P. J., R. J. Gurney, T. J. Schmugge. 1983. "A Soil and Atmospheric Boundary Layer Model for Evapotranspiration and Soil Moisture Studies," *Water Resources Research* 19(2):371.

Campbell, A. J., and R. C. Sidle, 1984. "Prediction of Peak Flows on Small Watersheds in Oregon for Use in Culvert Design," *Water Resources Bulletin* 20(1):9.

Campbell, C. J., 1970. "Ecological Implications of Riparian Vegetation Management," *Journal of Soil and Water Conservation* 25(2):49.

Carroll, S. S., and T. R. Carroll, 1989. "Effect of Uneven Snow Cover on Airborne Snow Water Equivalent Estimates Obtained by Measuring Terrestrial Gamma Radiation," *Water Resources Research* 25(7):1505.

Carter, L. J., 1979. "Uncontrolled SO_2 Emissions Bring Acid Rain," *Science* 204:1179.

Carter, V., M. S. Bedinger, R. P. Novitzki, and W. O. Wilen, 1979. "Water Resources and Wetlands," In *Wetland Functions and Values: The State of Our Understanding, Proceedings* of the National Symposium on Wetlands, American Water Resources Association, 1978. Orlando, FL, pp. 344-76.

Casey, D., P. N. Nemetz, and D. H. Uyeno, 1983. "Sampling Frequency for Water Quality Monitoring: Measures of Effectiveness," *Water Resources Research* 19(5):1107.

Chang, A. T. C., D. K. Hall, J. L. Foster, A. Rango, and J. C. Shiue, 1979. "Passive Microwave Sensing of Snow Characteristics Over Land," In *Satellite Hydrology* edited by Deutsch et al. 1981, pp. 213-17.

Chang, M., 1977. "An Evaluation of Precipitation Gage Density in a Mountainous Terrain," *Water Resources Bulletin* 13(1):39.

Chang, M., and D. C. Boyer, 1977. "Estimates of Low Flows Using Watershed and Climatic Parameters," *Water Resources Research* 13(6):997.

Changnon, S. A., and D. M. A. Jones, 1972. "Review of the Influences of the Great Lakes on Weather," *Water Resources Research* 4(2):360.

Changnon, S. A., and J. L. Vogel, 1981. "Hydroclimatological Characteristics of Isolated Severe Rainstorms," *Water Resources Research* 17(6):1694.

Cheng, J. D., 1989. "Streamflow Changes After Clear-Cut Logging of a Pine Beetle-Infested Watershed in Southern British Columbia, Canada," *Water Resources Research* 25(3):449.

Chery, D. E., 1968. *Output Response to Pulse Inputs of a 'Scaled' Laboratory Watershed System*, U.S. Department of Agriculture, Southwest Watershed Research Center, Tucson, AZ.

Chiang, S. L., and G. W. Petersen, 1970. "Soil Catena Concept for Hydrologic Interpretations," *Journal of Soil and Water Conservation* 25(6):225.

Chow, V. T. 1964. *Handbook of Applied Hydrology*. McGraw-Hill Book Co., Inc., New York, NY.

-------, 1967. "Laboratory Study of Watershed Hydrology," *Proceedings*, International Hydrology Symposium, Fort Collins, CO I:194.

Chow, V. T., and S. J. Kareliotis, 1970. "Analysis of Stochastic Hydrologic Systems," *Water Resources Research* 6(6):1569.

Clary, W. P., and P. F. Ffolliott, 1969. "Water Holding Capacity of Pondersoa Pine Forest Floor Layers," *Journal of Soil and Water Conservation* 24(1):22.

Cline, R. G., and R. L. Jeffers, 1975. "Installation of Neutron Probe Access Tubes in Stony and Bouldery Forest Soils," *Soil Science* 120(1):71.

Cole, D. W., 1968. "A System for Measuring Conductivity, Acidity, and Rate of Water Flow in a Forest Soil," *Water Resources Research* 4(5):1127.

Coleman, G., and D. G. DeCoursey, 1976. "Sensitivity and Model Variance Analysis Applied to Some Evaporation and Evapotranspiration Models," *Water Resources Research* 12(5):873.

Collins, M. R., and R. M. Myrick, 1966. *Effects of Juniper and Pinyon Eradication on Streamflow from Corduroy Creek Basin, Arizona* Professional Paper 491-B, Geological Survey, USDI, Washington, DC, 12 pp.

Colman, E. A., and T. M. Hendrix, 1949. "The Fiberglas Electrical Soil-Moisture Instrument," *Soil Science* 67(6):425.

-------, 1953. *Vegetation and Watershed Management*. The Ronald Press Company, New York, NY.

Cooper, C. F., 1967. "Rainfall Intensity and Elevation in Southwestern Idaho," *Water Resources Research* 3(1):181.

Corbett, E. S., and J. A. Lynch, 1989. "Hydrologic Production Zones in a Headwater Watershed," In *Headwaters Hydrology*, Proceedings of the Symposium, Woessner and Potts, editors. p. 573.

Corps of Engineers, n.d. *The Great Flood of 1972*. North Atlantic Division, New York, NY.

-------, 1976. *A Perspective on Flood Plain Regulations for Flood Plain Management*. Office of the Chief of Engineers, Department of the Army, Washington, DC, 20314.

-------, 1976. *A Perspective on Flood Plain Regulations for Flood Plain Management*. Department of the Army, Washington, DC.

Cotecchia, V., A. Inzaghi, E. Pirastru, and R. Ricchena, 1968. "Influence of the Physical and Chemical Properties of Soil on Measurements of Water Content Using Neutron Probes," *Water Resources Research* 4(5):1023.

Council on Environmental Quality, 1981. *Global Energy Futures and the Carbon Dioxide Problem*. United States Government Printing Office, Washington, DC.

Cowardin, L. M., V. Carter, F. C. Golet, and E. T. LaRoe, 1979. *Classification of Wetlands and Deepwater Habitats of the United States*, FWS/OBS/-79/31, USF&WS, USDI, Washington, DC, 103 pp.

Cox, L. M., W. J. Rawls, and J. F. Zuzel, 1975. "Snow: Nature's Reservoir," *Water Resources Bulletin* 11(5):1009.

Cox, M. B., 1952. "Recording the Intake of Water into the Soil," *Journal of Soil and Water Conservation* 7(2):79.

Crawford, N. H., and R. K. Linsley, Jr., 1966. *Digital Simulation in Hydrology: Stanford Watershed Model IV*, Technical Report 39, Department of Civil Engineering, Stanford University, Palo Alto, CA.

Cronan, C. S., and C. L. Schofield, 1979. "Aluminum Leaching Response to acid Precipitation: Effects on High-Elevation Watersheds in the Northeast," *Science* 204(4390):304.

Crouse, R. P., and L. W. Hill, 1962. "What's Happening at San Dimas?" Miscellaneous Paper No. 68, Pacific Southwest Forest and Range Experiment Station, Forest Service, USDA, Berkeley, CA.

Crouse, R. P., E. S. Corbett, and D. W. Seegrist, 1966. "Methods of Measuring and Analyzing Rainfall Interception by Grass," *Bulletin of the International Association of Scientific Hydrology* XI(2):110.

Curtis, W. R., 1966. "Forest Zone Helps Minimize Flooding in the Driftless Area," *Journal of Soil and Water Conservation* 21(3):101.

Dake, J. M. K., 1972. "Evaporative Cooling of a Body of Water," *Water Resources Research* 8(4):1087.

Dalrymple, T., 1960. *Flood-Frequency Analysis*, Water-Suply Paper 1543-A, Geological Survey, USDI, Government Printing Office, Washington, DC.

Datta, B., and M. H. Houck, 1984. "A Stochastic Optimization Model for Real-Time Operation of Reservoirs Using Uncertain Forecasts," *Water Resources Research* 20(8):1039.

Davenport, D. C., J. E. Anderson, L. W. Gay, B. E. Kynard, E. K. Bonde, and R. M. Hagan, 1979. "Phreatophyte Evapotranspiration and its Potential Reduction without Eradication," *Water Resources Bulletin* 15(5):1293.

Davenport, D. C., P. E. Martin, E. B. Roberts, and R. M. Hagan, 1976. "Conserving Water by Antitranspirant Treatment of Phreatophytes," *Water Resources Research* 12(5):985.

Davenport, D. C., P. E. Martin, and R. M. Hagan, 1976. "Aerial Spraying of Phreatophytes with Antitranspirant," *Water Resources Research* 12(5):991.

———, 1982a. "Evapotranspiration from Riparian Vegetation: Water Relations and Irrecoverable Losses for Saltcedar," *Journal of Soil and Water Conservation* 37(4):233.

———, 1982b. "Evapotranspiration from Riparian Vegetation: Conserving Water by Reducing Saltcedar Transpiration," *Journal of Soil and Water Conservation* 37(4):237.

Davis, E. A., 1984. "Conversion of Arizona Chaparral to Grass Increases Water Yield and Nitrate Loss," *Water Resources Research* 20(11):1643.

———, 1987. "Chaparral Conversion and Streamflow: Nitrate Increase is Balanced Mainly by a Decrease in Bicarbonate," *Water Resources Research* 23(1):215.

Dawdy, D. R., and W. B. Langbein, 1960. "Mapping Mean Areal Precipitation," *Bulletin* of the International Association of Scientific Hydrology, 19:16.

Day, F. P., and C. D. Monk, 1977. "Seasonal Nutrient Dynamics in the Vegetation on a Southern Appalachian Watershed," *American Journal of Botany* 64:1126.

DeBano, L. F., 1969. "Water Repellant Soils: A Worldwide Concern in Management of Soil and Vegetation," *Agricultural Science Review* 7(2):13.

-------, 1977. "Influence of Forest Practices on Water Yield, Channel Stability, Erosion, and Sedimentation in the Southwest," *Proceedings* of the National Convention of the Society of American Foresters, Albuquerque, NM, pp. 74-78.

DeBano, L. F., P. H. Dunn, and C. E. Conrad, 1977. "Fire's Effect on Physical and Chemical Properties of Chaparral Soils," Technical Report WO-3, Forest Service, USDA, Washington, DC.

DeBano, L. F., and L. F. Schmidt, 1989. *Improving Southwestern Riparian Areas Through Watershed Management*, General Technical Report RM-182, Rocky Mountain Forest and Range Experiment Station, Forest Service, USDA, Fort Collins, CO 33pp.

DeByle, N. V., and H. F. Haupt, 1965. *The Intermountain Precipitation Storage Gage*. Research Note INT-34, Forest Service, Intermountain Forest and Range Experiment Station, Ogden, UT.

DeByle, N. V., and P. E. Packer, 1972. "Plant Nutrient and Soil Losses in Overland Flow from Burned Forest Clearcuts," *Proceedings* of National Symposium of Watersheds in Transition, American Water Resources Association, Ft. Collins, CO, pp. 296-307.

Decker, J. P., 1963. "An Analysis and Simplification of the Blaney-Criddle Method for Estimating Evapotranspiration," Research Note RM-2, Forest Service, Rocky Mountain Forest and Range Experiment Station, Fort Collins, CO.

de la Cruz, A. A., 1978. "Primary Production Processes: Summary and Recommendations," in Good et al., p. 79.

Dennis, A. S., 1970. "Modifying Precipitation by Cloud Seeding," *Journal of Soil and Water Conservation* 25(3):88.

Dennis, H. W., and E. C. Griffin, 1971. "Some Effects of Trincheras on Small River Basin Hydrology," *Journal of Soil and Water Conservation* 26(6):240.

Department of Agriculture, 1938. *Soils and Men: the 1938 Yearbook of Agriculture*, Washington, DC.

Department of Agriculture, 1955. *Water: the 1955 Yearbook of Agriculture*, Washington, DC.

Department of Environmental Conservation, 1989. *Nonpoint Source Assessment Report*, Division of Water, Bureau of Water Quality Management, Albany, NY.

-------, 1990. *Nonpoint Source Management Program*. Division of Water, Bureau of Water Quality Management, Albany, NY.

Department of Natural Resources, 1955. "Soil-Vegetation Surveys in California," Division of Forestry, Sacramento, CA.

Deutsch, M., D. R. Wiesnet, and A. Rango, editors, 1981. *Satellite Hydrology*, Proceedings of the Fifth Annual Illiam T. Pecora Memorial Symposium on Remote Sensing, Sioux Falls, SD. American Water Resources Association, Bethesda, MD.

DeWalle, D. R., and J. R. Meiman, 1971. "Energy Exchange and Late Season Snowmelt in a Small Opening in Colorado Subalpine Forest," *Water Resources Research* 7(1):184.

DeWalle, D. R., and A. Rango, 1972. "Water Resources Applications of Stream Channel Characteristics on Small Forested Basins," *Water Resources Bulletin* 8(4):697.

Dickerson, R. R. et al., 1987. "Thunderstorms: An Important Mechanism in the Transport of Air Pollutants," *Science* 235:460.

Dils, R. E., 1953, "Influence of Forest Cutting and Mountain Farming on Some Vegetation, Surface Soil and Surface Runoff Characteristics," USDA, Forest Service, Southeastern Forest Experiment Station, Station Paper No. 24, Asheville, NC.

-------, 1957. *A Guide to the Coweeta Hydrologic Laboratory*, USDA, Forest Service, Southeastern Forest Experiment Station, Asheville, NC.

Dingman, S. L., 1978a. "Synthesis of Flow-Duration Curves for Unregulated Streams in New Hampshire," *Water Resources Bulletin* 14(6):1481.

-------, 1978b. "Drainage Density and Streamflow: a Closer Look," *Water Resources Research* 14(6):1183.

-------, 1984. *Fluvial Hydrology*. W. H. Freeman and Company, New York, NY, 383, pp.

Dissmeyer, G. E., E. S. Corbett, and W. T. Swank, 1975. "Summary of Municipal Watershed Management Surveys in the Eastern United States," in Sopper and Corbett, 1975. pp. 185-196.

Dortignac, E. J., 1951. *Design and Operation of Rocky Mountain Infiltrometer*, USDA, Forest Service, Station Paper No. 5, Rocky Mountain Forest and Range Experiment Station, Fort Collins, CO.

Doty, R. D., 1970. "Influence of Contour Trenching on Snow Accumulation," *Journal of Soil and Water Conservation* 25(3):102.

-------, 1983. "Stream Flow in Relation to Ohia Forest Decline on the Island of Hawaii," *Water Resources Bulletin* 19(2):217.

Douglass, J. E., 1962a. *Variance of Nuclear Moisture Measurements*. USDA, Forest Service, Southeastern Forest Experiment Station, Station Paper No. 143, Asheville, NC.

\-\-\-\-\-\-\-, 1962b. *A Method for Determining the Slope of Neutron Moisture Meter Calibration Curves*. USDA, Forest Service, Southeastern Forest Experiment Station, Station Paper No. 154, Asheville, NC.

\-\-\-\-\-\-\-, 1983. "The Potential for Water Yield Augmentation from Forest Management in the Eastern United States," *Water Resources Bulletin* 19(3):351.

Douglass, J. E., and W. T. Swank, 1975. "Effects of Management Practices on Water Quality and Quantity: Coweeta Hydrologic Laboratory, North Carolina," In *Proceedings*, Municipal Watershed Management Symposium, edited by Sopper and Corbett, pp. 1-13.

Dracup, J. A., K. S. Lee, and E. G. Paulson, Jr., 1980. "On the Definition of Droughts," *Water Resources Research* 16(2):297.

Dragoun, F. J., 1969. "Effects of Cultivation and Grass on Surface Runoff," *Water Resources Research* 5(5):1078.

Dress, P. E., and R. C. Field, 1987. *The 1985 Symposium on Systems Analysis in Forest Resources*. Georgia Center for Continuing Education, Athens, GA.

Duckstein, L., and C. C. Kisiel, 1971. "Efficiency of Hydrologic Data Collection Systems Role of Type I and II Errors," *Water Resources Bulletin* 7(3):592.

Dunford, E. G., 1966. "Forests-Users of Water," *Proceedings* of the Annual Meeting of the Society of American Foresters, Seattle, WA, pp.146-50.

Dunne, T., 1978. "Field Studies of Hillslope Flow Processes," Chapter 7 in Kirkby, 1978. pp. 227-93.

Dunne, T., and R. D. Black, 1970. "Partial Area Contributing to Storm Runoff in a Small New England Watershed," *Water Resources Research* 6(5):1296.

\-\-\-\-\-\-\-, 1971. "Runoff Processes during Snowmelt," *Water Resources Research* 7(5):1160.

Dunne, T., and L. B. Leopold, 1978. *Water in Environmental Planning*. W. H. Freeman and Company, San Francisco, CA.

Duvick, D. N., and T. J. Blasing, 1981. "A Dendroclimatic Reconstruction of Annual Precipitation Amounts in Iowa Since 1680," *Water Resources Research* 17(4):1183.

Eagleson, P. S., 1972. "Dynamics of Flood Frequency," *Water Resources Research* 8(4):878.

\-\-\-\-\-\-\-, 1978a. "Climate, Soil, and Vegetation: 3. A Simplified Model of Soil Moisture Movement in the Liquid Phase," *Water Resources Research* 14(5):722.

\-\-\-\-\-\-\-, 1978b. "Climate, Soil, and Vegetation: 4. The Expected Value of Annual Evapotranspiration," *Water Resources Research* 14(5):731.

-------, 1982. "Ecological Optimality in Water-Limited Natural Soil-Vegetation Systems: 1. Theory and Hypotheses," *Water Resources Research* 18(2):325.

Eagleson, P. S., and T. E. Tellers, 1982. "Ecological Optimality in Water-Limited Natural Soil-Vegetation Systems: 2. Tests and Applications," *Water Resources Research* 18(2):341.

Edgington, J. M., and G. L. Rolfe, 1974. "An Inexpensive Sampler for Monitoring Water Quality," *Forestry Research Report* 74-2, University of Illinois at Urbana-Champaign, IL.

Eisel, L. M., 1972. "Watershed Management: a Systems Approach," *Water Resources Research* 8(2):326.

Eisenlohr, W. S., 1966. "Water Loss from a Natural Pond through Transpiration by Hydrophytes," *Water Resources Research* 2(3):443.

-------, 1967. "Measuring Evapotranspiration from Vegetation-Filled Prairie Potholes in North Dakota," *Water Resources Bulletin* 3(1):59.

Eschner, A. R., 1965. *Forest Protection and Streamflow From an Adirondack Watershed.* Doctoral Dissertation, SUNY College of Environmental Science and Forestry, Syracuse, NY, 209 pp.

-------, 1966. "Interception and Soil Moisture Distribution," *Proceedings*, International Symposium on Forest Hydrology, The Pennsylvania State University, University Park, PA.

-------, 1980. "*De Facto* Snowpack Management in the Northeast," Paper presented at ASCE Watershed Management Symposium, Boise, ID 11 pp., mimeographed.

Eschner, A. R., and D. R. Satterlund, 1963. "Snow Deposition and Melt Under Different Vegetative Covers in Central New York," US Forest Service Research Note NE-13, Northeastern Forest Experiment Station, Upper Darby, PA, 6 pp.

-------, 1966. "Forest Protection and Streamflow from an Adirondack Mountain Watershed," *Water Resources Research* 2(4):765.

Escobar, L. A., and I. Rodriguez-Iturbe, 1982. "A Modeling Scheme for the Study of Drainage Density," *Water Resources Research* 18(4):1029.

Ettenheim, G. P., 1962. *Fog Studies at Arcata, California Under "Operation Pea Soup,"* Technical Note No. 571, AFCRL-62-804, Aeronautical Icing Research Laboratories for Office of Aerospace Research, United States Air Force, Bedford, MA.

Farmer, E. E., and J. E. Fletcher, 1972. "Rainfall Intensity-Duration-Frequency Relations for the Wasatch Mountains of Northern Utah," *Water Resources Research* 8(1):266.

Faust, S. D., and O. M. Ally, 1981. *Chemistry of Natural Waters*. Ann Arbor Science Publishers, Inc., Ann Arbor, MI, 400 pp.

Federer, C. A., 1973. "Forest Transpiration Greatly Speeds Streamflow Recession," *Water Resources Research* 9(6):1599.

-------, 1982. "Transpirational Supply and Demand: Plant, Soil, and Atmospheric Effecrts Evaluated by Simulation," *Water Resources Research* 18(2):355.

Federer, C. A., and D. Lash, 1978. "Simulated Streamflow Response to Possible Differences in Transpiration Among Species of Hardwood Trees," *Water Resources Research* 14(6):1089.

Feller, M. C., and J. P. Kimmins, 1984. "Effects of Clearcutting and Slash Burning on Streamwater Chemistry and Watershed Nutrient Budgets in Southwestern British Columbia," *Water Resources Research* 20(1):29.

Ferguson, E. R., and W. B. Duke, 1954. "Devices to Facilitate King-Tube Soil-Moisture Samples," In *Some Field, Laboratory, and Office Procedures for Soil-Moisture Measurement*, USDA, Forest Service, Southern Forest Experiment Station, Occasional Paper 135, pp. 26-29.

Ferguson, R. I., 1985. "Runoff From Glacierized Mountains: A Model For Annual Variation and Its Forecasting," *Water Resources Research* 21(5):702.

Ffolliott, P. F., G. J. Gottfried, and M. B. Baker, Jr., 1989. "Water Yield from Forest Snowpack Management: Research Findings in Arizona and New Mexico," *Water Resources Research* 25(9):1999.

Ffolliott, P. F., and D. B. Thorud, 1977. "Water Yield Improvement by Vegetation Management," *Water Resources Bulletin* 13(3):563.

Ffolliott, P. R., G. J. Gottfried, and M. B. Baker, 1989. "Water Yield From Fores Snowpack Management: Research Findings in Arizona and New Mexico," *Water Resources Research* 25(9):1999.

Field, R., 1985. "Urban Runoff: Pollution Sources, Control, and Treatment," *Water Resources Bulletin* 21(2):197.

Fisher, R. T., 1903. "A Study of the Redwood," In *The Redwood*, Bureau of Forestry, USDA Bull. 38, Government Printing Office, Washington, DC. pp 7-28.

Fisk, D., 1989. *Wetlands: Concerns and Successes*. Proceedings of the AWRA Symposium, Tampa, Florida. American Water Resources Association, 5410 Grosvenor Lane, Suite 220, Bethesda, MD 20814.

FitzPatrick, E. A., 1971. *Pedology: a Systematic Approach to Soil Science*. Oliver & Boyd, Edinburgh.

Fleming, P. M., 1970. "A Diurnal Distribution Function for Daily Evaporation," *Water Resources Research* 6(3):937.

Follett, R. F., G. A. Reichman, E. J. Doering, and L. C. Benz, 1973. "A Nomograph for Estimating Evapotranspiration," *Journal of Soil and Water Conservation* 28(2):90.

Forest Service, 1946. *Water and Our Forests*, Misc. Pub. No. 600, USDA< Washington, DC.

-------, 1961. *Handbook on Soils*, Department of Agriculture, Washington, DC, [Offset, looseleaf].

-------, 1980. *An Approach to Water Resources Evaluation of Non-Point Silvicultural Sources (A Procedural Handbook)*, EPA-600/8-80-012 (2 volumes), USDA, Washington, DC.

Foroud, N., R. S. Broughton, and G. L. Austin, 1984. "The Effects of a Moving Rainstorm on Direct Runoff Properties," *Water Resources Bulletin* 20(1):87.

Foster, I. D. L., I. C. Grieve, and A. D. Christmas, 1981. "The Use of Specific Conductance in Studies of Natural Waters and Soil Solutions," *Hydrological Sciences Bulletin* 26(3):257.

Fralish, J. S., 1977. "Upland Forest Ecosystems and their Relation to Water Quality," in Aubertin, 1977.

Francko, D. A., and R. G. Wetzel, 1986. *To Quench our Thirst*. The University of Michigan Press, Ann Arbor, MI.

Frank, B., and C. A. Betts, 1946. *Water and Our Forests*, USDA Miscellaneous Publication No. 600, Washington, DC, 29 pp.

Frank, E. C., and R. Lee, 1966. *Potential Solar Beam Irradiation on Slopes: Tables for 30% to 50% Latitude*. U.S. Forest Service Research Paper RM-18, Fort Collins, CO.

Freeze, R. A., 1972. "Role of Subsurface Flow in Generating Surface Runoff: 2. Upstream Source Areas," *Water Resources Research* 8(5):1272.

Freeze, R. A., and J. A. Cherry, 1979. *Groundwater*. Prentice-Hall, Inc., Englewood Cliffs, NJ.

Frere, M. H., D. A. Woolshiser, J. H. Caro, B. A. Stewart, and W. R. Wischmeier, 1977. "Control of Nonpoint Water Pollution from Agriculture: Some Concepts," *Journal of Soil and Water Conservation* 32(8):260.

Fritschen, L. J., and P. Doraiswamy, 1973. "Dew: and Addition to the Hydrologic Balance of Douglas Fir," *Water Resources Bulletin* 9(4):891.

Fritschen, L. J., J. Hsia, and P. Doraiswamy, 1977. "Evapotranspiration of a Douglas Fir Determined With a Weighing Lysimeter," *Water Resources Research* 13(1):145.

Frohliger, J. O., and R. Kane, 1975. "Precipitation: its Acidic Nature," *Science* 189:455.

Gardiner, V., 1979. "Estimation of Drainage Density From Topological Variables," *Water Resources Research* 15(4):909.

Gardner, W. H., 1968. "How Water Moves in Soil," *Crops and Soils Magazine* [volume, number, and page designations not included in reprint].

Garstka, W. U., L. D. Love, B. C. Goodell, and F. A. Bertle, 1958. *Factors Affecting Snowmelt and Streamflow*, Bureau of Reclamation, USDI, and Forest Service, USDA, Rocky Mountain Forest and Range Experiment Station, Fort Collins, CO 189 pp.

Gary, H. L., 1972. "Rime Contributes to Water Balance in High-Elevation Aspen Forests," *Journal of Forestry* 70(2):93.

-------, 1974. "Snow Accumulation and Snowmelt as Influenced by a Small Clearing in a Lodgepole Pine Forest," *Water Resources Research* 10(2):348.

-------, 1975. "Airflow Patterns and Snow Accumulation in a Forest Clearing," *Proceedings* Western Snow Conference, 43:106.

-------, 1980. "Patch Clearcuts to Manage Snow in Lodgepole Pine," *Proceedings*, ASC Watershed Management Symposium, Boise, ID pp 335-46.

-------, 1982. "Stream Water Quality in a Small Commercial Campground in Colorado," *Journal of Environmental Health* 45(1):5.

Gaskin, J. W., J. E. Douglass, and W. T. Swank, compilers. 1983. *Annotated Bibliography of Publications on Watershed Management and Ecological Studies at Coweeta Hydrologic Laboratory, 1934, 1984*. General Technical Report SE-30, Southeastern Forest Experiment Station, Forest Service, USDA, Asheville, NC, 140 pp.

Geiger, R., 1957. *The Climate Near the Ground*. Harvard University Press, Cambridge, MA. 494 pp.

Gentry, R. C., 1970. "Hurricane Debbie Experiments, August, 1969," *Science* 168:473.

Geological Survey, 1967. *Magnitude and Frequency of Floods in the United States: Part 11. Pacific Slope Basins in California.* Water-Supply Paper 1685. Department of The Interior, Government Printing Office, Washington, DC.

-------, 1976. *Surface Water Supply of the United States, 1966-70: Part 1. North Atlantic Slope Basins, Volume 3, Basins from Maryland to York River*. Water-Supply Paper 2103. Department of the Interior, Government Printing Office, Wasington, DC.

Gibbs, R. J., 1970. "Mechanisms Controlling World Water Chemistry," *Science* 170:1088.

Gillette, R., 1972. "Stream Channelization: Conflict between Ditchers, Conservationists," *Science* 176:890.

References 331

Gleick, J., 1987. *Chaos: Making a New Science*. Viking Penguin, Inc., New York, NY, 352 pp.

Gleick, P. H., 1989. "Climate Change, Hydrology, and Water Resources," *Reviews of Geophysics* 27(3):329.

Golding, D. L., and R. H. Swanson, 1986. "Snow Distribution Patterns in Clearings and Adjacent Forest," *Water Resources Research* 22(13):1931.

Good, R. E., D. F. Whigham, R. L. Simpson, and C. G. Jackson, Jr., 1978. *Freshwater Wetlands: Ecological Processes and Management Potential*. Academic Press, New York, NY.

Goodell, B. C., 1958. "Watershed Studies at Fraser, Colorado," *Proceedings*, Society of American Foresters, pp. 42-5.

-------, 1963. "A Reappraisal of Precipitation Interception by Plants and Attendant Water Loss," *Journal of Soil and Water Conservation* 18(6):231.

-------, 1966. "Watershed Treatment Effects on Evapotranspiration," in *International Symposium on Forest Hydrology*, Pennsylvania State University, University Park, PA, Pergamon Press, NY, pp. 477-82.

Goodman, J., 1985. "The Collection of Fog Drip," *Water Resources Research* 21(3):392.

Gosselink, J. G., and R. E. Turner, 1978. "The Role of Hydrology in Freshwater Wetland Ecosystems," in Good et al., p. 63.

Government Accounting Office, 1988. *Public Rangelands*. GAO/RCED-88-105, Washington, DC, 85 pp.

Gray, D. M., and D. H. Male, 1981. *Handbook of Snow*. Pergamon Press, Toronto, Canada, 776 pp.

Greeson, P. E., 1978. "Microbiological Monitoring for Water-Quality Assessment," *Journal of Food Protection* 41(4):309.

Greeson, P. E., J. F. Clark, and J. E. Clark, 1978. Wetland Functions and Values: the State of Our Understanding. American Water Resources Association, 5410 Grosvenor Lane, Suite 220, Bethesda, MD 20814.

Hage, K. D., 1975. "Averaging Errors in Monthly Evaporation Estimates," *Water Resources Research* 11(3):359.

Haines, B. L., 1984. "Kinetics of Nutrient Uptake," paper presented at Fiftieth Anniversary Symposium on Long Range Research on Forested Watersheds at the Coweeta Hydrologic Laboratory, University of Georgia, Athens, GA.

Hall, F. R., 1968. "Base-Flow Recession — A Review," *Water Resources Research* 4(5):973.

Hamilton, C. E., Editor, 1978. *Manual on Water.* STP 442A, American Society of Testing Materials, 1916 Race Street, Philadelphia, PA 19103.

Hamilton, E. L., 1954. *Rainfall Sampling on Rugged Terrain.* Technical Bulletin No. 1096, U.S. Department of Agriculture, Government Printing Office, Washington, DC.

-------, and L. F. Reimann, 1958. *Simplified Method of Sampling Rainfall on the San Dimas Experimental Forest,* Technical Paper No. 26, Forest Service, California Forest and Range Experiment Station, Berkeley, CA.

Hammond, A. L., 1971. "Weather Modification: a Technology Coming of Age," *Science* 172:548.

-------, 1973a. "Weather and Climate Modification: Progress and Problems," *Science* 181:644.

-------, 1973b. "Hurricane Prediction and Control: Impact of Large Computers" *Science* 181:643.

Hannaford, J. F., and A. J. Brown, 1979. "Application of Snow Covered Area to Runoff Forecasting in the Sierra Nevada, California," In *Satellite Hydrology* edited by Deutsch et al. 1981, pp. 165-72.

Harbaugh, T. E., 1966. *Time Distribution of Runoff from Watersheds,* Ph.D. Dissertation, University of Illinois, Urbana, IL.

Harr, R. D., 1982. "Fog Drip in the Bull Run Municipal Watershed, Oregon," *Water Resources Bulletin* 18(5):785.

-------, 1983. "Potential for Augmenting Water Yield Through Forest Practices in Western Washington and Western Oregon," *Water Resources Bulletin* 19(3):383.

Harr, R. D., and R. L. Fredriksen, 1988. "Water Quality After Logging Small Watersheds Within the bull Run Watershed, Oregon," *Water Resources Bulletin* 24(5):1103.

Harr, R. D., A. Levno, and R. Mersereau, 1982. "Streamflow Changes after Logging 130-year-old Douglas Fir on Two Small watersheds," *Water Resources Research* 18(3):637.

Harr, R. D., and F. M. McCorison, 1979. "Initial Effects of Clearcut Logging on Size and Timing of Peak Flows in a Small Watershed in Western Oregon," *Water Resources Research* 15(1):90.

Harris, H. J. H., K. Cartwright, and T. Torii, 1979. "Dynamic Chemical Equilibrium in a Polar Desert Pond: A Sensitive Index of Meteorological Cycles," *Science* 204:301.

Harrison, W. D., and T. E. Osterkamp, 1981. "A Probe Method for Soil Water Sampling and Subsurface Measurements," Water Resources Research 17(6):1731.

References 333

Hart, G., 1966. "Forest Cutting to Increase Streamflow in the White Mountains," *N. H. Forest Notes*, 4 pp.

Hart, G. E., and D. A. Lomas, 1979. "Effects of Clearcutting on Soil Water Depletion in an Engelmann Spruce Stand," *Water Resources Research* 15(6):1598.

Hasler, A. D., and B. Ingersoll, 1968. "Dwindling Lakes," *Natural History* 77(9):8.

Haupt, H. F., 1970. "Relation of Wind Exposure and Forest Cutting to Changes in Snow Accumulation," *Proceedings*, Symposium on Modification of Snowfall, Snowcover, and Ice Cover, Banff, AL, 12 pp.

-------, 1972. "Relation of Wind Exposure and Forest Cutting to Changes in Snow Accumulation," *Proceedings*, International Symposium on the Role of Snow and Ice in Hydrology, Banff, AL, Canada.

-------, 1979a. *Local Climatic and Hydrologic Consequences of Creating Openings in Climax Timber of North Idaho*, Research Paper INT-223, Intermountain Forest and Range Experiment Station, Ogden, UT, 43 pp.

-------, 1979b. *Effects of Timber Cutting and Revegetation on Snow Accumulation and Melt in North Idaho*. Research Paper INT-224, Intermountain Forest and Range Experiment Station, Ogden, UT, 14 pp.

Hawkins, R. H., 1970. "Effect of Streamflow Regimen on Reservoir Yield," *Water Resources Research* 5(5):1115.

Hayes, G. L., 1944. "A Method of Measuring Rainfall on Windy Slopes," *Monthly Weather Review* 72:111.

Heaney, J. P., and W. C. Huber, 1984. "Nationwide Assessment of Urban Runoff Impact on Receiving Water Quality," *Water Resources Bulletin* 20(1):35.

Heede, B. H., 1960. *A Study of Early Gully-Control Structures in the Colorado Front Range*, Station Paper No. 55, Rocky Mountain Forest and Range Experiment Station, Forest Service, USDA, Fort Collins, CO. 42 pp.

-------, 1967. "The Fusion of Discontiuous Gullies — a Case Study," *Bulletin of the International Association of Scientific Hydrology* 12:42 .

-------, 1972. *Flow and Channel Characteristics of Two High Mountain Streams*, Research Paper RM-96, Rocky Mountain Forest and Range Experiment Station, Forest Service, USDA, Ft. Collins, CO.

-------, 1976. *Gully Development and Control: the Status of our Knowledge*, Research Paper RM-169, Rocky Mountain Forest and Range Experiment Station, Forest Service, USDA, Ft. Collins, CO.

-------, 1980. *Stream Dynamics: an Overview for Land Managers*, General Technical Report RM-72, Rocky Mountain Forest and Range Experiment Station, Forest Service, USDA, Ft. Collins, CO.

-------, 1985. "Interactions Between Stream-side Vegetation and Stream Dynamics," In *Riparian Ecosystems and their Management: Reconciling Conflicting Uses, Proceedings*, General Technical Report RM-120, Rocky Mountain Forest and Range Experiment Station, Forest Service, USDA, Ft. Collins, CO, pp. 54-58.

-------, 1987. "Overland Flow and Sediment Delivery Five Years After Timber Harvest in a Mixed Conifer Forest (Arizona)," *Journal of Hydrology* 91(1987):205.

Helmers, A. E., 1954. "Precipitation Measurements on Wind-Swept Slopes," *Transactions*, American Geophysical Union 35(3):471.

Helvey, J. D., 1964. *Rainfall Interception by Hardwood Forest Litter in the Southern Appalachians*, U.S. Forest Service Research Paper SE-8, Southeast Forest Experiment Station, Asheville, NC, 8 pp.

-------, 1967. "Interception by Eastern White Pine," *Water Resources Research* 3(3):723.

Helvey, J. D., and W. B. Fowler, 1980. "A New Method for Sampling Snow Melt and Rainfall in Forests", *Water Resources Bulletin* 16(5):938.

Helvey, J. D., and J. H. Patric, 1965. "Canopy and Litter Interception of Rainfall by Hardwoods of Eastern United States," *Water Resources Research* 1(2):193.

Hem, J. D., 1971. *Study and Interpretation of the Chemical Characteristics of Natural Water*. Second Edition. Geological Survey Water-Supply Paper No. 1473, U.S. Government Printing Office, Washington, DC.

Hendricks, E. L., 1962. "Hydrology," *Science* 135(3505):699.

Hershfield, D. M., 1961. *Rainfall Frequency Atlas of the United States*. Technical Bulletin No. 40, National Weather Service, Government Printing Office, Washington, DC.

-------, 1969. "A Note on Areal Rainfall Definition," *Water Resources Bulletin* 5(3):49.

Hewlett, J. D., 1958. "Pine and Hardwood Forest Water Yield," *Journal of Soil and Water Conservation* 13(3):106.

-------, 1961. *Soil Moisture as a Source of Base Flow from Steep Mountain Watersheds*, USDA, Forest Service, Station Paper No. 132, Southeastern Forest Experiment Station, Asheville, NC.

-------, 1971. "Comments on the Catchment Experiment to Determine Vegetal Effects on Water Yield," *Water Resources Bulletin* 7(2):376.

References

------, 1982. *Principles of Hydrology*. University of Georgia Press, Athens, Ga.

Hewlett, J. D., and J. E. Douglass, 1961. "A Method for Calculating Error of Soil Moisture Volumes in Gravimetric Sampling," *Forest Science* 7:265.

Hewlett, J. D., and A. R. Hibbert, 1967. "Factors Affecting the Response of Small Watersheds to Precipitation in Humid Areas," In *Proceedings of International Symposium on Forest Hydrology*, Sopper and Lull, editors, p. 275.

Hewlett, J. D., and J. D. Helvey, 1970. "Effects of Forest Clear-Felling on the Storm Hydrograph," *Water Resources Research* 6(3):768.

Hewlett, J. D., and W. L. Nutter, 1969. *An Outline of Forest Hydrology*, University of Georgia Press, Athens, GA.

Hibbert, A. R., 1967. "Forest Treatment Effects on Water Yield," In Sopper and Lull (1967), pp. 527-43.

------, 1969. "Water Yield Changes after Converting a Forested Catchment to Grass," *Water Resources Research* 5(3):634.

------, 1971. "Increases in Streamflow after Converting Chaparral to Grass," *Water Resources Research* 7(1):71.

------, 1983. "Water Yield Improvement Potential by Vetation Management on Western Rangelands," *Water Resources Bulletin* 19(3):375.

Hibbert, A. R., and E. A. Davis, 1986. "Streamflow Response to Converting Arizona Chaparral in a Mosaic Pattern," In *Hydrology and Water Resources in Arizona and the Southwest*, American Water Resources Association, Tucson, AZ, pp. 123-31.

Hibbert, A. R., E. A. Davis, and D. G. Scholl, 1974. *Chaparral Conversion Potential in Arizona, Part I: Water Yield Response and Effects on Other Resources*, Research Paper RM-126, Forest Service, USDA, Rocky Mountain Forest and Range Experiment Station, Fort Collins, CO, 36 pp.

Higgins, D. A., A. R. Tiedemann, T. M. Quigley, and D. B. Marx, 1989. "Streamflow Characteristics of Small Watersheds in the Blue Mountains of Oregon," *Water Resources Bulletin* 25(6):1131.

Higgins, R. J., 1981. "Use and Modification of a Simple Rainfall-Runoff Model for Wet Tropical Catchments," *Water Resources Research* 17(2):423.

Hidore, J. J., 1971. "Annual, Seasonal, and Monthly Changes in Runoff in the United States," *Water Resources Bulletin* 7(3):554.

Hill, A. R., 1986. "Stream Nitrate-N Loads in Relation to Variations in Annual and Seasonal Runoff Regimes," *Water Resources Bulletin* 22(5):829.

Hill, R. W., A. L. Huber, E. K. Israelsen, and J. P. Riley, 1972. "A Self-Verifying Hybrid Computer Model of River-Basin Hydrology," *Water Resources Bulletin* 8(5):909.

Hillel, D., 1980. *Fundamentals of Soil Physics*, Academic Press, Inc., New York, NY.

Hindman, E. E., R. D. Borys, and P. J. DeMott, 1983. "Hydrometeorological Significance of Rime Ice Deposits in the Colorado Rockies," *Water Resources Bulletin* 19(4):619.

Hirsch, R. M., J. R. Slack, and R. A. Smith, 1982. "Techniques of Trend Analysis for Monthly Water Quality Data," *Water Resources Research* 18(1):107.

Hofmann, W., and S. E. Rantz, 1968. "What is Drought?," *Journal of Soil and Water Conservation* 23(3):105.

Hollis, G. E., 1975. "The Effect of Urbanization on Floods of Different Recurrence Interval," *Water Resources Research* 11(3):431.

Holtan, H. N., 1966. "A Model for Computing Watershed Retention from Soil Parameters," *Journal of Soil and Water Conservation* 20(3):91.

Hoover, M. D., 1953. *Interception of Rainfall in a Young Loblolly Pine Plantation*. U.S. Forest Service Station Paper No. 21, Southeastern Forest Experiment Station, Asheville, NC, 13 pp.

-------, 1969. "Vegetation Management for Water Yield," *Proceedings*, Symposium on Water Balance in North America, Banff, AL, pp. 191-95.

Hoover, M. D., and C. F. Leaf, 1966. "Process and Significance of Interception in Colorado Subalpine Forest," International Symposium on Forest Hydrology *Proceedings*, Pergamon Press, Oxford, England, pp. 213-24.

Hornbeck, J. W., 1964. *The Importance of Dew in Watershed Management Research*, NE-24, Northeastern Forest Experiment Station, Forest Service, Parsons, WV.

-------, 1973. "Storm Flow from Hardwood-Forested and Cleared Watersheds in New Hampshire," *Water Resources Research* 9(2):346.

Hornbeck, J. W., R. S. Pierce, and C. A. Federer, 1970. "Streamflow Changes after Forest Clearing in New England," *Water Resources Research* 6(4):1124.

Hornbeck, J. W., and K. G. Reinhart, 1964. "Water Quality and Soil Erosion as Affected by Logging in Steep Terrain," *Journal of Soil and Water Conservation* 19(4):23.

Horton, R. E., 1933. "The Role of Infiltration in the Hydrologic Cycle," *Transactions* of the American Geophysical Union 14:446.

References

-------, 1939. "Analysis of Runoff-Plat Experiments with Varying Infiltration-Capacity," *Transactions, American Geophysical Union* 20:693.

Horton, J. S., 1972. "Management Problems in Phreatophyte and Riparian Zones," *Journal of Soil and Water Conservation* 27(2):57.

Horton, J. S., T. W. Robinson, and H. R. McDonald, 1964. *Guide for Surveying Phreatophyte Vegetation*, Agriculture Handbook No. 266, Forest Service, USDA, Government Printing Office, Washington, DC.

Huff, F. A., 1970. "Time Distribution Characteristics of Rainfall Rates," *Water Resources Research* 6(2):447.

-------, 1975. "Urban Effects on the Distribution of Heavy Convective Rainfall," *Water Resources Research* 11(6):889.

Hughes, E. E., 1968. "Phreatophytes: Problems and Perspectives," *Water Resources Bulletin* 4(4):50.

Hunt, C. B., 1967. *Physiography of the United States*, W. H. Freeman and Company, San Francisco, CA.

Hussain, S. B., C. M. Skau, and R. O. Meeuwig, 1968. "Infiltrometer Studies on Non-Wettable Soils on East Sierra Nevada," Project Report No. 11, Center for Water Resources Research, Desert Research Institute, University of Nevada, Reno, NV, pp. 19-28.

Hutchinson, G. E., *A Treatise on Limnology, Volume I: Geography, Physics, and Chemistry*. John Wiley & Sons, Inc., London, England.

Hydrology Committee, Ed., 1957. *Hydrology Handbook*. American Society of Civil Engineers, New York, NY.

Idso, S. B., 1981. "Relative Rates of Evaporative Water Losses from Open and Vegetation Covered Water Bodies," *Water Resources Bulletin* 17(1):46.

Ingwersen, J. B., 1985. "Fog Drip, Water Yield, and Timber Harvesting in the Bull Run Municipal Watershed, Oregon," *Water Resources Bulletin* 21(3):469.

Jarboe, J. E., and C. T. Haan, 1974. "Calibrating a Water Yield Model for Small Ungaged Watersheds," *Water Resources Research* 10(2):256.

Jenny, H., 1941. *Factors of Soil Formation*. McGraw-Hill Book Co., Inc., New York, NY.

Johannessen, M., and A. Henriksen, 1978. "Chemistry of Snow Meltwater: Changes in Concentration During Melting," *Water Resources Research* 14(4):615.

Johnson, E. A., 1952. "Effect of Farm Woodland Grazing on Watershed Values in the Southern Appalachian Mountains," *Journal of Forestry* 50:109.

Johnson, P. L., and W. T. Swank, 1973. "Studies of Cation Budgets in the Southern Appalachians on Four Experimental Watersheds with Contrasting Vegetation," *Ecology* 54:70.

Johnston, R. S., 1970. "Evapotranspiration from Bare, Herbaceous, and Aspen Plots: A Check on a Former Study," *Water Resources Research* 6(71):324.

Junge, C. E., 1958. "The Distribution of Ammonia and Nitrate in Rain Water over the United States," American Geophysical Union *Transactions* 39(2):241.

Kadlec, R. H., and H. Alvord, 1989. "Mechanisms of Water Quality Improvement in Wetland Treatment Systems," in Fisk (1989), p. 489.

Kane, D. L, and J. Stein, 1983. "Water Movement into Seasonally Frozen Soils," *Water Resources Research* 19(6):1547.

Kattelmann, R., 1989a. "Hydrology of Four Headwater Basins in the Sierra Nevada," In *Headwaters Hydrology*, Proceedings of the Symposium, Woessner and Potts, editors, p. 141.

-------, 1989b. "Groundwater Contributions in an Alpine Basin in the Sierra Nevada," In *Headwaters Hydrology*, Proceedings of the Symposium, Woessner and Potts, editors, p. 361.

Kattelmann, R. C., N. H. Berg, and R. Rector, 1983. "The Potential for Increasing Streamflow from Sierra Nevada Watersheds," *Water Resources Bulletin* 19(3):395.

Kazman, R. G., 1987. "Mathematical Models and the Real World," *Environmental Geology Water Science* 10(3):125.

Kennedy, V. C., 1971. "Silica Variation in Stream Water with Time and Discharge," In *Advances in Chemistry Series* No. 106:94.

Kennedy, V. C., G. W. Zellweger, and R. J. Avanzino, 1979. "Variation of Rain Chemistry During Storms at Two Sites in Northern California," *Water Resources Research* 15(3):687.

Kerr, R. A., 1981. "There is More to 'Acid Rain' Than Rain," *Science* 211:692.

-------, 1982. "Tracing Sources of Acid Rain Causes Big Stir," *Science* 215:881.

-------, 1983a. "The Carbon Cycle and Climate Warming," *Science* 222:1107.

-------, 1983b. "Carbon Dioxide and a Changing Climate," *Science* 222:491.

-------, 1984. "The Moon Influences Western U.S. Drought," *Science* 224:587.

-------, 1986. "Chinook Winds Resemble Water Flowing Over a Rock," *Science* 231:1244.

-------, 1987. "Winds, Pollutants Drive Ozone Hole," *Science* 238:156.

-------, 1988. "Linking Earth, Ocean, and Air at the AGU," *Science* 239:259.

-------, 1989a. "Volcanoes Can Muddle the Greenhouse," *Science* 245:127.

-------, 1989b. "How to Fix the Clouds in Greenhouse Models," *Science* 243:28.

King, J. G., and L. C. Tennyson, 1984. "Alteration of Streamflow Characteristics Following Road Construction in North Central Idaho," *Water Resources Research* 20(8):1159.

King, W. W., C. O. Brater, and G. Woodburn, 1960. *Hydraulics*. 5th Edition, John Wiley & Sons, Inc., New York, NY.

Kirkby, M. J., Editor, 1978. *Hillslope Hydrology*. John Wiley & Sons, Inc., Chichester, England.

Kittredge, J., 1948. *Forest Influences*. McGraw-Hill Book Company, Inc., New York, NY.

Kleiss, B. A., E. E. Morris, J. F. Nix, and J. W. Barko, 1989. "Modification of Riverine Water Quality by an Adjacent Bottomland Hardwood wetland," in Fisk (1989), p. 429.

Klock, G. O., 1972. "Snowmelt Temperature Influence on Infiltration and Soil Water Retention," *Journal of Soil and Water Conservation* 27(1):12.

Knapp, B. J., 1979. *Elements of Geographical Hydrology*. George Allen & Unwin, London, England.

Knapp, R., M., D. W. Green, E. C. Pogge, and C. Stanford, 1975. "Development and Field Testing of a Basin Hydrology Simulator," *Water Resources Research* 11(6):879.

Koshi, P. T., 1966. "Soil-Moisture Measurement by the Neutron Method in Rocky Wildland Soils," *Proceedings of the Soil Science Society of America* 30(2):282.

Kovner, J. L., 1957. "Evapotranspiration and Water Yields Following Forest Cutting and Natural Regrowth," *Proceedings*, 1956 Annual Meeting in Memphis, TN, Society of American Foresters, Washington, DC, pp. 106-10.

Kramer, P. J., and T. T. Kozlowski, 1979. *Physiology of Woody Plants*. Academic Press, Inc., New York, NY.

Krammes, J. S., and L. F. DeBano, 1965. "Soil Wettability: a Neglected Factor in Watershed Management," *Water Resources Research* 1(2):283.

Krishnan, K. P. R., J. J. Lizcano, L. E. Erikson, and L. T. Fan, 1974. "Evaluation of Methods for Estimating Stream Water Quality Parameters in a Transient Model from Stochastic Data," *Water Resources Bulletin* 10(5):899.

Krug, E. C., and C. R. Frink, 1983. "Acid Rain on Acid Soil: a New Perspective," *Science* 221:520.

Krumholz, L. A., and S. E. Neff, 1970, "The Freshwater Stream, a Complex Ecosystem," *Water Resources Bulletin* 6(1):163.

Krutilla, J. V., M. D. Bowes, and P. Sherman, 1983. "Watershed Management for Joint Production of Water and Timber: A Provisional Assessment," *Water Resources Bulletin* 19(3):403.

Kunkle, S. H., and J. R. Meiman, 1968. *Sampling Bacteria in a Mountain Stream*, Hydrology Papers No. 28, Colorado State University, Ft. Collins, CO.

LaBaugh, J. W., 1986. "Wetland Ecosystem Studies from a Hydrologic Perspective," *Water Resources Bulletin* 22(1):1.

Lambert, J. L., W. R. Gardner, and J. R. Boyle, 1971. "Hydrologic Response of a Young Pine Plantation to Weed Removal," *Water Resources Research* 7(4):1013.

Lane, L. J., M. H. Diskin, D. E. Wallace, and R. M. Dixon, 1978. "Partial Area Response on Small Semiarid Watersheds," *Water Resources Bulletin* 14(5):1143.

Langbein, W. B. et al., 1947. "Topographic Characteristics of Drainage Basins," U.S. Geological Water-Supply Paper No. 968-C, U.S. Department of the Interior, Washington, DC.

Langbein, W. B., and J. V. B. Wells, 1955. "The Water in the Rivers and Creeks," In *Water, the 1955 Yearbook of Agriculture*, United States Government Printing Office, Washington, DC.

Langford, R. H., and F. P. Kapinos, 1979. "The National Water Data Network: A Case History," *Water Resources Research* 15(6):1687.

Larson, L. W., and E. L. Peck, 1974. "Accuracy of Precipitation Measurements for Hydrologic Forecasting," *Water Resources Research* 10(4):857.

Law, A. M., and W. D. Kelton, 1982. *Simulation Modeling and Analysis*. McGraw-Hill Book Co., Inc., New York, NY.

Lawrence, G. B., C. T. Driscoll, and R. D. Fuller, 1988. "Hydrologic Control of Aluminum Chemistry in an Acidic Headwater Stream," *Water Resources Research* 24(5):659.

Lawson, E. R., 1967. "Throughfall and Stemflow in a Pine-Hardwood Stand in the Ouachita Mountains of Arkansas," *Water Resources Research* 3(3):731.

Leaf, A. L., Editor, 1979. *Impact of Intensive Harvesting on Forest Nutrient Cycling, Proceedings*, SUNY College of Environmental Science and Forestry, Syracuse, NY, 421 pp.

Leaf, C. F., 1974. *A Model for Predicting Erosion and Sediment Yield from Secondary Forest Road Construction*, Research Note RM-274, Rocky Mountain Forest and Range Experiment Station, Forest Service, USDA, Ft. Collins, CO.

--------, 1975a. *Watershed Management in the Rocky Mountain Subalpine Zone: the Status of Our Knowledge*, Research Paper RM-137, Rocky Mountain Forest and Range Experiment Station, USDA, Fort Collins, CO, 31 pp.

--------, 1975b. *Watershed Management in the Central and Southern Rocky Mountains: A Summary of the Status of Our Knowledge by Vegetation Types*, Research Paper RM-142, Rocky Mountain Forest and Range Experiment Station, USDA, Fort Collins, CO, 28 pp.

Leaf, C. F., and R. R. Alexander, 1975. *Simulating Timber Yields and Hydrologic Impacts Resulting from Timber Harvest on Subalpine Watersheds*, Research Paper RM-137, Rocky Mountain Forest and Range Experiment Station, Forest Service, USDA, Fort Collins, CO, 31 pp.

Leaf, C. F., and G. E. Brink, 1973. *Hydrologic Simulation Model of Colorado Subalpine Forest*, Research Paper RM-107, Rocky Mountain Forest and Range Experiment Station, Forest Service, USDA, Fort Collins, CO, 23 pp.

Leaf, C. F., and J. L. Kovner, 1972. "Sampling Requirements for Areal Water Equivalent Estimates in Forested Sub-alpine Watersheds," *Water Resources Research* 8(3):713.

Lee, R., 1963. "The Topographic Sampler," *Journal of Forestry* 61:922.

--------, 1967. "The Hydrologic Importance of Transpiration Control by Stomata," *Water Resources Research* 3(3):737.

--------, 1970. "Theoretical Estimates versus Forest Water Yield," *Water Resources Research* 6(5):1327.

--------, 1980. *Forest Hydrology*. Columbia University Press, New York, NY. 349 pp.

Lee, S., 1985. *Weekly Hydrologic Drought Index — Development and Application in Two Watersheds in New York State*. Ph.D Dissertation, SUNY College of Environmental Science and Forestry, Syracuse, NY.

Leonard, R. E., and K. G. Reinhart, 1963. *Some Observations on Precipitation Measurement on Forested Experimental Watersheds*, Research Note NE-6, Forest Service, Northeastern Forest Experiment Station, Upper Darby, PA.

--------, and A. R. Eschner, 1968. "Albedo of Intercepted Snow," *Water Resources Research* 4(5):931.

Leopold, L. B., M. G. Wolman, and J. P. Miller, 1964. *Fluvial Processes in Geomorphology*. W. H. Freeman and Company, San Franciso, CA.

Lettenmaier, D. P., 1979. "Dimensionality Problems in Water Quality Network Design," *Water Resources Research* 15(6):1692.

Lettenmaier, D. P., and S. J. Burges, 1978. "Climatic Change: Detection and Its Impact on Hydrologic Design," *Water Resources Research* 14(4):679.

Leupold & Stevens, Inc., 1987. *Stevens Water Resources Data Book.* P. O. Box 688, Beaverton, OR, 190 pp.

Lewis, D. C., 1968. "Annual Hydrologic Reponse to Watershed Conversion from Oak Woodland to Annual Grassland," *Water Resources Research* 4(1):59.

Lewis, W. M, and M. C. Grant, 1980. "Acid Precipitation in the Western United States," *Science* 207:176.

Liebetrau, A. M., 1979. "Water Quality Sampling: Some Statistical Considerations," *Water Resources Research* 15(6):1717.

Likens, G. E., 1970. "Effects of Deforestation on Water Quality," In *Proceedings* of the ASCE Symposium on Interdisciplinary Aspects of Watershed Management, Bozeman, MT.

Lindroth, A., 1985. "Canopy Conductance of Coniferous Forests Related to Climate," *Water Resources Research* 21(3):297.

Linsley, R. K., and J. B. Franzini, 1964. *Water-Resources Engineering.* McGraw-Hill Book Co., Inc., New York, NY.

Linsley, R. K., M. A. Kohler, and J. L. H. Paulhus, 1949. *Applied Hydrology.* McGraw-Hill Book Company, Inc., New York, NY.

Litten, S., 1990. "Limits of Laboratory Data," *Water Bulletin*, Division of Water, NYS Department of Environmental Conservation 51:14.

Lull, H. W., and K. G. Reinhart, 1955. *Soil-Moisture Measurement.* USDA, Occasional Paper No. 140, Forest Service, Southern Forest Experiment Station, New Orleans, LA.

Lull, H. W., and F. M. Rushmore, 1960. *Snow Accumulation and Melt under Certain Forest Conditions in the Adirondacks*, Station Paper No. 138, Northeastern Forest Experiment Station, Forest Service, USDA, Upper Darby, PA, 16 pp.

Lynch, J. A., C. M. Hanna, and E. S. Corbett, 1986. "Predicting pH, Alkalinity, and Total Acidity in Stream Water During Episode Events," *Water Resources Research* 22(6):905.

Macan, T. T., 1974. *Freshwater Ecology.* Second Edition. John Wiley & Sons, Inc., New York, NY.

Macan, T. T., 1974. *Freshwater Ecology.* John Wiley & Sons, NY.

Mace, A. C., 1966. "Accuracy of Soil Moisture Readings with Unsealed Access Tubes," Research Note 61, Rocky Mountain Forest and Range Experiment Station, Forest Service, USDA, Fort Collins, CO.

Mace, A. C., and J. R. Thompson, 1969. "Modifications and Evaluations of the Evapotranspiration Tent," Research Paper RM-50, Rocky Mountain Forest and Range Experiment Station, Forest Service, USDA, Forest Service, USDA, Fort Collins, CO.

Malcolm Pirnie Engineers, 1968. Herkimer and Oneida Counties Comprehensive Public Water Supply Study. 226 Westchester Avenue, White Plains, NY. 119 pp.

Male, D. H., and R. J. Granger, 1981. "Snow Surface Energy Exchange," *Water Resources Research* 17(3):609.

Manning, J. C., 1987. *Applied Principles of Hydrology*. Merrill Publishing Company, Columbus, OH.

March, W. J., J. R. Wallace, and L. W. Swift, 1979. "An Investigation into the Effect of Storm Type on Precipitation in a Small Mountain Watershed," *Water Resources Research* 15(2):298.

Markowitz, E. M., 1971. "The Chance a Flood Will be Exceeded in a Period of Years," *Water Resources Bulletin* 7(1):40.

Marsh, G. P., 1874. *The Earth as Modified by Human Action*. Scribners, NY.

Martin, C. W., and R. S. Pierce, 1979. "Clearcutting Configurations Affect the Magnitude and Duration of Nutrient Losses in Northern Hardwood Forests" [abstract], In Leaf, 1979; p. 406.

Martin, S. C., and D. E. Ward, 1970. "Rotating Access to Water to Improve Semidesert Cattle Range Near Water," *Journal of Range Management* 23(1):22.

Martinec, J., 1982. "Transfer of Results on Snowmelt Runoff from Small to Big Basins," *Proceedings*, Symposium on Hydrology of Research Basins, Berne, SW, pp. 801-09.

Martinec, J., and A. Rango, 1981. "Areal Distribution of Snow Water Equivalent Evaluated by Snow Cover Monitoring," *Water Resources Research* 17(5):1480.

Martinelli Jr., M., 1964. *Watershed Management in the Rocky Mountain Alpine and Subalpine Zones*, Research Note RM-36, Rocky Mountain Forest and Range Experiment Station, Forest Service, USDA, Fort Collins, CO, 7 pp.

-------, 1967. "Possibilities of Snowpack Management in Alpine Areas," In *Proceedings, International Symposium on Forest Hydrology* edited by Sopper and Lull, 1967, pp. 225-31.

-------, 1965. "An Estimate of Summer Runoff from Alpine Snowfields," *Journal of Soil and Water Conservation* 20(1):24.

------, 1975. *Water-Yield Improvement from Alpine Areas: The Status of Our Knowledge*, Research Paper RM-138, Rocky Mountain Forest and Range Experiment Station, Forest Service, USDA, Fort Collins, CO, 16 pp.

Massman, W. J., 1980. "Water Storage on Forest Foliage: a General Model," *Water Resources Research* 16(1):210.

Matalas, N. C., and J. R. Wallis, 1973. "Eureka! It Fits a Pearson Type 3 Distribution," *Water Resources Research* 9(2):281.

Mather, J. R., 1978. *The Climatic Water Budget in Environmental Analysis*. Lexington Books, D. C. Heath and Company, Lexington, MA, 239 pp.

Maugh, T. H., 1979. "The Threat to Ozone is Real, Increasing," *Science* 206:1167.

Mawdsley, J. A., and M. F. Ali, 1985. "Estimating Nonpotential Evapotranspiration by Means of the Equilibrium Concept," *Water Resources Research* 21(3):383.

McCabe, G. J., and M. A. Ayers, 1989. "Hydrologic Effects of Climate Change in the Delaware River Basin," *Water Resources Bulletin* 25(6):1231.

McClimans, R. J., G. F. Taylor, A. Huggins, and A. F. Bowen, editors. 1978. *Annotated Bibliography of Forest Practices in Relation to Water Quality*, Research Report No. 37, Applied Forestry Research Institute, SUNY College of Environmental Science and Forestry, Syracuse, NY, 146 pp.

McColl, J. G., and D. F. Grigal, 1975. "Forest Fire: Effects on Phosphorous Movement to Lakes," *Science* 188:1109.

McCuen, R. H., and W. J. Rawls, 1979. "Classification of Evaluation of Flood Flow Frequency Estimation Techniques," *Water Resources Bulletin* 15(1):88.

McGuiness, J. L., 1963. "Accuracy of Estimating Watershed Mean Rainfall," *Journal of Geophysical Research* 68(16):4763.

McGuinness, J. L., and L. L. Harrold, 1971. "Reforestation Influences on Small Watershed Streamflow," *Water Resources Research* 7(4):845.

McGuinness, J. L., L. L. Harrold, and W. M. Edwards, 1971. "Relation of Rainfall Energy Streamflow to Sediment Yield from Small and Large Watersheds," *Journal of Soil and Water Conservation* 26(6):233.

McNaughton, K. G., and T. A. Black, 1973. "A Study of Evapotranspiration from a Douglas Fir Forest Using the Energy Balance Approach," *Water Resources Research* 9(6):1579.

References

Megahan, W. F., 1980. *Nonpoint Source Pollution from Forestry Activities in the Western United States: Results of Recent Research and Research Needs*. Forest Service, Intermountain Forest and Range Experiment Station, USDA, Ogden, UT.

-------, 1983. "Hydrologic Effects of Clearcutting and Wildfire on Steep Granitic Slopes in Idaho," *Water Resources Research* 19(3):811.

Meiman, J. R., 1987. "Influence of Forests on Snowpack Accumulation," In Troendle et al., 1987, pp. 61-67.

Meinzer, O. E., 1942. *Hydrology* Dover Publications, Ne York, NY.

Melton, M. A., 1957. *An Analysis of the Relations Among Elements of Climate, Surface Properties, and Geomorphology*, Technical Report No. 11, Department of Geology, Columbia University, New York, NY.

Merriam, R. A., 1973. "Fog Drip from Artificial Leaves in a Fog Wind Tunnel," *Water Resources Research* 9(6):1591.

Merriam, R. A., and K. R. Knoerr, 1961. "Counting Times Required with Neutron Soil-Moisture Probes," *Soil Science* 92(6):394.

Merry, C. J., H. L. McKim, R. E. Bates, S. G. Ungar, S. Cooper, and J. M. Power, 1979. "Snow Cover Mapping in Northern Maine Using LANDSAT Digital Processing Techniques," In *Satellite Hydrology* edited by Deutsch et al. 1981, pp. 197-98.

Metropolitan Council, 1982. *Water Resources Management: Nonpoint Source Pollution Control Technical Report*, 300 Metro Square Bldg., St. Paul, MN 55101.

Meyers, P. C., 1986. "Nonpoint-Source Pollution Control: the USDA Position," *Journal of Soil and Water Conservation* 41(3):156.

Miller, A., and J. C. Thompson, 1970. *Elements of Meteorology*. Charles E. Merrill Publishing Co., Columbus, OH.

Miller, D. E., 1964. "Estimating Moisture Retained by Layered Soils," *Journal of Soil and Water Conservation* 19(6):235.

-------, 1966. *Transport of Intercepted Snow from Trees During Snow Storms*, Research Paper PSW-33, Pacific Southwest Forest and Range Experiment Station, Forest Service, USDA, Berkeley, CA 30 pp.

-------, 1977. *Water at the Surface of the Earth*. Second Edition. Academic Press, New York, NY.

Miller, D. F., 1978. "Ozone Formation Related to Power Plan Emissions," *Science* 202:1186.

Miller, D. H., 1957. "Coastal Fogs and Clouds," *The Geographical Review* XLVII(4):591.

Miller, E. L., 1984. "Sediment Yield and Storm Flow Response to Clear-Cut Harvest and Site Preparation in the Ouachita Mountains," *Water Resources Research* 20(4):471.

Miller, W., and A. Rango, 1985. "Lake Evaporation Studies Using Satellite Thrmal Infrared Data," *Water Resources Bulletin* 21(6):1029.

Mitchell, J. K., and B. A. Jones, Jr., 1978. "Micro-Relief Surface Depression Storage: Changes During Rainfall Events and Their Application to Rainfall-Runoff Models," *Water Resources Bulletin* 14(4):777.

Mitsch, W. J., and J. G. Gosselink, 1986. *Wetlands*. Van Nostrand Reinhold Company, New York, NY.

Molz, F. J., and I. Remson, 1970. "Extraction Term Models of Soil Moisture Use by Transpiring Plants," *Water Resources Research* 6(5):1346.

Monk, C. D., D. A. Crossley, R. L. Todd, W. T. Swank, J. B. Waide, and J. R. Webster, 1977. "An Overview of Nutrient Cycling Research at Coweeta Hydrologic Laboratory," In Correll, 1977, pp. 35-50.

Moss, M. E., 1979. "Some Basic Considerations in the Design of Hydrologic Data Networks," *Water Resources Research* 15(6):1673.

Mount, A. B., 1972. *The Derivation and Testing of a Soil Dryness Index Using Run-Off Data*. Forestry Commission, Tasmania, Australia.

Murphy, C. E., and K. R. Knoerr, 1975. "The Evaporation of Intecepted Rainfall from a Forest Stand: An Analysis by Simulation," *Water Resources Research* 11(2):273.

Musgrave, G. W., 1956. "Soils and Watershed Protection," *Journal of Soil and Water Conservation* 11(3):125.

Mustonen, S. E., and J. L. McGuiness, 1967. "Lysimeter and Watershed Evapotranspiration," *Water Resources Research* 3(4):898.

Meyers, P. C., 1986. "Nonpoint-Source Pollution Control: the USDA Position," *Journal of Soil and Water Conservation* 41(3):156.

Myrup, L. O., T. N. Powell, D. A. Godden, and C. R. Goldman, 1979. "Climatological Estimate of the Average Monthly Energy and Water Budgets of Lake Tahoe, California-Nevada," *Water Resources Research* 15(6):1499.

Nace, R. L., 1974. "Pierre Perrault: The Man and His Contribution to Modern Hydrology," *Water Resources Bulletin* 10(4):633.

National Council of the Paper Industry for Air and Stream Improvement, Inc., 1988. *Procedures for Assessing the Effectiveness of Best Management Practices in Protecting Water and Stream Quality Associated with Managed Forests*, Technical Bulletin No. 538, 260 Madison Avenue, New York, NY 10016.

National Water Commission, 1973. *Water Policies for the Future*, Final Report to the President and the Congress, Water Information Center, Port Washington, NY 579 pp.

National Weather Service, 1977. *Five- to 60-Minute Precipitation Frequency for the Eastern and Central United States*. Technical Memorandum NWS HYDRO-35, National Oceanic and Atmospheric Administration, Silver Spring, MD.

Needham, J. G., and P. R. Needham, 1962. *A Guide to the Study of Fresh-Water Biology*. Fifth Edition. Hoden-Day, San Francisco, CA.

Nicolson, J. A., D. B. Thorud, and E. I. Sucoff, 1968. "The Interception-Transpiration Relationship of White Spruce and White Pine," *Journal of Soil and Water Conservation* 23(5):181.

Nielsen, D. R., M. Th. van Genuchten, and J. W. Biggar, 1986. "Water Flow and Solute Transport Processes in the Unsaturated Zone," *Water Resources Research* 22(9):89S.

Nierenberg, W. A., 1990. "Global Warming Report," Letter to the Editor, *Science* 247:14.

Nik, A. R. H., R. Lee, and J. D. Helvey, 1983. "Climatological Watershed Calibration," *Water Resources Bulletin* 19(1):47.

Novotny, V., Editor, 1988. *Nonpoint Pollution: 1988, Policy, Economy, Management, and Appropriate Technology, Symposium Proceedings*, American Water Resources Association, Bethesda, MD.

Oberlander, G. T., 1956. "Summer Fog Precipitation on the San Francisco Peninsula," *Ecology* 37(4):851.

Oberts, G. L., 1981. "Impact of Wetlands on Watershed Water Quality," in Richardson (1981), p. 213.

Odum, E. P., 1969. "Air-Land-Water=An Ecological Whole," *Journal of Soil and Water Conservation* 24(1):4.

Olson, D. F., and M. D. Hoover, 1954. *Methods of Soil Moisture Determination Under Field Conditions*. Station Paper No. 38, Southeastern Forest Experiment Station, Forest Service, USDA, Asheville, NC.

Osborn, H. B., 1983. "Timing and Duration of High Rainfall Rates in the Southwestern United States," *Water Resources Research* 19(4):1036.

Osborn, H. B., and R. B. Hickok, 1968. "Variability of Rainfall Affecting Runoff from a Semiarid Rangeland Watershed," *Water Resources Research* 4(1):199.

Osborn, H. B., K. G. Renard, and J. R. Simanton, 1979. "Dense Networks to Measure Convective Rainfall in the Southwestern United States," *Water Resources Research* 15(6):1701.

Osteen, C., W. D. Seitz, and J. B. Stall, 1981. "Managing Land to Meet Water Quality Goals," *Journal of Soil and Water Conservation* 36(3):138.

Packer, P. E., 1971. "Terrain and Cover Effects on Snowmelt in a Western White Pine Forest," *Forest Science* 17(1):125.

Palmer, W. C., 1965. *Meteorologic Drought*, U.S. Weather Bureau Research Paper No. 45, Government Printing Office, Washington, DC.

-------, 1968. "Keeping Track of Crop Moisture Conditions Nationwide: The New Crop Moisture Index," *Weatherwise* 21(4):156.

Pani, E. A., and D. R. Haragan, 1985. "Storm Characteristics of Convective-Scale Precipitation," *Water Resources Bulletin* 21(3):393.

Parker, G. G., 1955. "The Encroachment of Salt Water into Fresh," In *WATER: the 1955 Yearbook of Agriculture*, USDA, Government Printing Office, Washington, DC.

Parkes, M. W., and J. R. O'Callaghan, 1980. "Modeling Soil Water Changes in a Well-Structured Freely Draining Soil," *Water Resources Research* 16(4):755.

Parmele, L. H., 1972. "Errors in Output of Hydrologic Models Due to Errors in Input Potential Evapotranspiration," *Water Resources Research* 8(2):348.

Patric, J. H., 1961. "The San Dimas Large Lysimeters," *Journal of Soil and Water Conservation* 16(1):13.

Patric, J. H., and K. G. Reinhart, 1971. "Hydrologic Effects of Deforesting Two Mountain Watersheds in West Virginia," *Water Resources Research* 7(5):1182.

Patrick, R., 1970. "Benthic Stream Communities," *American Scientist* 58:546.

Patrick, R., V. P. Binetti, and S. G. Halterman, 1981. "Acid Lakes from Natural and Anthropogenic Causes," *Science* 211:446.

Pearce, A. J., L. K. Rowe, and J. B. Stewart, 1980. "Nighttime, Wet Canopy Evaporation Rates and the Water Balance of an Evergreen Mixed Forest," *Water Resources Research* 16(5):955.

Peirce, J. J., 1980. "Strategies to Control Nonpoint Source Water Pollution," *Water Resources Bulletin* 16(2):220.

Penman, H. L., 1948. "Natural Evaporation from Open Water, Bare Soil, and Grass," *Proceedings*, Royal Society of London, Series A 193:120.

-------, 1963. *Vegetation and Hydrology*. Commonweath Agricultural Bureaux, Farnham Royal, Bucks, England, 124 pp.

Penton, V. E., and A. C. Robertson, 1967. "Experience with the Pressure Pillow as a Snow Measuring Device," *Water Resources Research* 3(2):405.

Petts, G. E., 1984. *Impounded Rivers: Perspectives for Ecological Management*. John Wiley & Sons, Inc., Chichester, England.

Philander, G., 1989. "El Niño and La Niña," *American Scientist* 77(5):451.

Pierce, R. S., J. W. Hornbeck, G. E. Likens, and F. H. Bormann, 1970. "Effect of Elimination of Vegetation on Stream Water Quantity and Quality," *Proceedings of the Symposium on the Results of Research on Representative and Experimental Basins*, Wellington, NZ, pp. 311-28.

Pilgrim, D. H., D. D. Huff, and T. D. Steele, 1979. "Use of Specific Conductance and Contact Time Relations for Separating Flow Components in Storm Runoff," *Water Resources Research* 15(2):329.

Pisano, M., 1978. "Nonpoint Pollution: an EPA View of Areawide Water Quality Management," *Journal of Soil and Water Conservation* 31(3):94.

Pitlick, J., 1988. "Variability of Bed Load Measurement," *Water Resources Research* 24(1):173.

Ponce, S. L., Editor, 1983. *The Potential for Water Yield Augmentation Through Forest and Range Management*. Collection of articles from the *Water Resources Bulletin* 19(3):351.

Ponce, S. L., and J. R. Meiman, 1983. "Water Yield Augmentation Through Forest and Range Management, Issues for the Future," *Water Resources Bulletin* 19(3):415.

Potts, D. F., 1984. "Hydrologic Impacts of a Large-Scale Mountain Pine Beetle Epidemic," *Water Resources Bulletin* 20(3):373.

Pritsker, A. A. B., and C. D. Pegden, 1979. *Introduction to Simulation and SLAM*. Halsted Press, New York, NY.

Quinn, M. L., 1988. "Tennessee's Copper Basin: a Case for Preserving an Abused Landscape," *Journal of Soil and Water Conservation* 43(2):140.

Ramanthan, V., 1988. "The Greenhouse Theory of Climate Change: a Test by an Inadvertent Global Experiment" *Science* 240:293.

Rango, A., 1970. "Possible Effects of Precipitation Modification on Stream Channel Geometry and Sediment Yield," *Water Resources Research* 6(6):1765.

-------, 1979. "Snow and Ice," In *Satellite Hydrology*, edited by Deutsch et al. 1981, p. 157.

-------, 1983. "Application of a Simple Snowmelt-Runoff Model to Large River Basins," *Proceedings*, Western Snow Conference, pp. 89-99.

-------, 1985. "Assessment of Remote Sensing Input to Hydrologic Models," *Water Resources Bulletin* 21(3):423.

-------, 1987. "New Technology for Hydrological Data Acquisition and Applications," In *Water for the Future: Hydrology in Perspective*, Proceedings of the Rome Symposium, April, p. 511.

-------, 1988. "Progress in Developing and Operational Snowmelt-Runoff Forecast Model with Remote Sensing Input," *Nordic Hydrology* 19:65.

Rango, A., E. T., Engman, T. J. Jackson, J. C. Ritchie, and R. F. Paetzold, 1983. "Hydrological Research in the AgRISTARS Programme," *Proceedings*, Hamburg Symposium on Hydrological Application of Remote Sensing and Remote Data Transmission, IAHS Publication No. 145, pp. 579-89.

Rango, A., and J. Martinec, 1982. "Snow Accumulation Derived from Modified Depletion Curves of Snow Coverage," *Proceedings* of the Exeter Symposium on "Hydrological Aspects of Alpine and High Mountain Areas, IAHS Publ. No. 138, pp. 83-90.

Rango, A., and R. Roberts, 1987. "Snowmelt-Runoff Modeling in the Microcomputer Environment," Paper presented at the Western Snow Conference, Vancouver, BC.

Rasmussen, J. L., 1970. "Atmospheric Water Balance and Hydrology of the Upper Colorado River Basin," *Water Resources Research* 6(1):62.

Rasmussen, W. O., and P. F. Ffolliott, 1979. "Prediction of Water Yield using Satellite Imagery and a Snowmelt Simulation Model," In *Satellite Hydrology*, edited by Deutsch et al. 1981, pp. 193-96.

Reich, B. M., 1970. "Flood Series Compared to Rainfall Extremes," *Water Resources Research* 6(6):1655.

-------, 1971. "Seasonal Occurrence of Annual Floods on Small Pennsylvania Streams," *Water Resources Bulletin* 7(6):1153.

-------, 1973. *Effect of Agnes Floods on Annual Series in Pennsylvania*, Research Publication Number 74, Department of Civil Engineering, College of Engineering, The Pennsylvania State University, University Park, PA.

Reich, B. M., and L. A. V. Hiemstra, 1967. "Purpose and Performance of Peak Predictions," *Proceedings* International Hydrology Symposium, Fort Collins, CO.

Reid, L. M., and T. Dunne, 1984. "Sediment Production from Forest Road Surfaces," *Water Resources Research* 20(11):1753.

Reifsnyder, W. E., and H. W. Lull, 1965. *Radiant Energy in Relation to Forests*. Technical Bulletin No. 1344, Forest Service, USDA, United States Government Printing Office, Washington, DC.

Reigner, I. C., 1964. *Evaluation of the Trough-Type Rain Gage*, NE-20, Northeastern Forest Experiment Station, Forest Service, USDA, New Lisbon, NJ.

Reinhart, K. G., 1954. "Relation of Soil Bulk Density to Moisture Content as it Affects Soil-Moisture Records," In *Some Field, Laboratory, and Office Procedures for Soil-Moisture Measurement*, USDA, Forest Service, Southern Forest Experiment Station, Occasional Paper 135, pp. 12-21.

Reinhart, K. G., and R. S. Pierce, 1964. *Stream-Gaging Stations for Research on Small Watersheds*, Agriculture Handbook No. 268, Forest Service, USDA, Washington, DC. 37 pp.

Reinhart, K. G., A. R. Eschner, and G. R. Trimble, Jr., 1963. *Effect on Streamflow of Four Forest Practices*, USFS Research Paper NE-1, Northeastern Forest Experiment Station, USDA, Upper Darby, PA.

Rich, L. R., 1972. "Managing a Ponderosa Pine Forest to Increase Water Yield," *Water Resources Research* 8(2):422.

Richards, K. S., 1979. "Prediction of Drainage Density from Surrogate Measures," *Water Resources Research* 15(2):435.

Richardson, B. Z., 1966. "Installation of Soil Moisture Access Tubes in Rocky Soils," *Journal of Soil and Water Conservation* 21(4):143.

Richardson, B., editor. 1981. *Wetland Values and Management*. Minnesota Freshwater Society, 2500 Shadywood Road, Box 90, Navarre, MN.

Richardson, C. J., 1979. "Primary Productivity Values in Fresh Water Wetlands," in *Wetland Functions and Values: The State of Our Understanding, Proceedings* of the National Symposium on Wetlands, American Water Resources Association, 1978, Orlando, FL, pp. 131-45.

Richter, D. D., C. W. Ralston, and W. R. Harms, 1982. "Prescribed Fire: Effects on Water Quality and Forest Nutrient Cycling," *Science* 215:661.

Riekirk, H., 1983. "Impacts of Silviculture on Flatwoods Runoff, Water Quality, and Nutrient Budgets," *Water Resources Bulletin* 19(1):73.

Riggs, H. C., 1968a. *Low Flow Investigations*. Geological Survey, Department of The Interior, Washington, DC.

-------, 1968b. "Frequency Curves," Book 4, Chapter A2, Geological Survey, Department of The Interior, Washington, DC.

Robertson, J. K., T. W. Dolzine, and R. C. Graham, 1979. *Chemistry of Precipitation From Sequentially Sampled Storms*. The Science Research Laboratory, United States Military Academy, West Point, NY. Lithographed report.

Rodenhauser, K., 1989. "War in the Wilderness, New York State's Aerial Battle Against Acid Rain in the Adirondacks," Presentation at the Annual Meeting of the NYS Section of the American Water Resources Association, Albany, NY Nov. 15th. Unpublished talk. .

Rodriguez-Iturbe, I., J. M. Mejia, 1974. "On the Transformation of Point Rainfall to Areal Rainfall," *Water Resources Research* 10(4):729.

Rodriguez-Iturbe, I., B. F. de Power, M. B. Sharifi, and K. P. Georgakakos, 1989. "Chaos in Rainfall," *Water Resources Research* 25(7):1667.

Rogerson, T. L., 1968. "Thinning Increases Throughfall in Loblolly Pine Plantations," *Journal of Soil and Water Conservation* 23(4):141.

Rogerson, T. L., and W. R. Byrnes, 1968. "Net Rainfall Under Hardwoods and Red Pine in Central Pennsylvania," *Water Resources Research* 4(1):55.

Roth, F. A., and M. Chang, 1981. "Throughfall in Planted Stands of Four Southern Pine Species in East Texas," *Water Resources Bulletin* 17(5):880.

Rothacher, J., 1965. "Streamflow from Small Watershed on the Western Slope of the Cascade Range of Oregon," *Water Resources Research* 1(1):125.

-------, 1970. "Increases in Water Yield Following Clear-Cut Logging in the Pacific Northwest," *Water Resources Research* 6(2):653.

Rouse, W. R., P. F. Mills, and R. B. Stewart, 1977. "Evaporation in High Latitudes," *Water Resources Research* 13(6):909.

Rowe, P. B., 1940. *The Construction, Operation, and Use of the North Fork Infiltrometer*, Miscellaneous Publication No. 1, California Forest and Range Experiment Station, Forest Service, USDA, Berkeley, CA.

Roy, D. F., 1966. *Silvical Characteristics of Redwood*. U.S. Forest Service, Research Paper No. PSW-28, Pacifc Southwest Forest and Range Experiment Station, Berkeley, CA.

Rudel, R. K., H. J. Stockwell, and R. G. Walsh, 1973. "Weather Modification: an Economic Alternative for Augmenting Water Supplies," *Water Resources Bulletin* 9(1):116.

Ryan, J. A., I. G. Morison, and J. S. Bethel, 1974. "Ecosystem Modeling of a Forested River Basin," *Water Resources Bulletin* 10(4):703.

Sadeghipour, J. and J. A. Dracup, 1985. "Regional Frequency Analysis of Hydrologic Multiyear Droughts," *Water Resources Bulletin* 21(3):481.

Sanders, T. G., and D. D. Adrian, 1978. "Sampling Frequency for River Quality Monitoring," *Water Resources Research* 14(4):569.

Santeford, H. S., G. R. Alger, and J. G. Meier, 1972. "Snowmelt Energy Exchange in the Lake Superior Region," *Water Resources Research* 8(2):390.

Sartz, R. S., and D. N. Tolsted, 1974. "Effect of Grazing on Runoff from Two Small Watersheds in Southwestern Wisconsin," *Water Resources Research* 10(2):354.

Sartz, R. S., and W. R. Curtis, 1961. *Field Calibration of a Neutron-Scattering Soil Moisture Meter*, Station Paper No. 91, Lake States Forest Experiment Station, Forest Service, USDA, La Crosse, WI.

Sather, J. H., 1984. "Research Gaps in Assessing Wetland Functions," *National Wetlands Newsletter* 6(2):2.

Satterlund, D. R., 1972. *Wildland Watershed Management*. Ronald Press Company, Inc., New York, NY.

Satterlund, D. R., and A. R. Eschner, 1965. "Land Use, Snow, and Streamflow Regimen in Central New York," *Water Resources Research* 1(3):397.

Satterlund, D. R., and H. F. Haupt, 1970. "The Disposition of Snow Caught by Conifer Crowns," *Water Resources Research* 6(2):649.

-------, 1972. "Vegetation Management to Control Snow Accumulation and Melt in the Northern Rocky Mountains," *Proceedings*, National Symposium on Watersheds in Transition, Fort Collins, CO, pp. 200-05.

Scheidegger, A. E., 1970. "Stochastic Models in Hydrology," *Water Resources Research* 6(3):750.

Schmugge, T. J., T. J. Jackson, and H. L. McKim, 1980. "Survey of Methods for Soil Moisture Determination," *Water Resources Research* 16(6):961.

Schneider, S. H., 1989. "The Greenhouse Effect: Science and Policy," *Science* 243:771.

Schneider, W. J., 1961. "Flood Frequencies as Related to Land Use," *Bulletin* of the International Association of Scientific Hydrology VI(4):36.

Schneider, W. J., and G. R. Ayer, 1961. *Effect of Reforestation on Streamflow in Central New York*, Water-Supply Paper 1602, Geological Survey, USDI, Washington, DC, 61 pp.

Schermerhorn, V. P., 1967. "Relations Between Topography and Annual Precipitation in Western Oregon and Washington," *Water Resources Research* 3(3):707.

Shafer, R. A., C. F. Leaf, and J. K. Marron, 1979. "LANDSAT Derived Snow Cover as an Input Variable for Snow Melt Runoff Forecasting in South Central Colorado," In *Satellite Hydrology*, edited by Deutsch et al. 1981, pp. 218-24.

Sharp, A. L, J. J. Bond, J. W. Neuberger, A. R. Kuhlman, and J. K. Lewis, 1964. "Runoff as Affected by Intensity of Grazing on Rangeland," *Journal of Soil and Water Conservation* 19(3):103.

Sherman, 1932. "The Relation of Hydrographs of Runoff to Size and Character of Drainage Basin," *Transactions, American Geophysical Union* 13:332-9.

Sherman, L. K., and G. W. Musgrave, 1942. *Chapter VII - Infiltration*, In *Hydrology* edited by O. E. Meinzer, Dover Publications, Inc., New York, NY pp. 244-58.

Sherman, L. K., 1932. "The Relation of Hydrographs of Runoff to Size and Character of Drainage Basins," *Transactions* of the American Geophysical Union 19(2):447.

Schindler, D. W., 1988. "Effects of Acid Rain on Freshwater Ecosystems," *Science* 239:149.

Siegel, D. I., 1981. "The Effect of Snowmelt on the Water Quality of Filson Creek and Omaday Lake, Northeastern Minnesota," *Water Resources Research* 17(1):238.

Simpson, R. H., and J. S. Malkus, 1964. "Experiments in Hurricane Modification" *Scientific American* 211(6):27.

Simpson, R. L., and D. F. Whigham, 1978. "Seasonal Patterns of Nutrient Movement in a Freshwater Tidal Marsh," in in Good et al., p. 243.

Singh, B., and G. Szeicz, 1979. "The Effect of Intercepted Rainfall on the Water Balance of a Hardwood Forest," *Water Resources Research* 15(1):131.

Singh, K. P., 1968. "Some Factors Affecting Base Flow," *Water Resources Research* 4(5):985.

-------, 1971. "Model Flow Duration and Streamflow Variability," *Water Resources Research* 7(4):1031.

-------, 1976. "Unit Hydrographs, A Comparative Study," *Water Resources Bulletin* 12(2):381.

Singh, V. P., and P. K. Chowdhury, 1986. "Comparing Some Methods of Estimating Mean Areal Rainfall," *Water Resources Bulletin* 22(2):275.

Smith, W. L. et al., 1986. "The Meteorological Satellite: Overview of 25 Years of Operation," *Science* 231:455.

Smith, W., and C. F. Hains, 1961. *Flow-Duration and High- and Low-Flow Tables for California Streams*. Geological Survey and California Department of Water Resources, Sacramento, CA.

Snyder, G. G., H. F. Haupt, and G. H. Belt, Jr, 1975. *Clearcutting and Burning Slash Alter Quality of Stream Water in Northern Idaho*, Research Paper INT-168, Intermountain Forest and Range Experiment Station, Forest Service, USDA, Ogden, UT.

Sofia, S., P. Demarque, and A. Endal, 1985. "From Solar Dynamo to Terrestrial Climate," *American Scientist* 73:326.

Soil Conservation Society of America, 1976. *Resource Conservation Glossary*. 7515 Northeast Ankeny Road, Ankeny, IA 50021.

Soil Conservation Service, 1975. *Soil Taxonomy, A Basic System of Soil Classification for Making and Interpreting Soil Surveys*, Agriculture Handbook 436, USDA, Washington, DC, 754 pp.

Soons, J. M., and D. E. Greenland, 1970. "Observations on the Growth of Needle Ice," *Water Resources Research* 6(2):579.

Sopper, W. E., and E. S. Corbett, 1975. *Municipal Watershed Management Symposium Proceedings*, Forest Service, General Technical Report NE-13, Northeastern Forest Experiment Station, Upper Darby, PA.

Sopper, W. E., and L. A. V. Hiemstra, 1970. "Effects of Simulated Cloud Seeding on Streamflow of Selected Watersheds in Pennsylvania," *Water Resources Bulletin* 6(5):754.

Sopper, W. E., and H. W. Lull, 1964. "Streamflow Characteristics of Physiographic Units in the Northeast," *Water Resources Research* 1(1):115.

-------, Editors, 1967. *International Symposium on Forest Hydrology*, Proceedings of a National Science Foundation Advanced Science Seminar held at The Pennsylvania State University, University Park, PA, Aug. 29-Sept. 10, 1965.

Spittlehouse, D. L., and T. A. Black, 1981. "A Growing Season Water Balance Model Applied to Two Douglas Fir Stands," *Water Resources Research* 17(6):1651.

Squires, P., 1971. "Possibilities of Increased Water Supply from Cloud Seeding," *Water Resources Bulletin* 7(5):951.

Srinilta, S., D. R. Nielsen, and D. Kirkham, 1969. "Steady Flow of Water through a Two-Layer Soil," *Water Resources Research* 5(5):1053.

Stallings, J. H., 1952. "Raindrops Puddle Surface Soil," *Journal of Soil and Water Conservation* 7(2):70.

Stalnaker, C. B., 1979. "The Use of Habitat Structure Preferenda for Establishing Flow Regimes Necessary for Maintenance of Fish Habitat," In Ward and Stanford (1979):321.

Stark, N., 1980. "Changes in Soil Water Quality Resulting From Three Timber Cutting Methods and Three Levels of Fiber Utilization," *Journal of Soil and Water Conservation* 35(4):183.

Stein, R. A., 1965. "Laboratory Studies of Total Load and Apparent Bed Load," *Journal of Geophysical Research* 70(8):1831.

Storey, H. C., 1959. "Effects of Forest on Runoff," *Journal of Soil and Water Conservation* 14(4):152.

Storr, D., J. Tomlain, H. F. Cork, and R. E. Munn, 1970. "An Energy Budget Study above the Forest Canopy at Marmot Creek, Alberta, 1967," *Water Resources Research* 6(3):705.

Strahler, A. N., 1957. "Quantitative Analysis of Watershed Geomorphology," *Transactions* of the American Geophysical Union 38:913.

Strahler, A. N., and A. H. Strahler, 1973. *Environmental Geoscience*. Hamilton Publishing Co., Santa Barbara, CA, 5511 pp.

Striffler, W. D., 1959. "Effects of Forest Cover on Soil Freezing in Northern Lower Michigan," USDA, Forest Service, Lake States Experiment Station, Staion Paper No. 76.

Stroud, R. H., and R. G. Martin, 1973. "Influence of Reservoir Discharge Location on the Water Quality, Biology, and Sport Fisheries of Reservoirs and Tail Waters," In Ackerman et al. (1973):540.

Stumm, W., and J. J. Morgan, 1970. *Aquatic Chemistry*. Wiley-Interscience, New York, NY.

Sturges, D. L., 1968. "Evapotranspiration at a Wyoming Mountain Bog," *Journal of Soil and Water Conservation* 23(8):23.

-------, 1975. *Hydrologic Relations on Undisturbed and Converted Big Sagebrush Lands: The Status of Our Knowledge*, Research Paper RM-140, Forest Service, Rocky Mountain Forest and Range Experiment Station, Fort Collins, CO 23 pp.

-------, 1977. *Snow Accumulation and Melt in Sprayed and Undisturbed Big Sagebrush Vegetation*, Research Note RM-348, Rocky Mountain Forest and Range Experiment Station, Forest Service, USDA, Fort Collins, CO 6 pp.

Sun, M., 1986. "Ground Water Ills: Many Diagnoses, Few Remedies," *Science* 232:1490.

Sundeen, K. D., C. F. Leaf, and G. M. Bostrum, 1989. "Hydrologic Functions of Sub-Alpine Wetlands in Colorado," in Fisk (1981), p. 401.

Sundquist, E. T., 1985. "Geological Perspectives on Carbon Dioxide and the Carbon Cycle," in Sundquist and Broecker, 1985. pp. 5-60.

Sundquist, E. T., and W. S. Broecker, editors, 1985. *The Carbon Cycle and Atmospheric CO_2: Natural Variations Archean to Present*, Geophysical Monograph 32, American Geophysical Union, Washington, DC.

Swank, G. W., and R. W. Booth, 1970. "Snow Fencing to Redistribute Snow Accumulation," *Journal of Soil and Water Conservation* 25(5):197.

Swank, W. T., and W. H. Caskey, 1982. "Nitrate Depletion in a Second-Order Mountain Stream," *Journal of Environmental Quality* 11:581.

Swank, W. T., and D. A. Crossley, Jr., editors, 1988. *Forest Hydrology and Ecology at Coweeta*, Ecological Studies 66, Springer-Verlag, New York, NY.

Swank, W. T., and J. B. Waide, 1988. "Characterization of Baseling Precipitation and Stream Chemistry and Nutrient Budgets for Control Watersheds," in Swank and Crossley (1988), pp. 57-79.

Swank, W. T., and J. E. Douglass, 1974. "Streamflow Greatly Reduced by Converting Deciduous Hardwood Stands to Pine," *Science* 185:857.

Swank, W. T., and N. H. Miner, 1968. "Conversion of Hardwood-Covered Watersheds to White Pine Reduces Water Yield," *Water Resources Research* 4(5):947.

Swank, W. T., N. R. Goebel, and J. D. Helvey, 1972. "Interception Loss in Loblolly Pine Stands of the South Carolina Piedmont," *Journal of Soil and Water Conservation* 27(11):160.

Swanson, F. J., and G. W. Lienkaemper, 1978. *Physical Consequences of Large Organic Debris in Pacific Northwest Streams*, General Technical Report PNW-69, Pacific Northwest Forest and Range Experiment Station, Forest Service, USDA, Portland, OR.

Swanson, R. H., 1972. "Water Transpired by Trees is Indicated by Heat Pulse Velocity," *Agricultural Meteorology* 10:277.

Swift, L. W., and W. T. Swank, 1981. "Long Term Responses of Streamflow Following Clearcutting and Regrowth," *Hydrological Sciences Bulletin* 26:245.

Swift, L. W., W. T. Swank, J. B. Manking, R. J. Luxmoore, and R. A. Goldstein, 1975. "Simulation of Evapotranspiration and Drainage from Mature and Clear-Cut Deciduous Forests and Young Pine Plantation," *Water Resources Research* 11(5):667.

Tabler, R. D., and D. L. Sturges, 1986. "Watershed Test of a Snow Fence to Increase Streamflow: Preliminary Results," *Proceedings*, Symposium on Cold Regions Hydrology, Anchorage, AK, pp. 53-61.

Tajchman, S. J., 1971. "Evapotranspiration and Energy Balances of Forest and Field," *Water Resources Research* 7(3):511.

Tangborn, W. V., 1980. "A Model to Forecast Short-Term Snowmelt Runoff Using Synoptic Observations of Streamflow, Temperature, and Precipitation," *Water Resources Research* 16(4):778.

Tanner, C. B., 1957. "Factors Affecting Evaporation from Plants and Soils," *Journal of Soil and Water Conservation* 12(5):221.

Thomas, H. E., 1955. "Underground Sources of our Water," In *WATER: the 1955 Yearbook of Agriculture*, USDA, Government Printing Office, Washington, DC.

Thomas, R. B., 1985. "Estimating Total Suspended Sediment Yield With Probability Sampling," *Water Resources Research* 21(9):1381.

Thompson, A. E., 1960. "Timber and Water, Twin Harvests on Seattle's Cedar River Watershed," *Journal of Forestry* 58(4):299.

Thompson, J. R., 1968. "Effect of Grazing on Infiltration in a Western Watershed," *Journal of Soil and Water Conservation* 23(2):63.

-------, 1974. "Energy Budget Measurements Over Three cover Types in Eastern Arizona," *Water Resources Research* 10(5):1045.

Thompson, L. M., 1973. "Cyclical Weather Patterns in the Middle Latitudes," *Journal of Soil and Water Conservation* 28(2):87.

Thornthwaite, C. W. et al., 1944. "Report of the Committee on Transpiration and Evaporation," *Transactions* of the American Geophysical Union 25(V):683.

-------, and J. R. Mather, 1957. "Instructions and Tables for Computing Potential Evapotranspiration and the Water Balance," *Publications in Climatology* X(3):181.

Thorud, D. B., 1967. "The Effect of Applied Interception on Transpiration Rates of Potted Ponderosa Pine," *Water Resources Research* 3(2):443.

Thorud, D. B., and D. A. Anderson, 1969. *Freezing in Forest Soil as Influenced by Soil Properties, Litter, and Snow*. Water Resources Research Center, Graduate School, University of Minnesota, Minneapolis, MN.

Tice, R. H., 1968. *Magnitude and Frequency of Floods in the United States, Part 1-B, North Atlantic Slope Basins, New York to York River*. Government Printing Office, Washington, DC.

Tiedeman, A. R. et al., 1979. *Effects of Fire on Water*, General Technical Report WO-10, Forest Service, USDA, Washington, DC.

Tiedemann, A. R., J. D. Helvey, and T. D. Anderson, 1978. "Stream Chemistry and Watershed Nutrient Economy Following Wildfire and Fertilization in Eastern Washington," *Journal of Environmental Quality* 7(4):580.

Todd, D. K., 1959. *Ground Water Hydrology*. John Wiley and Sons, Inc., New York, NY.

Todorovic, P., and D. A. Woolhiser, 1972. "On The Time When the Extreme Flood Occurs," *Water Resources Research* 8(6):1433.

Trimble, S. W., and F. H. Weirich, 1987. "Reforestation Reduces Streamflow in the Southeastern United States," *Journal of Soil and Water Conservation* 42(7):274.

Trimble, G. R., R. S. Sartz, and R. S. Pierce, 1958. "How Type of Soil Frost Affects Infiltration," *Journal of Soil and Water Conservation* 13(2):81.

Troendle, C. A., 1970. "The Flow Interval Method for Analyzing Timber Harvesting Effects on Streamflow Regimen," *Water Resources Research* 6(1):328.

-------, 1983. "The Potential for Water Yield Augmentation From Forest Management in the Rocky Mountain Region," *Water Resources Bulletin* 19(3):359.

-------, 1987. *The Potential Effect of Partial Cutting and Thinning on Streamflow from the Subalpine Forest*, Research Paper RM-274, Rocky Mountain Forest and Range Experiment Station, Forest Service, USDA, Fort Collins, CO, 7 pp.

Troendle, C. A., and R. M. King, 1985. "The Effect of Timber Harvest on the Fool Creek Watershed, 30 Years Later," *Water Resources Research* 21(12):1915.

-------, 1987. "The Effect of Partial and Clearcutting on Streamflow at Deadhorse Creek, Colorado," *Journal of Hydrology* 90:145.

Troendle, C. A. et al., Coordinators, 1987. *Management of Subalpine Forests: Building on 50 Years of Research, Proceedings of a Technical Conference*. General Technical Report RM-149, USDA Forest Service, Fort Collins, CO.

Troendle, C. A., and M. R. Kaufman, 1987. "Influence of Forests on the Hydrology of the Subalpine Forest," In Troendle et al. 1987, pp. 68-78.

Troendle, C. A., and C. F. Leaf, 1981. "Effects of Timber Harvesting in the Snow Zone on Volume and Timing of Water Yield," in Baumgartner, 1980. pp. 231-243.

-------, 1981. "Effects of Timber Harvest in the Snow Zone on Volume and Timing of Water Yield," In *Interior West Watershed Management*, Cooperative Extension, University of Washington,, Pullman, WA, pp. 231-43.

Troendle, C. A., and M. A. Nilles, 1987. "The Effect of Clearcutting on Chemical Exports in Lateral Flow from Differing Soil Depths on a Subalpine Forested Slope," *Proceedings of the IASH-AISH Symposium on Forest Hydrology and Watershed Management*, Vancouver, BC.

Tryon, C. P., 1972. "Partial Cutting and Increased Water Yields, a New Multiresource Approach," *Journal of Soil and Water Conservation* 27(2):66.

Turco, R. P., O. B. Toon, T. P. Ackerman, J. B. Pollack, and C. Sagan, 1983. "Nuclear Winter: Global Consequences of Multiple Nuclear Explosions," *Science* 222(4630):1283.

Turk, J. T., and D. H. Campbell, 1987. "Estimates of Acidification of Lakes in the Mt. Zirkel Wilderness Area, Colorado," *Water Resources Research* 23(9):1757.

Tyler, S. W., and S. W. Wheatcraft, 1990. "Fractal Processes in Soil Water Retention," *Water Resources Research* 26(5):1047.

Ulaby, F. T., G. A. Bradley, and M. C. Dobson, 1979. "Potential Application of Satellite Radar to Monitor Soil Moisture," In *Satellite Hydrology* edited by Deutsch et al. 1981, pp. 363-70.

Urie, D. H., 1971. "Estimated Groundwater Yield Following Strip Cutting in Pine Plantations," *Water Resources Research* 7(6):1497.

Van Bavel, C. H. M., 1966. "Potential Evaporation: The Combination Concept and Its Experimental Verification," *Water Resources Research* 2(3):455.

van der Valk, A. G., C. B. Davis, J. L. Baker, and C. E. Beer, 1978. "Natural Fresh Water Wetlands as Nitrogen and Phosphorous Traps for Land Runoff," in Greeson et al. 1978, p. 457.

van't Woudt, B. D., J. Whittaker, and K. Nicolle, 1979. "Ground Water Replenishment from Riverflow," *Water Resources Bulletin* 15(4):1016.

van Hylckama, T. E. A., 1970. "Water, Something Peculiar," *Hydrological Sciences* 24(4):499.

Veihmeyer, F. J., 1929. "An Improved Soil-Sampling Tube," *Soil Science* 27:147.

Verry, E. S., 1986. "Forest Harvesting and Water: the Lake States Experience," *Water Resources Bulletin* 22(60:1039.

Verry, E. S., J. R. Lewis, and K. N. Brooks, 1983. "Aspen Clearcutting Increases Snowmelt and Storm Flow Peaks in North Central Minnesota," *Water Resources Bulletin* 19(1):59.

Viessman, W., J. W. Knapp, G. L. Lewis, and T. E. Harbaugh, 1977. *Introduction to Hydrology*. Second Edition. Harper & Row, Publishers, New York, NY.

Vigon, B. W., 1985. "The Status of Nonpoint Source Pollution: its Nature, Extent, and Control," *Water Resources Bulletin* 21(2):179.

Viskanta, R., and J. S. Toor, 1972. "Radiant Energy Transfer in Waters," *Water Resources Research* 8(3):595.

Vitousek, P. M., J. R. Gosz, C. C. Grier, J. M. Melillo, W. A. Reiners, and R. L. Todd, 1979. "Nitrate Losses from Disturbed Ecosystems," *Science* 204:469.

Vogelmann, H. W., 1976. "Rain-Making Forests," *Natural History*, March.

Wagle, R. F., 1971. "Understanding a Watershed from the Biological Viewpoint," *Water Resources Bulletin* 7(2):244.

Walling, D. E., 1977. "Assessing the Accuracy of Suspended Sediment Rating Curves for a Small Basin," *Water Resources Research* 13(3):531.

Wallis, J. R., 1977. "Climate, Climatic Change, and Water Supply," EOS 58(11):1012.

Ward, J. V., and J. A. Stanford, Editors, 1979. *The Ecology of Regulated Streams*. Plenum Press, New York, NY.

Warfvinge, P., and H. Sverdrup, 1988. "Soil Liming as a Measure to Mitigate Acid Runoff," *Water Resources Research* 24(5):701.

Water Quality Group, 1988. *Annotated Bibliography of Nonpoint Source Literature*, North Carolina Agricultural Extension Service, North Caorlina State University, Raleigh, NC.

Water Resources Council, 1978. *The Second National Water Assessment*, 2120 L Street, Washington, DC.

------, 1978. *Floodplain Management Guidelines*. 43 FR 6030.

Weather Bureau, 1958. "The World's Heaviest Rains," *Daily Weather Map* April 14.

------, Weather Bureau, 1961. *Rainfall Frequency Atlas of the United States*. Technical Publication No. 40, Government Printing Office, Washington, DC.

------, 1962. *Instructions for Climatological Observers*. Circular B, Revised. U.S. Department of Commerce, Government Printing Office, Washington, DC.

Webb, M. S., and D. W. Phillips, 1973. "An Estimate of the Role of Lake Effect Snowstorms in the Hydrology of the Lake Erie Basin," *Water Resources Research* 9(1):103.

Webster, J. R., 1984. "Forest Disturbance and Stream Ecosystem Stability," paper presented at Fiftieth Anniversary Symposium on Long Range Research on Forested Watersheds at the Coweeta Hydrologic Laboratory, University of Georgia, Athens, GA.

Wei, T. C., and J. L. McGuinness, 1973. "Reciprocal Distance Squared Method: A Computer Technique for Estimating Areal Precipitation," Agricultural Research Service, ARS-NC-8, USDA, Washington, DC.

Weitzman, S., and K. G. Reinhart, 1957. "Water Yields from Small Forested Watersheds," *Journal of Soil and Water Conservation* 12(2):56.

Wilde, S. A., and D. M. Spyridakis, 1961. "Determination of Soil Moisture by the Immersion Method," *Soil Science* 94(2):132.

Willett, H. C., and F. Sanders, 1959. *Descriptive Meteorology*. Academic Press, Inc., New York, NY.

Williams, G. P., 1978. "Bank-Full Discharge of Rivers," *Water Resources Research* 14(6):1141.

Wilm, H. G., 1950. *Statistical Control in Hydrologic Forecasting*, Reprint from Research Note 61, Pacific Northwest Forest and Range Experiment station, Forest Service, USDA, Portland, OR; SUNY College of Environmental Science and Forestry, Syracuse, NY.

-------, 1955. "Foresters in the Watershed Movement," *American Forests* 61(2):28.

Wilm, H. G., and E. G. Dunford, 1948. *Effect of Timber Cutting on Water Available for Streamflow from a Lodgepole Pine Forest*, Technical Bulletin 968, U.S. Department of Agriculture, Washington, DC.

Winter, T. C., 1977. "Classification of the Hydrologic Settings of Lakes in the North Central United States," *Water Resources Research* 13(4):753.

-------, 1981. "Uncertainties in Estimating the Water Balance of Lakes," *Water Resources Bulletin* 17(1):82.

Wisler, C. O., and E. F. Brater, 1949. *Hydrology*, John Wiley & Sons, Inc., New York, NY.

-------, 1959. *Hydrology*. Second Edition. John Wiley & Sons, Inc., New York, NY.

Winger, P. V., P. J. Lasier, M. Hudy, D. Fowler, and M. J. Van Den Avyle, 1988. "Sensitivity of High-Elevation Streams in the Southern Blue Ridge Province to Acidic Deposition," *Water Resources Bulletin* 23(3):379.

Woessner, W. W., and D. F. Potts, Editors, 1989. *Headwaters Hydrology*, Proceedings of the American Water Resources Association Symposium, held at the University of Montana, Missoula, MT, June 27-29.

Wooding, R. A., 1966. "A Hydraulic Model for the Catchment-Stream Problem," *Journal of Hydrology* 3:254; 3:268; 4:21 (two earlier parts in 1965).

Woodley, W. L., 1970. "Rainfall Enhancement by Dynamic Cloud Modification," *Science* 170(3954):127.

Woodwell, G. M., J. E. Hobbie, R. A. Houghton, J. M. Melillo, B. Moore, B. J. Peterson, and G. R. Shaver, 1983. "Global Deforestation: Contribution to Atmospheric Carbon Dioxide," *Science* 222(4628):1081.

Woods, L. G., 1966. "Increasing Watershed Yield Through Management," *Journal of Soil and Water Conservation* 21(3):95.

WQED, and National Academy of Sciences, 1986. "The Solar Sea," Episode Number 6 of "The Planet Earth," PBS/TV, Pittsburgh, PA.

Yang, C. T., 1971. "Potential Energy and Stream Morphology," *Water Resources Research* 7(2):311.

Yarnell, D. L., 1935. *Rainfall Intensity-Frequency Data*. Miscellaneous Publication No. 204, U.S. Department of Agriculture, Government Printing Office, Washington, DC.

Yates, H., A. Strong, D. McGinnis Jr., and D. Tarpley, 1986. "Terrestrial Observations from NOAA Operational Satellites," *Science* 231:463.

Yen, B. C., and V. T. Chow, 1969. "A Laboratory Study of Surface Runoff Due to Moving Rainstorms," *Water Resources Research* 5(5):989.

Yevjevich, V., 1968. "Misconceptions in Hydrology and Their Consequences," *Water Resources Research* 4(2):225.

Young, C. E., R. A. Klawitter, and J. E. Henderson, 1972. "Hydrologic Model of a Wetland Forest," *Journal of Soil and Water Conservation* 27(3):122.

Young, C. T., and J. B. Stall, 1971. "Note on the Map Scale Effect in the Study of Stream Morphology," *Water Resources Research* 7(3):712.

Young, I. G., 1973. "Conductivity: Danger, Handle With Care!" Honeywell, Inc., Fort Washington, PA.

Zebuhr, R. H., 1968. *Studies in Water Quality: a Small River in the White Mountains of New Hampshire*. MS Thesis, SUNY College of Environmental Science and Forestry, Syracuse, NY.

Ziemer, R. R., 1981. "Storm Flow Response to Road Building And Partial Cutting in Small Streams of Northern California," *Water Resources Research* 17(4):907.

Ziemer, R. R., I. Goldberg, and N. A. MacGillivray, 1967. *Measuring Moisture Near Soil Surface*. USDA, Forest Service, Pacific Southwest Forest and Range Experiment Station, Research Note PSW-158.

Zinke, P. J., 1958. "The Soil-Vegetation Survey as a Means of Classifying Land for Multiple-Use Forestry," *Proceedings of the Fifth World Forestry Congress*, Reprint, State Division of Natural Resources, Sacramento, CA.

-------, 1967. "Forest Interception Studies in the United States," *Proceedings of the International Symposium on Forest Hydrology*, edited by W. E. Sopper and H. W. Lull, Pergamon Press, New York NY, pp. 137-61.

Appendices

A **Conversion Tables**
   (Courtesy of Leupold & Stevens Inc., Beaverton, Oregon) **366**

B **Map Scales** **370**

C **Watershed Eccentricity** **371**

D **Glossary** **372**

Table A-1 English/Metric Length Conversions

Name	mm.	cm.	in.	dm.	ft.	yd.	m.	rd.	ch.	hm.	fur.	km.	mi.	naut.mi*
millimetres	1	10	25.4	100	304.8	914.4	1000	5029.2	20116.8	100,000	201,168	1,000,000	1,609,347	1,853,248
centimetres	.1	1	2.54	10	30.48	91.44	100	502.9	2011.68	10,000	20116.8	100,000	160,935	185,325
inches	.03937	.3937	1	3.937	12	36	39.37	198	792	3937	7920	39,370	63,360	72,963
decimetres	.01	.1	.254	1	3.048	9.144	10	50.29	201.17	1000	2011.7	10,000	16,093	18,532
feet	.00328	.0328	.08333	.32808	1	3	3.2808	16.5	66	328.08	660	3280.8	5280	6080.2
yards	.00109	.01093	.0278	.10936	.33333	1	1.0936	5.5	22	109.36	220	1093.6	1760	2026.8
metres	.001	.01	.02540	.1	.30480	.91440	1	5.0262	20.116	100	201.17	1000	1609.3	1853.2
rods	1.99x10⁻⁴	.00199	.00505	.01988	.0606	.18181	.1988	1	4	19.883	40	198.83	320	368.85
chains	4.97x10⁻⁵	4.97x10⁻⁴	.00126	.00497	.01515	.04545	.04971	.25	1	4.9708	10	49.708	80	92.23
hectometres	10⁻⁵	10⁻⁴	.00025	.001	.00305	.00914	.01	.05029	.20117	1	2.0117	10	16.093	18.53
furlongs	4.97x10⁻⁶	4.97x10⁻⁵	1.26x10⁻⁴	4.97x10⁻⁴	.00151	.00454	.00497	.025	.1	.49078	1	4.9708	8	9.223
kilometres	10⁻⁶	10⁻⁵	2.54x10⁻⁵	10⁻⁴	3.05x10⁻⁴	9.15x10⁻⁴	.001	.00503	.02012	.1	.20117	1	1.6093	1.853
miles	6.21x10⁻⁷	6.21x10⁻⁶	1.58x10⁻⁵	6.21x10⁻⁴	1.89x10⁻⁴	5.68x10⁻⁴	6.21x10⁻⁴	.00312	.0125	.06213	.125	.62137	1	1.151
nautical miles	5.39x10⁻⁷	5.39x10⁻⁶	1.37x10⁻⁵	5.39x10⁻⁴	1.64x10⁻⁴	4.92x10⁻⁴	5.39x10⁻⁴	.00271	.01084	.05396	.10844	.5396	.8684	1

*Value adopted by the U. S. Coast and Geodesic Survey. A speed of 1 nautical mile per hour is called a knot.

Table A-2 English/Metric Area Conversions

Name	cm².	sq. in.	sq. ft.	sq. yd.	m².	sq. rd.	sq. ch.	a.	ha.	km².	sq. mi.
square centimetres	1	6.452	929	8361	10⁴	252,908	4,046.528	40,465.284	10⁸	10¹⁰	2.59x10¹⁰
square inches	.155	1	144	1296	1550	39,204	627,264	6,272,640	155x10⁷	155x10⁷	4,014,489,600
square feet	1.076x10⁻³	.00694	1	9	10.76	272.25	4356	43,560	107,639	10,763,500	27,878,400
square yards	1.196x10⁻⁴	7.716x10⁻⁴	.1111	1	1.196	30.25	484	4840	11.960	1,195,900	3,097,600
square metres	10⁻⁴	6.452x10⁻⁴	.0929	.8361	1	25.29	404.7	4047	10⁴	10⁶	2,589,998
square rods	3.953x10⁻⁶	2.551x10⁻⁵	3.673x10⁻³	.03306	.0395	1	16	160	395.37	39,537	102,400
square chains	2.471x10⁻⁷	1.594x10⁻⁶	2.296x10⁻⁴	.00207	.00247	.0625	1	10	24.71	2471	6400
acres	2.471x10⁻⁸	1.594x10⁻⁷	2.296x10⁻⁵	2.066x10⁻⁴	2.471x10⁻⁴	.00625	.10	1	2.47104	247.1	640
hectares	10⁻⁸	6.452x10⁻⁹	9.29x10⁻⁶	8.361x10⁻⁵	10⁻⁴	.00253	.04047	.4047	1	100	259
square kilometres	10⁻¹⁰	6.452x10⁻¹⁰	9.29x10⁻⁸	8.361x10⁻⁷	10⁻⁶	2.52x10⁻⁵	4.047x10⁻⁴	4.047x10⁻³	.01	1	2.59
square miles	3.861x10⁻¹¹	2.491x10⁻¹⁰	3.587x10⁻⁸	3.228x10⁻⁷	3.861x10⁻⁷	9.766x10⁻⁶	1.563x10⁻⁴	1.563x10⁻³	3.861x10⁻³	.3861	1

Table A-3 English/Metric Volume Conversions

Name	cm³	cu. in.	L.	U.S. gal.	Imp. gal.	cu. ft.	cu. yds.	m³	ac. ft.	s.f.d.	m.g.
cubic centimetres	1	16.39	1000	3785.4	4542.5	28,317	764,560	10^6	1.233×10^9	2.451×10^9	3.785×10^9
cubic inches	.06102	1	61.0234	231	277.274	1728	46,656	61,023	75,271,680	149,299,200	231×10^6
liters	.001	.016387	1	3.7854	4.5425	28.317	764.56	1000	1,233,490	2,451,250	$3,785,430$
U.S. gallons	2.642×10^{-4}	.004329	.26417	1	1.200	7.4805	201.974	264.17	325,851	646,317	10^6
imperial gallons	2.201×10^{-4}	.003607	.22008	.83311	1	6.2321	168.267	220.83	271,472	538,453	833,111
cubic feet	3.531×10^{-5}	5.787×10^{-4}	.03631	.13368	.16046	1	27	35.3145	43,560	86,400	133.681
cubic yards	1.308×10^{-6}	2.143×10^{-5}	.001308	.00495	.00594	.03704	1	1.30794	1613.33	3200	4951
cubic metres	10^{-6}	1.639×10^{-5}	.001	.003785	.00464	.02832	.76456	1	1233.49	2451.25	3785
acre feet	8.107×10^{-10}	1.328×10^{-8}	8.107×10^{-7}	3.069×10^{-6}	3.684×10^{-6}	2.296×10^{-5}	6.199×10^{-4}	8.107×10^{-4}	1	1.9835	3.0684
second-foot-day	4.081×10^{-10}	6.698×10^{-9}	4.081×10^{-7}	1.547×10^{-6}	1.588×10^{-6}	1.157×10^{-5}	3.125×10^{-4}	4.081×10^{-4}	.5042	1	1.5472
million U.S. gallons	2.644×10^{-10}	4.329×10^{-9}	2.644×10^{-7}	10^{-6}	1.2×10^{-6}	7.481×10^{-6}	2.02×10^{-4}	2.643×10^{-4}	.3260	.6463	1

Table A-4 English/Metric Volume per Time Conversions

Name	gal. per day U.S.	gal. per day imp.	cu.ft. per day	m³ per day	gal. per min. U.S.	gal. per min. imp.	l/sec.	gal. per sec. U.S.	gal. per sec. imp.	a.c.ft. per day	sec.ft. or cusec.	m³/sec.
U.S. gallons per day	1	1.200	7.4805	264.17	1440	1728	22,824	86,400	103,680	325,850	646,317	22,824,288
imp. gallons per day	.8333	1	6.233	220.14	1200	1440	19,020	72,000	86,400	271,542	538,860	19,020,240
cubic feet per day	.1337	.1605	1	35.314	192.50	231.12	3051.2	11,550	13,860	43,560	86,400	3,051,173
cubic metres per day	.003785	.004544	.02832	1	5.451	6.541	86.4	327.06	392.47	1233.5	2446.6	86,400
U.S. gals. per minute	6.944x10⁻⁴	8.333x10⁻⁴	5.195x10⁻³	.1836	1	1.200	15.860	60	72	226.28	448.83	15,850
imp. gals. per minute	5.787x10⁻⁴	6.944x10⁻⁴	4.327x10⁻³	.1528	.8333	1	13.208	50	60	188.57	374.03	13,208
litres per second	4.382x10⁻⁵	5.258x10⁻⁵	3.278x10⁻⁴	.0116	.0631	.0767	1	3.785	4.542	14.276	28.317	1000
U.S. gals. per second	1.157x10⁻⁵	1.389x10⁻⁵	8.658x10⁻⁵	3.067x10⁻³	.01667	.02	.2648	1	1.2	3.771	7.480	264.2
imp. gals. per second	9.647x10⁻⁶	1.157x10⁻⁵	7.215x10⁻⁵	2.548x10⁻³	.01389	.01667	.2201	.8333	1	3.142	6.232	220.1
acre feet per day	3.069x10⁻⁶	3.683x10⁻⁶	2.296x10⁻⁵	8.106x10⁻⁴	.00442	.00530	.0700	.2652	.3183	1	1.9835	70.046
cubic feet per second	1.548x10⁻⁶	1.856x10⁻⁶	1.157x10⁻⁵	4.086x10⁻⁴	.00223	.00267	.0383	.1337	.1606	.5042	1	35.314
cubic metres per second	4.381x10⁻⁸	5.258x10⁻⁸	3.278x10⁻⁷	1.157x10⁻⁵	6.309x10⁻⁵	7.572x10⁻⁵	.001	.00378	.00454	.0142	.0283	1

Table B-1 Map Scales for Selected Representative Fractions

Fractional Scale (1)	Feet per Inch (2)	Inches per 1000 Feet (3)	Inches per Mile (4)	Miles per Inch (5)	Acres per Squinch (6)	Squinches per Acre (7)	Sqmiles per Squinch (8)
1: 600	50.00	20.00	105.60	0.009	0.057	17.4240	0.0001
1: 1,200	100.00	10.00	52.80	0.019	0.230	4.3560	0.0004
1: 1,500	125.00	8.00	42.24	0.024	0.359	2.7878	0.0006
1: 2,000	166.67	6.00	31.68	0.032	0.638	1.5682	0.0010
1: 2,400	200.00	5.00	26.40	0.038	0.918	1.0890	0.0014
1: 3,000	250.00	4.00	21.12	0.047	1.435	0.6970	0.0022
1: 3,600	300.00	3.33	17.60	0.057	2.066	0.4840	0.0032
1: 4,800	400.00	2.50	13.20	0.076	3.673	0.2722	0.0057
1: 6,000	500.00	2.00	10.56	0.095	5.739	0.1742	0.0090
1: 7,200	600.00	1.67	8.80	0.114	8.264	0.1210	0.0129
1: 7,920	660.00	1.52	8.00	0.125	10.000	0.1000	0.0156
1: 9,600	800.00	1.25	6.60	0.152	14.692	0.0681	0.0230
1: 10,000	833.33	1.20	6.34	0.158	15.942	0.0627	0.0249
1: 12,000	1,000.00	1.00	5.28	0.189	22.957	0.0436	0.0359
1: 19,200	1,600.00	0.63	3.30	0.303	58.770	0.0170	0.0918
1: 24,000	2,000.00	0.50	2.64	0.379	91.827	0.0109	0.1435
1: 25,000	2,083.33	0.48	2.53	0.395	99.639	0.0100	0.1557
1: 62,500	5,208.33	0.19	1.01	0.986	622.744	0.0016	0.9730
1: 125,000	10,416.67	0.10	0.51	1.973	2,490.977	0.0004	3.8922
1: 250,000	20,833.33	0.05	0.25	3.946	9,963.907	0.0001	15.5686

Column Derivations

(2) = Column (1) / 12

(3) = 12,000 / Column (1)

(4) = 5280 ft/mi x 12 in/ft / Column (1)

(5) = 1 / Column (4)

(6) = (Column (1))^2 / 43,560 ft^2/acre x 144 in^2/ft^2

(7) = 1 / Column (6)

(8) = (Column (2))^2 / (5280)^2 ft^2/mi^2

C Watershed Eccentricity

The concept of watershed eccentricity is derived from the formula for the eccentricity of an ellipse, but with appropriate modification to render the parameter directional with respect to the mouth of the watershed and responsive primarily to the shape of the lower half of the drainage area. The reasons for this are (1) the impact of shape on peak flow is an optimization process as one progresses from long, narrow watersheds with the outlet at one end to wide, narrow watersheds with the outlet at one side; (2) for storms that uniformly and completely cover the watershed and that cease prior to storage medium saturation (as is natural under most conditions), the flood peak has been found to be generated in the lower portion of the drainage basin in laboratory model studies under a rainfall simulator; and (3) the concept that a circle is the shape that will produce the maximum peak has been found to be erroneous.

Requirements for a useful shape variate, then, include that it should (1) be dimensionless, (2) be easily measured, (3) primarily involve the runoff contribution from the lower half of the drainage, and (4) optimize with compactness of the basin. Determined from a carboard cut-out of the watershed, which may be suspended and plumbed from a straight pin so as to locate the center of mass, watershed eccentricity meets these requirements, and may be calculated and expressed as follows:

$$\tau = (|L_C^2 - W_L^2|)^{0.5} / W_L$$

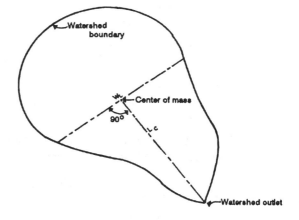

where τ = watershed eccentricity, a dimensionless parameter;

L_C = length from the outlet to the center of mass of the watershed; and

W_L = width of the watershed perpendicular to L_C and at the basin's center of mass, both in the same units.

It can be seen that, as L_C approaches the value of W_L, the value of τ approaches zero: thus, lower values of τ are found to associated with high flood peak potential and, as either L_C or W_L gets large, τ approaches infinity, associating high values of τ with low flood peaks under the experiental assumptions that the storm uniformly, instantaneously, and completely covered the watershed model.

D Glossary

Acre-foot - the amount of water necessary to cover an acre (43,560 square feet) to a depth of one foot, or 43,560 cubic feet, which is equivalent to 325,828 gallons

Antecedent Moisture Conditions - a description of the amount of water in storage at some point in time (usually the start of an hydrologic event) that is relevant to the event

Best Management Practices (BMP) - a set of field activities that provide the most effective means for reducing pollution from a nonpoint source

Cold front - the interface between an advancing mass of air that is colder than the one it is replacing, usually at the point of contact with the ground surface

Convectional storm - a rain event that results from unequal heating of the land surface such that a rising column of air cools beyond the dew point and becomes unstable, producing a cumulonimbus cloud, typically exhibiting violent local winds, high intensity rainfall over a small area and for a short duration, hail, thunder, and lightning

Cubic feet per second (cfs) - the basic unit of stream flow in the English system, a flow of volume equal to one cubic foot passing through a vertical plane in one second

Cyclonic storm - a rain or snow event that results from air that is forced to rise over (1) a wedge of cold air that is slipping beneath it (cold front), or (2) a wedge of warmer air that it is overriding (warm front)

Depression Storage - water temporarily detained on the surface of the Earth in puddles and cavities that have little or no surface outlet

Detention Storage - water temporarily detained in the non-capillary pores of the soil, free to move by gravity, which it generally does within about 24 hours of the event that filled the pores

Dew Point - the temperature to which the air must be cooled in order to reach 100 percent relative humidity, or saturation

Evapotranspiration - the combined evaporative-type processes, including evaporation, interception, and transpiration, usually applied to biological systems

Falling limb - the portion of the hydrograph trace immediately following the peak and reflecting the decreasing production of storm flow

Flow interval (quarter-, half-) - the shortest period of time in days (usually) during which one-half (-quarter) of the annual runoff occurs

Fog drip - water that is collected on the surface of vegetation and falls to the ground, as warm, moist air is advected over the vegetation

Freezing rain - water that freezes upon reaching a surface, the temperature of which is below freezing

Graupel - a snow or ice crystal heavily coated with rime

Gross precipitation - the amount of precipitation measured in the open, that is, before the interception process depletes the amount reaching the ground

Guard cell - the plant cell(s) bounding the stomate

Hail - solid ice precipitation that has resulted from repeated cycling through the freezing level within a cumulonimbus cloud

Hydrograph - the trace of stage (height) or discharge of a stream over time, sometimes restricted to the short period during storm flow

Hygroscopic coefficient - the level of tension at which water is considered to be "bound" to the soil particles, 31 atmospheres

Infiltration - the rate of movement of water from the atmosphere into the soil

Interception - (1) the process whereby the downward movement of precipitation is interrupted and redistributed, or (2) the amount of water lost to soil moisture by this process, often expressed as a percent

Leaf drip - water that is intercepted and rerouted to the ground via collection on and drip from leaf surfaces

Litter - the vegetative material on the surface of the soil, referred to as the *Oi* horizon

Lysimeter - an artificial device for evaluating the water budget by enclosing a block of soil, often on a scale, with equipment for monitoring inputs and outputs

Net precipitation - the amount of precipitation reaching the ground under a vegetative cover, thus, gross precipitation minus interception loss, corrected for stemflow

Occluded front - the widespread (cloudiness and precipitation) that results when the faster-moving cold front overtakes the warm front and lifts it aloft, often stalling in the process

Orographic storm - a rain or snow event that results from lifting (and consequent cooling) of an air mass over a mountain barrier, usually restricted to the windward side of the mountains, and often producing hot, dry winds on the lee side

Overland flow - surface runoff

Phreatophyte - literally a water-loving plant, one that thrives in wet sites and/or has the ability to tap deep saturation zones

Phytometer - a device for controlling the input of water to a plant, usually simply a sealed pot, that can be weighed so as to evaluate water use and growth

Precipitation - the downward movement of water in liquid or solid form from the atmosphere following condensation in the atmosphere due to cooling of the air below the dew point

Precipitation intensity - the rate of precipitation, in units of depth per unit time

Rain - the liquid form of precipitation

Recession curve - the decreasing portion of the (storm) hydrograph that represents base flow

Retention storage - water retained in the capillary pores of the soil, not free to move by gravity, and in large part available to plants

Rising limb - the increasing portion of the storm hydrograph

Second-feet - (see Cubic feet per second)

Snow - one of the common solid forms of precipitation

Soil aggregates - conglomeration of inorganic and organic matter in the A horizon which, if present, is referred to as the "mull" layer

Stemflow - water that flows down the trunks of trees following interception

Stomates - the minute openings in the underside (generally) of leaves that permits passage of gases, including oxygen, carbon dioxide, and water vapor

Stream order - a system of numbering streams according to sequence of tributary size

Sublimation - transformation of a solid to the gaseous phase without passing through the normally intermediate liquid phase

Subsurface runoff - water that moves through the aerated portion of the soil to the stream and behaves more like overland flow than base flow

Surface runoff - water that runs across the top of the soil without infiltrating the soil

Throughfall - precipitation that falls directly through the vegetative cover to the soil

Transpiration - the movement of water from the soil or ground water reservoir via the stomata in plant cells to the atmosphere

Variable source area - the flexible zone near to and extending the stream that contributes runoff to the channel during a runoff-producing event

Warm front - the interface between an advancing mass of air that is warmer than the one it is replacing, usually at the point of contact with the ground surface

Water table - the upper surface of the ground water reservoir

Wilting point - the tension at which water is held in the soil beyond which plants (normally) cannot withdraw soil moisture

Author Index

(Initials are used only where authors of the same last name have citations in the same year)

Abdul 200
Abelson 279
Achuthan 10
Adams 64
Adrian 178, 242
Aguado 64
Alexander 126, 300
Alger 27, 64
Ali 127
Alley 126
Alvord 136
Aly 184
American Society of Civil
 Engineers 203
Anderson 135, 154, 200, 289, 293, 294,
 297, 298, 299
Anderson, E. A. 27
Anderson, H. W. 297, 299
Aron 232
Aubertin 118
Austin 124, 364
Avanzino 245
Ayer 113, 120, 297
Ayers 278

Bailey 287
Baker 113, 114, 126, 287, 300

Bakun 277
Barge 64
Barker 87
Barnes 300
Barnhart 227
Bastin 63
Bates 95, 109, 296
Beaumont 63
Belt 283
Ben-Asher 125
Benton 128
Benz 99
Berg 122, 296
Berndt 64
Berris 300
Bertle 298
Beschta 118
Bethel 126
Bethlahmy 69, 119, 282, 289
Betson 199
Betters 126
Beven 149, 230
Biggar 150, 179
Binetti 278
Birtles 230
Bissell 64
Biswas 179

Black 32, 85, 87, 91, 94, 100, 101, 104,
    105, 106, 115, 121, 123, 125, 126,
    130, 133, 200, 211, 218, 219, 222,
    238, 255, 256, 261, 264, 265, 266,
    267, 268, 269, 270, 275, 291, 307
Black, T. A. 109
Blackburn 115
Blaney 99
Blasing 223
Blumenstock 79, 80
Bodhaine 205
Booth 64
Booty 278
Bormann 116
Borys 91
Bostrum 136
Boughton 217
Bowes 296
Bowley 300
Boyd 243, 245
Boyer 231
Boyle 114
Bradley 300
Brady 143, 146, 150, 151, 154, 155,
    162, 168, 177, 180
Bras 63
Brater 201, 225, 226, 254, 264, 266
Brater 255
Brezonik 278
Brink 300
Broadfoot 171
Brooks 110
Broughton 264
Brown 87, 115, 118, 246, 255, 269
Brown, G. W. 287
Brussock 269
Brutsaert 149
Buchanan 190, 191
Buckman 143, 150, 151, 154, 155, 168,
    180
Bureau of Reclamation 275
Burges 231
Burt 200
Byrnes 87

Camillo 125
Campbell 128, 205, 278
Carroll 300
Carter 133, 278
Cartwright 246
Casey 243, 245
Caskey 117
Chang 69, 85, 86, 231, 300
Changnon 55, 65
Cheng 289
Cherry 9, 139, 140, 141, 181
Chery 256
Chiang 165
Chow 9, 12, 73, 76, 78, 99, 124, 127,
    128, 139, 142, 176, 185, 187, 202,
    205, 216, 232, 235, 236, 251, 256,
    264
Chowdhury 71
Christmas 238
Clark 115, 255
Clary 84
Cline 174
Cole 181
Coleman 124
Collins 292
Colman 173
Connaughton 287
Conrad 119
Cooper 55
Corbett 201, 245, 290
Cork 25
Corps of Engineers 221
Cotecchia 174
Council on Environmental Quality 25
Cowardin 135
Cox 62, 179
Crawford 232
Criddle 99
Cronan 278
Cronn 222
Crouse 119
Curtis 109, 174

Dake   26, 97
Dalrymple   222
Datta   232
Davenport   127, 128
Davis   113, 115, 283, 292
Dawdy   69
De la Cruz   132
DeBano   119, 149, 286, 292
DeByle   60
Decker   100
DeCoursey   124
Demarque   27
DeMott   91
Dennis   43, 285
Department of Agriculture   3
Department of Environmental
   Conservation   307, 308
Department of Natural Resources   155
DePinto   278
DeWalle   64, 200
Dickerson   245
Dils   111, 114
Dingman   43, 45, 148, 185, 203, 212,
   227, 229, 234, 267
Diskin   200
Dissmeyer   301, 302
Dixon   200, 269
Dobson   300
Doering   99
Dolzine   246
Doraiswamy   93, 103
Dortignac   178
Doty   119, 299
Douglass   102, 113, 123, 172, 174, 290,
   291, 295
Dracup   223
Dragoun   114
Dress   126
Driscoll   278
Duckstein   232
Duke   171
Dunford   109, 112, 297, 311
Dunn   119

Dunne   96, 100, 179, 200, 201, 202,
   217, 222, 223, 287
Duque   63
Duvick   223

Eagleson   97, 109, 212, 230
Edgerton   278
Edgington   242, 243
Eisel   311
Eisenlohr   108, 128
Endal   27
Eschner   17, 85, 113, 121, 298, 300, 302
Escobar   267
Ettenheim   93

Farmer   55
Faust   184
Federer   125
Feller   289
Ferguson   171, 230
Ferguson   230
Ffolliott   84, 121, 297, 300, 301
Field   126, 308
Fink   278
FitzPatrick   143, 155
Fleming   78
Fletcher   55
Follett   99
Forest Service   144, 145, 146, 147, 153,
   168, 169, 290, 308
Foroud   264
Foster   238
Fowler   60
Fralish   118
Francko   140
Frank   29
Franzini   178, 187, 204
Fredriksen   118
Freeze   9, 139, 140, 141, 181, 198
Frere   282, 283
Fritschen   93, 103
Frohliger   278
Fuller   278

Gardiner 267
Gardner 114, 149
Garstka 297
Gary 62, 86, 287, 300
Gaskin 290
Geiger 42
Gentry 275
Geological Survey 188, 189, 218, 242, 243
Germann 149
Getty 255
Gevers 63
Geyer 255
Gibbs 237
Gillette 302, 303
Gillham 200
Gleason 298
Gleick 48, 233, 255, 277, 301
Goldberg 174
Golding 300
Good 136
Goodell 96, 101, 102, 297
Goodman 93
Gosselink 133, 134, 241
Gottfried 297, 300
Government Accounting Office 308
Graham 246
Granger 64
Grant 278
Greenland 154
Greeson 242
Grieve 238
Griffin 285
Grigal 289

Haan 230
Hagan 127, 128
Hage 79
Haines 116
Hains 212
Hall 203, 209
Halterman 278
Hamilton 59, 60

Hammond 275
Hanna 245
Haragan 56
Harbaugh 264
Harms 289
Harr 91, 111, 114, 118, 122, 295, 300
Harris 246
Harrison 175
Harrold 102
Hart 114, 298
Hasler 246
Hathaway 255
Haupt 60, 86, 283, 300
Hawkins 276
Hayes 59
Heaney 306
Heede 115, 233, 283, 284, 285
Helmers 59
Helvey 60, 85, 86, 110, 112, 124, 289
Hem 30, 32, 234, 239
Henderson 135
Hendrix 173
Hendry 278
Henriksen 245
Henry 109, 296
Hershfield 66, 67
Hertzler 255
Hewlett 12, 102, 104, 109, 110, 112, 172, 175, 198, 240
Hibbert 101, 113, 123
Hibbert 128, 198, 283, 290, 291, 292, 295
Hickock 65
Hidore 196, 275
Hiemstra 222, 275
Higgins 231, 282
Hill 119, 232, 233, 243
Hillel 177
Hindman 91
Hirsch 245
Hofmann 223
Hollis 220
Holtan 230

Author Index

Hoover 86, 87, 171, 172, 173, 293, 294, 298
Hornbeck 93, 102, 110, 112, 116, 287, 300
Horton 14, 128, 179
Houck 232
Hsia 103
Huber 232, 233, 306
Huff 56, 65
Hughes 127
Humphries 64
Hussain 149
Hutchinson 184
Hydrology Committee 202, 205

Idso 108
Ingersoll 246
Ingwersen 93
Inzaghi 174
Israelsen 232, 233

Jackson 174
James 128
Jarboe 230
Jeffers 174
Jenny 151
Johannessen 245
Johnson 118, 288
Johnston 102
Jones 55
Junge 238

Kadlec 136
Kane 176, 177, 278
Kapinos 242
Kareliotis 232
Kattelmann 122, 196, 198, 291, 296
Kaufman 122
Kazman 232
Kelton 229
Kennard 135
Kennedy 237, 243, 244, 245

Kerr 54, 223, 277, 278
Kimmins 289
King 111, 112, 185, 256, 287, 300
Kirkby 142, 177
Kirkham 180
Kisiel 232
Kittredge 91, 208
Klock 64
Knapp 231
Knoerr 98, 125, 174
Kohler 202, 203, 204, 205, 206, 207, 208, 209, 215
Kohler 28, 266
Koshi 174
Kovner 64, 1112
Kowall 267
Kozlowski 95, 96, 130
Kramer 95, 96, 130
Krammes 149
Krishnan 232
Krug 278
Krumholz 237
Krutilla 296
Krygier 115
Kuhnke 64
Kunkle 115

LaBaugh 135
Laden 255
Lambert 114
Lane 200
Langbein 69, 267
Langford 242
Larsen 67
Lash 125
Law 229
Lawrence 278
Lawson 87
Leaf 64, 86, 118, 126, 136, 290, 300, 301
Lee 29, 124, 223, 224, 261, 262, 289
Lee, R. 95, 104
Lee, S. 223

Lefor 135
Leonard 17, 60
Leopold 96, 100, 179, 202, 217, 222, 223, 267
Lettenmaier 231, 245
Levno 114
Lewis 110, 114
Lewis 278
Liebetrau 244
Lienkaemper 283
Likens 115, 116
Lindroth 20
Linsley 28, 187, 202, 203, 204, 205, 206, 207, 208, 209, 215, 232, 266
Litten 242
Lomas 114
Lorent 63
Love 297
Lull 15, 16, 17, 18, 19, 20, 21, 171, 172, 218, 290, 298
Lynch 201, 245

Macan 134, 241
Mace 99, 174
MacGillivray 174
Mah 64
Male 64
Malkus 275
Manning 75
March 55
Marius 199
Markowitz 217
Marron 301
Marsh 109, 271
Martin 116
Martin 127, 128, 241, 287
Martinec 63, 64, 300, 301
Martinelli 298, 300
Marx 282
Massman 93
Matalas 217
Mather 79, 101, 103
Maugh 278

Mawdsley 127
McCabe 278
McClimans 308
McColl 289
McCorisin 111
McCuen 217
McDonald 128
McGuiness 103, 246
McGuinness 60, 71, 102
McHughs 255
McKim 174
McNaughton 32, 100
Meeuwig 149
Megahan 110
Meier 27, 64
Meiman 64, 115, 121, 123, 296, 297
Meinzer 181
Mejia 67
Melton 267
Merriam 93, 174
Merry 301
Mersereau 114
Miller 19, 36, 38, 43, 45, 46, 55, 57, 58, 60, 62, 73, 91, 127, 132, 174, 267, 279
Miller, D. H. 91, 93, 95, 127, 128, 175, 299, 300
Miller, E. L. 111
Mills 79
Miner 86, 113
Minshall 255
Mitsch 133, 134, 241
Molz 104
Morgan 184
Morison 126
Moss 65
Mount 175
Munn 25
Murphy 98, 125
Musgrave 155, 178, 179
Mustonen 103
Myers 306
Myrick 292
Myrup 78

Author Index

Nace 4
National Council of the Paper Industry for Air and Stream Improvement 308
National Water Commission 306
National Weather Service 66
Needham 241
Neff 237
Nemetz 245
Nemetze 243
Nicolle 269
Nicolson 94
Nielsen 150, 179, 180
Nierenberg 277
Nik 124
Nilles 115
Novotny 308
Nutter 240

O'Callaghan 229
Oberlander 91, 93
Oberts 136
Odum 311
Olson 171, 172, 173
Osborn 65
Osteen 307
Osterkamp 175

Palmer 223, 300
Pani 56
Parker 140
Parkes 229
Parmele 232
Patric 85, 86, 103, 111, 114, 118
Patrick 240, 278
Paulhus 28, 202, 203, 204, 205, 206, 207, 208, 209, 215, 266
Paulson 223
Pearce 108
Peck 64
Peck 67
Pegden 227
Peirce 307

Penman 99, 100, 101
Penton 63
Perrault 4
Petersen 165
Petts 241
Philander 277
Phillips 55
Pierce 116, 154, 193, 194, 195
Pilgrim 246
Pirastru 174
Pisano 307
Pitlick 243
Ponce 121, 123, 295, 296
Potts 119, 289
Prévost 154
Pritsker 227

Quigley 282
Quinn 280

Rachford 232
Ralston 289
Ramanthan 277
Rango 63, 64, 174, 200, 231, 276, 300
Rantz 223
Rasmussen 126, 301
Rawls 62, 217
Rector 122, 296
Reich 212, 222
Reichman 99
Reid 287
Reifsnyder 15, 16, 17, 18, 19, 20, 21
Reigner 85
Reimann 60
Reinhart 60, 109, 111, 113, 114, 171, 172, 193, 194, 195, 287, 293, 294
Remson 104
Renard 65
Ricchena 174
Rice 297
Rich 102
Richards 267
Richardson, B. Z. 174

Richter 289
Riekirk 114, 116
Riggs 211, 216, 231
Riley 232, 233
Roberts 231
Robertson 246
Robertson 63, 246
Rodenhauser 278
Rodriguez-Iturbe 63, 67, 233, 267
Rogers 126
Rogerson 86, 87
Rolfe 242, 243
Roth 85, 86
Rothacher 111, 113
Rouse 79, 128
Rowe 108, 178
Rudel 275
Rushmore 298
Ryan 126

Sadeghipour 223
Sanders 35, 36, 242
Santeford 27, 64
Sartz 154, 174, 288
Sather 134
Satterlund 78, 86, 121, 298, 300
Scheffe 278
Scheidegger 232
Schermerhorn 55
Schindler 278
Schmidt 286
Schmuggee 174
Schneider 113, 220, 277, 297
Schofield 278
Scholl 292
Schwartz 255
Seitz 307
Shafer 301
Sharp 287
Sheridan 275
Sherman 14, 178, 179, 201, 254, 296
Sidle 205
Siegel 300

Simanton 65
Simpson 136, 275
Singh 202, 230
Singh 71, 97
Skau 149
Slack 245
Smith 212, 245
Smith 50
Snyder 283
Sofia 27
Soil Conservation Service 155
Soil Conservation Society of
    America 141, 176
Somers 190, 191
Soons 154
Sopper 218, 275, 290
Spyridakis 174, 175
Srinilta 180
Stall 267, 307
Stallings 176
Stalnaker 241
Stark 116
Stein 176, 177, 236
Stewart 79, 108
Stockwell 275
Storey 110
Storr 25, 78
Strahler 48, 266, 268
Striffler 154, 177
Stroud 241
Stumm 184
Sturges 107, 108, 292, 300
Sun 141
Sundeen 136
Sundquist 277
Sverdrup 278
Swank 64, 86, 112, 113, 117, 118, 238, 290, 291
Swanson 99, 283, 300
Swift 112, 125
Swift 55
Szeicz 97

Author Index

Tabler 300
Tajchman 26, 98
Tangborn 232
Tanner 89
Taylor 118, 255
Tellers 97
Tennyson 287
Thomas 140, 245
Thompson 20, 36, 38, 43, 45, 46, 57, 58, 60, 62, 73, 99, 301
Thompson, J. R. 282, 287
Thompson, L. M. 223
Thornthwaite 79, 80, 100, 101, 103
Thorud 94, 121, 154
Tice 220
Tiedemann 1191, 282, 289
Tjachman 26
Todd 139
Todorovic 217
Tolsted 288
Tomlain 25
Toor 27
Torii 246
Trimble 113, 154, 291
Troendle 111, 112, 113, 115, 122, 290, 295, 300
Tryon 110
Turco 25
Turk 278
Turner 134

Ulaby 300
Urie 114
USDA 197
Uyeno 243
Uyeno 245

Van Bavel 79
Van Genuchten 150
Van Hylckama 128, 234
Van't Woudt 269
Veihmeyer 171
Verry 110, 114, 300

Viessman 204, 233, 255
Vigon 308
Vitousek 116
Vogel 65
Vogelmann 91

Wagle 10
Waide 238
Wallace 55, 200
Walling 245
Wallis 217, 277
Walsh 275
Ward 287
Warvinge 278
Water Quality Group 308
Water Resources Council 140, 217
Weather Bureau 40, 41, 57, 59, 66, 67, 75, 221
Webb 55
Webster 117
Wei 71
Weirich 291
Weitzman 109
West 297
Wetzel 140
Whittaker 269
Wilde 174, 175
Willet 35, 36
Williams 216
Wilm 109, 112, 229, 297, 311
Winger 278
Winter 108, 184, 185
Wisler 201, 225, 226, 255, 264, 266
Wolman 267
Wooding 256
Woodley 275
Woods 110
Woodwell 276
Woolhiser 217
WQED 27

Yang 32
Yarnell 66

Yates 300
Yen 264
Yevjevich 2
Young 238
Young 267

Zebuhr 115
Zellweger 245
Ziemer 111, 174
Zinke 84, 181
Zuzel 62

Subject Index

(* indicates term is defined in Glossary, starting on page 372)

(Software names are in CAPS)

Absolute humidity 45
Accuracy 242
Acid deposition 233, 246, 278
Acid rain xxi, 238, 278, 280
Acid runoff 278
Acidic precipitation 154
Actual evapotranspiration 30
Adhesion 7, 143, 166
Adhesive and cohesive forces of water 234
Adiabatic cooling 45, 50, 56
Adirondack Forest Preserve 109, 121
Adirondack Mountains 238
Adirondack Park 301
Adirondack Snow Tube 62
Advective fog 47, 91
Aeolian Soils 163
Aeronautical Icing Research Laboratory 93
Aggradation 249
Aggregates 83
Agricultural drought 223
Agricultural lands 282
   Best Management Pratices on 308
Agricultural Research Service 60, 97, 170, 296
Agriculture 248
Air mass weather 55
Air masses 48
Air, molecular weight of 40

Air-soil interface xix
Alaska 31
Albedo 17, 27
   of forest canopy 17
Aliasing 228
Alkalinity 245
   and acid rain 278
All-terrain vehicles 281
Allegheny Plateau, vegetation and snow deposition and melt 297
Alluvial soils 1217, 149, 163
Alpine snow zone, and snowpack management 298
Alpine zone 196, 260
Alsea Watershed Study 115
Aluminum 153, 162, 165
   and acid rain 278
American Elm 86
American Society of Testing Materials 242
Ammonia 238
Amount of water on Earth 8
Aneroid barometer 36
Animals, in soil 150
Annual hydrograph 110
   effect of forest cutting on 110
Annual precipitation 270, 274
Annual series, ignores the second highest event in any year 215
   randomly distributed events 216

underestimates flood magnitude   215
Annual water budgets, timber harvesting, impacts of   119
Annual yield   113, 115, 253
   conflicting results following herbicide treatments in different locations   112
   decrease in during protection   121
   decreases during regrowth   112
   Increase in annual yield barely detectable on south-facing watershed   112
   increased by partial cutting   112
   and watershed shape   265
Antecedent moisture conditions*   xix, 7, 13, 14, 89, 95, 175, 201, 202, 206, 219, 224
Antitranspirant, and phreatophytes   128
Ants   118
Appalachian Mountains   55, 85, 112, 114, 118, 123
Appropriation Doctrine of water rights   275
   ecosystem   235, 304
   environment   xxi, 240
   flora and fauna, and acid rain   278
   habitat modification, and phreatophytes   128
   system   118
Aquiclude   141, 152
Aquifer   141
   exchange rates, in models   230
Aquifers, and wetlands   133
Aquifuge   141
Arapaho National Forest   296
Area-elevation curve   257
Area-inches   29
Areal precipitation   63
Areawide Waste Treatment Management   305
Arid area rainfall determination   65
Aridity   222
   index   105
Arithmetic average   67

Arizona   91, 101, 113, 114, 115, 121, 291, 296, 300, 301
   and phreatophytes   127
Arkansas-Red-White River Basin   140
Artesian aquifer   141
ARZMLT   126
Aspect   22, 42, 104, 164, 205, 250, 252, 260, 270
   watershed   212
Aspen   110
Atmosphere   4, 9, 137, 176, 183, 237, 238, 276
   density of   34
   evaporative potential of the   97
Atmospheric particulates   25
   pressure   74, 95
      and ground water   140
   water   34
Augite   151
Australia   175
Average annual flow   254
Average annual rainfall   16
Average temperature   42
Azonal soils   163

Back radiation   24
   and the greenhouse effect   276
Bacteria   165
   and wetlands   136
Bank storage   5, 110, 115
Bank-full stage   216
Bark beetle   119, 289
Bark, water content of   129
Barometer   35
Basalt   151
Base flow   5, 122, 142, 175, 185, 198, 200, 203, 209
   recession   121, 205, 209
   and wetlands   133
Base level   249, 271
Basin characteristics   266
Basin lag   203
Bedrock   250

Subject Index

Best Management Practices* xxi, 109, 241, 248, 279, 282, 307
  basic principles 286
  more art than science 287
Bias 228
Bible, floods and droughts in 223
Bicarbonate 239, 241
Biological agents 151
Biosphere 9, 82, 183
Black body 16, 23, 25
Black box models 232
Blaney-Criddle method 99
  and models 99,
  shortcoming of 100
Bogs 134
Bound water 170
Boundary layer model 125
Bountiful, Utah 105
Bowen Ratio 19, 78
  in models 124, 227
Boyles' Law 38
Branching habit of vegetation, and albedo 19
British Columbia 289
Broad-crested weirs 191
BROOK 125
Budget water, effect of forest cutting on 110
Buffering capacity, and acid rain 278
Bulk density 148, 165, 170
  measurement of 171
Bull Run (Oregon) Municipal Watershed 93
Bureau of Reclamation 63, 275
BURP 126

Calcium 116, 162, 238
Calcium carbonate 239
Calibration, of experiental watersheds 111
  of models 228
California 54, 103, 114, 122, 129, 155, 178, 181, 212, 281
  Current 31
  Water Project 301

Camel, and straw 274
Canada 48
Canopy, depth/roughness of and albedo 19
Capillarity 148, 234
Capillary action 165
  conductivity 149, 179
  fringe 127
    and phreatophytes 128
  pores 7, 166
  rise 150
Carbon dioxide 19, 25, 30, 43, 95, 177, 231, 238, 239, 244, 246
  and the greenhouse effect 276
  and water quality 241
Carbonate 153, 246
  in rock 152
Carbonic acid 30, 239
Cascade Range 113, 118, 155
Catchment 249
Catena 164
Catskill Forest Preserve 109, 301
Cattle, compaction of soil by 177
Cedar River watershed 301
Cental Sierra Snow Lab 297
Channel 11, 251
  interception 186, 200, 226
  stabilization 283
  storage 5, 7, 57, 110, 185, 250, 251
  width, correlated with annual runoff 200
Channelization 302
Chaos xix, 233
Chaparral 101, 113
Charles' Law 40
Chatahoochie National Forest 115
Chemical bonds, in soil 148
Chemung River, NY 222
Cherrapunji, India 57
Chinook 54, 235
Chlorofluorocarbons, and ozone depletion 278
Chow, Ven Te 251
Clay 141, 143, 170
  and wetlands 132

Clean Water Act 279
Clear-cutting 110, 113, 118, 240, 301
   in model 125
Climate xix, 10, 27, 34, 97, 122, 137, 151,
   152, 182, 223, 271, 279
   change 231
   classification 79, 155
Climatogram 104
Climatology 311
Climax vegetation 166
Cloud cover 25, 27, 29, 30
   and the greenhouse effect 276
   seeding 43, 56, 235
Clouds 22, 38, 46, 50, 238, 240
Coast Range 115
Coast Redwoods 93, 94, 109, 111, 281
   water stored in 129
Coastal Plain 116
   upwelling, and the greenhouse
      effect 277
   zone 132
Coefficient of permeability 4
Cohesion 7, 30, 143, 166
Cold front* 52, 57
Collection efficiencies 93
Colloid action 153
Colloidal suspensions 148
Colluvial Soils 163
Colman Meter for monitoring soil
      moisture 173
Colman, E. A. 103, 173
Color, of water 236
Colorado 54, 109, 111, 112, 119, 136, 249,
   275, 296, 301, 311
Colorado River 296
   and phreatophytes 128
Colorado River Basin 140
   in model 126
Colorado Rockies 91, 238, 289
Colorado Springs 301
Columbia River Basin 149
Commerce 9
Commercial forest harvest operations 115

Community growth, long range planning
   of 211
Community planning 220
Compaction 56, 171, 177
Compactness coefficient 265, 270
Comprehensive planning 312
Comprehensive Soil Classification
      System 155
Computers 100, 104, 124, 202, 231, 241,
   254, 300
Concrete frost (*See* Frost, concrete)
Condensation xix, 30, 50, 53, 56, 62, 64,
   235
   nuclei 30, 43, 46, 238
Conductivity 181
   meter 238
Connate water 140
Connecticut 135, 212
Conservation practices 307
Continental Divide 91, 112, 296
Continuous gullies (*See* Gully,
      continuous 284
Contour lines 252
   plowing 303
Convectional storms* 55, 56, 60, 65, 219
   classification of 56
Conversion of units 191, Appendix A
Copper 116
Copper Basin 280
Coriolis Force 35, 48, 50
Corning, NY 222
Corps of Engineers 63, 220, 251, 296, 302
Cottonwood 127
Coweeta Hydrologic Laboratory 55, 87,
   101, 110, 111, 115, 117, 166, 175, 199,
   280, 288, 290, 291
   in model 125
Crown density, and albedo 19
Crown, water content of 129
Cubic foot per second per square mile 192
Culverts 202, 205, 286
Cyclone 48
Cyclonic precipitation 216
Cyclonic Storms* 48, 50, 56, 65, 67, 219

Subject Index

Dalton's Law 74, 79, 229
   and models 227
Dams 10, 256, 310
Darcy's Law 141, 180
   in models 227, 230
Death Valley 73
Debris in stream channels 283
Deciduous swamp 133
Deep seepage 102, 142, 1165
Defense 9
Defoliation by insects and diseases 289
Deforestation 109
Degradation 272
Degree-days 99
Density of water 234
Denver 54, 301
Depression storage* 6, 57, 139, 166, 183, 204, 224, 250, 273
Desalinization 274
Deserts 57, 162
   precipitation on 56
Detention storage* 7, 139, 166, 175, 250, 273
   and models 227
Deterministic models 228, 229
Dew formation 93
Dew point* 43, 50
   in models 231
Diameter limit cutting 113
Diffusion 94
Diffusion pressure gradients 130
Digital computer models 233
Discharge 5, 186, 203
Discontinuous gullies 284
Disproportionate Percentages, principle of 273
Dissolved gases 237
Dissolved oxygen 242, 244, 246
   and water quality 241
   in model 126
Dissolved salts, and soil moisture sampling relations 172
Dissolved solids 237
Distribution graphs 202

Diversion 10, 275
Doldrums 48
Douglasfir 32, 86, 92, 103, 109, 111, 114
   in model 125
Downstream basin 311
Downstream flood control measures 303
Drain tile 180
Drainage 163, 270
Drainage area 212
   related to flood peak 217
Drainage basin 184, 249, 267, 270
   network 252, 266
      in models 232
      pattern 250, 257, 267, 270
   to lake area ratio 185
   size 205
Drainage, and wetlands 132
   water 166
Drought 27, 149, 154, 196, 222, 275
   in model 127
   regional extreme method 223
Dry ice 43
Dryfall 64, 238
Ducktown, TN 280
Dunes 163
Dust 30, 43
Dyerville Flat, CA 130
Dynamic models 229

Earthworms 165
East Branch of Oneida Creek, NY 214
Eastern hardwood forests 123
Eastern White Pine 85, 113, 166
Ecological succession, and albedo 19
Ecosystem model 126
   and wetlands 134
ECOWAT 126
Eel River, CA 218, 212
Effective precipitation 91
Effects on Average and Extremes of Flow of
   aspect and orientation 261
   drainage network 269
   elevation and slope 259
   watershed shape 265

watershed size 254
Effluent standards 242
   stream 268
El Niño 91
   and the greenhouse effect 277
Electrical conductivity of water 234, 242
Electrical Resistivity Methods for measuring soil moisture 172
Electromagnetic spectrum 16
Electronic models 228
Electronic stream/watershed systems, in models 232
Elephants 118
Elevation 32, 64, 71, 74, 91, 228, 238, 250, 252, 257, 270, 273
   and precipitation 55
Elevation-weighting method for determining areal precipitation 67
Eluviation, zone of 153
Emergent vegetation 108
Emission from the forest field crops 26
Energy 73, 90, 94, 240
Energy balance, and ozone depletion 278
   importance of water in 235
Energy budget 22, 29, 77, 98, 229
   budget, and models 227
      Geometry of 20
      in model 125
   laws, and models 227
Engelmann spruce 114, 119, 289, 300
Engineering designs 216
   hydrology xxi, 8
England 230
Enhancement, a phase of watershed management 279, 289
Environment, large-scale changes in 274
Environmental chemistry 233
   impact analysis 246, 285
   impacts 10
   water xx, 95, 129
Environmental Impact Statement xxii
Environmental Protection Agency 242, 305
Ephemeral stream 268

Epply Thermoelectric Pyroheliometer 27, 28
Erosion 2, 32, 56, 83, 109, 113, 118, 122, 152, 184, 236, 249, 271, 274, 276, 280, 281, 282, 295, 304
   pavements 283
   and wetlands 132
   effects of wildfire on 120
   resistance to 152
Errors, in model 126
Estuarine wetlands 132, 135
Eucalyptus 93
Eureka, CA 181
Eutrophic lakes 184
Eutrophication 246
Evaporate water at Earth's surface 26
Evaporation xix, xx, 4, 31, 73, 80, 84, 94, 96, 100, 108, 137
   factors that affect 73, 90
   index, EVI 78
   pan 75
   reduction 275
   and soil moisture sampling 172
Evapotranspiration* 2, 5, 8, 19, 25, 26, 27, 30, 32, 77, 80, 88, 96, 101, 108, 115, 119, 123, 124, 154, 164, 168, 175, 196, 205, 224, 229, 250, 254, 257, 260, 270, 273, 290, 297, 303
   Research 107
      Field Investigations of Hydrologic Impacts 119
      General Studies 107
      Vegetation Manipulation 109
   tent 99
   Measuring 98
   in models 125, 232
   and wetlands 133
Exfiltration, in models 230
Experimental watersheds 100, 175
   first 296

FA type infiltrometer 178
Facet 22

Subject Index

Falling limb* 12, 202
   analysis of 206
Farm Bill of 1985 307
Farmland 120, 210
Feldspar 151
Fencing 287
Fens 134
Fernow Experimental Forest 93, 113, 114
Fertilizer 306
Field (water quality) kits 245
Field analysis of water quality 242
Field capacity 96, 168
Field moisture, used for drought determination 224
Fire 54, 55, 119, 240, 2811, 288
   damage 288
Fish and Wildlife Service 135
Fish Creek, NY 214
Fish demands for oxygen 237
   habitat 241
   metabolism 32
   and wetlands 133
Fisheries 9
Fishery protection 302
Flashy streamflow (runoff) 113, 213, 255, 260, 261, 270
   from high elevation watersheds 260
Flood control 110, 202, 215, 276, 302, 303
   and phreatophytes 127
   and wetlands 132
Flood forecasting, and remote sensing 174
Flood frequency 273
Flood frequency analysis 205, 210, 211, 216, 222
   elaborate means of 217
   stratifying the data 217
Flood frequency curve 215
   in models, misleading 229
Flood fringe 221, 302
Flood gates 256
Flood insurance 220
   regulations 216
Flood peaks 120, 184, 263, 270, 286, 295
   generated in lower half of watershed model 200
   and watershed shape 265
Flood prevention 215, 303
   protection 215
   retarding basins 214
   walls 10
Flood, defined 216
Mean Annual 217
   100-year 216
   Standard Project 221
Floodplain 185, 215, 220, 221, 273
   management 216
   regulations 220
   and phreatophytes 128
   and wetlands 132
Floodplain Information Series 220
Floods xvi, 2, 8, 27, 63, 196, 215, 241, 253, 280
Floodway 221, 302, 304
Florida 56, 114, 116, 278
Flow duration analysis 210, 222
   curves 205, 211, 213, 217, 273, 295
Flows, minimum daily 119, 121, 184, 213, 273
Flumes 191, 256
Flushing action 141, 184
   and wetlands 136
Flushing stream system 115, 117
Fluvioglacial Soils 163
Foehn 45, 54
Fog 46, 47, 55
   in Redwood stand 130
Fog drip* 4, 91, 94
Foliage, water stored on 132
Fool Creek Watersheds 111, 112
   cutting studies 296, 300
Forecasting weather 46
Forest canopy 83
Forest communities and water quality 240
Forest cutting 274
   and streamflow of 290
   effect on water available for runoff 296

Forest destruction, and the greenhouse
    effect   276
Forest fire   56, 275
    control   109
    danger   175
Forest floor   58
Forest hydrology   112
Forest Influences   xvi
Forest lands   177
    Best Management Practices on   308
    management   109, 123
Forest Service   60, 97, 102, 115, 128, 170, 280, 290, 296, 302
Forest soils   83, 152
Forest types in the United States   292
Forest vegetation   252
    water stored on   132
Forest, natural, undisturbed, and infiltration capacity   177
Forested watersheds for municipal supplies   302
Forestry   248
FORPLAN   126
Fossil fuels   25, 30, 276
Founders Grove   130
Four-wheel drive vehicles   281
Fox Creek Experimental Watersheds   93
Fractals   255
Fragipan   153, 179
Fraser Experimental Forest   112, 296
Freeze/thaw cycles   164, 177, 271
Freezing and boiling points of water   234, 250
Freezing rain*   4, 60
Fresh waters   9, 10, 271
Freshwater wetlands   132, 134
Frost   7, 154, 177
    and wetlands   134
    concrete   154, 200
    honeycomb   154
    stalactite   154
Froude's model law   256
Frozen soils   43
Fungi   154

Gaging station   191
Gas constant   40
Gas law   38
Gases, exchange between air and leaf   95
    exchange between air and soil   177
General Land Office   252
Geochemist   xxi
Geohydrology   139
Geological Survey   181, 187, 249, 252, 296
Geologically bound water   139
Geology   151, 163, 311
Geomorphic characteristics, stratification of flood data by   212
Geomorphology   249
Georgia   115
Geothermal heat   64
Glacial Soils   163
Glaciers   9
Gley   153
Global climate change   277, 278
    warming   276
Global climate models, and the greenhouse effect   277
Gneiss   152
Granites   151
Granular frost   154
Gravel   141
Gravimetric Method for measuring soil moisture   170
Gravitational water   170, 179
Gravity   34, 139, 166, 179, 248, 250, 271
    and similitude   256
Grazing   282, 303
Great Lakes   10, 55, 73
Great Plains   140
    aquifers, recharge of   238
Great Soil Groups   155, 181
Greenhouse Effect   276
Graupel*   60
Gross precipitation*   85
Ground water   276
Ground water   xvi, 3, 5, 7, 13, 110, 127, 139, 175, 183, 185, 200, 203, 205, 250
    flow   209

Subject Index

accretion to   227
pollution of   305
recharge   105
　　and wetlands   133
storage   213
Ground Water Recharge, Season of   105
Guard cells*   95
Gulf Coast   57
Gulf of Mexico   53
Gulf Stream   31
Gullies   282
Gully, continuous   284
　　discontinuous   284
　　plugs   283, 304
Gully control   283
Gunn-Bellani radiometer   28

H. J. Andrews Experimental Forest   113
Hail*   4, 56, 60
　　suppression   275
Half-flow interval*   113, 210, 273
　　and land use   297
Hardness, and acid rain   278
Harvesting   240
Hawaii   111, 119
Head, hydraulic   141
Head cuts, gully   284
Heat pulse   98
Heavy metals, and wetlands   135
Herbicides   112, 223, 292
High lead logging systems   301
High Plains   281
　　and phreatophytes   127
　　precipitation on   56, 60
High pressure   50
　　systems   35
Highly erodible lands   283
Honeycomb frost   (*See* Frost, honeycomb)
Hook gage   75
Horizon nomenclature   152
Horizonation   152, 164
Horizons in soil   152
Horse latitudes   48
Horton, R. E.   198, 225

Horton-Thiessen polygon method   67, 69
Hubbard Brook Experimental Forest   298, 110, 112, 115, 116
Humidity   42
Humidity Index   105
Humus   84, 153, 165, 182
Hurricane Agnes   222
Hurricane control   10, 275
Hurricanes   31, 66, 216, 218, 219, 245
Hydraulic conductivity   141, 149, 175
Hydraulic gradient   141, 181
　　head   141, 180
　　jump   187
　　similitude   256
Hydraulics   185
Hydric soils   132
Hydrogen bonding   234, 244
Hydrogeology   139
Hydrograph*   xvi, 12, 115, 142, 184, 186, 191, 201, 243, 248, 250, 254, 261
　　and infiltration determination   179
　　behavior   225
　　separation   142, 198
　　simulation   227
Hydrographic region   210
　　boundaries, in models   230
　　units   273
Hydrologic behavior   196, 254
Hydrologic Characteristics, of Major Forest Types   293
　　at the time of sampling,   245
　　at time of water quality sampling   243
Hydrologic cycle   xxi, 1, 15, 18, 25, 27, 29, 31, 57, 77, 137, 142, 175, 210, 230, 248, 311
　　and models   227
　　and ozone depletion   278
Hydrologic data networks, in models   232
Hydrologic drought   223
　　environment, effects of wildfire on   120
　　forecasting   229
　　index   224
　　integrity   295
　　models, and remote sensing   174

records, filling gaps in, in models 232
scientist 312
seasons 104, 205, 217, 219, 222, 224, 243
Hydrologic similitude 256
Hydrologic Year (*See* Water Year)
Hydrologist 233
  and acid rain 278
  and ozone depletion 278
Hydrology 34, 139, 196, 198, 201, 246, 285, 3111
  science of xviii, 233
  and wetlands 133, 134, 135
Hydroperiod 134, 241
Hydrophobic soils 149
Hydrophytic vegetation 132
Hydropower 123, 275
Hydrosphere 9, 183, 223, 237, 271
Hydrospheric systems 132
Hydrostatic force 94
Hygroscopic coefficient* 144, 170
Hygroscopic storage 7, 170
Hygrothermograph 41, 46, 59
Hypolimnion 244
Hypsometric curve 257

Ice 4, 60, 131, 163, 235, 249, 250, 271
  crystals 30, 43, 235
  in raingage 58
  vapor pressure over 43
Idaho 55
Igneous rocks 151
  poor aquifers 238
Illinois State Water Survey 65
Illuviation, zone of 153
Immersion Method of soil moisture determination 174
Impoundment 304
India 57
Industrial watersheds 123
Infiltration* xix, xviii, 2, 4, 13, 56, 83, 95, 149, 154, 164, 166, 176, 198, 210, 214, 225, 227, 234, 250, 257, 271, 273, 287, 303

Infiltration capacity 176
  effects of wildfire on 120
  in models 230
  measurement of 177
  model 230
Infiltrometer 177
  concentric ring, 179
Influent stream 268
Inorganic material 150, 165
Insecticides 233
Insects 150, 154, 165
Instantaneous flow values, used for flood frequency analysis 215
Instantaneous peaks 217
Instrumentation 290
Instruments xxii
  for measuring atmospheric humidity 45
  for measuring energy 27
  for measuring evaporation 75
  for measuring percolation 180
  for measuring rainfall 58
  for measuring runoff 186
  for measuring snow 62
  for measuring soil moisture 171, 174
  for measuring water quality 242
Intensity-duration-frequency curves 72
Intensity-duration-frequency relations 80
Interception* xx, 4, 77, 80, 82, 96, 98, 108, 112, 122, 132, 137, 224, 229
  Conservational Effects of 91
  effects of 82
  effects of wildfire on 120
  Mechanical Effects of 83
  in model 125
  Quantitative Effects of 83
  storage 139, 166
Interflow 200
Intermittent stream 268
Interstices 144, 178
Intrazonal soils 163
Inverse Influence, principle of 272
Inwashing of fine particles 176
Iron 153, 162, 165, 181, 239
Irregular runoff variations 196

Irrigation 100
Isohyetal method 67, 69
Isopleths 194

Jack Creek 119
Jack pine 114
Jet stream 35, 48, 53
Johnstown Flood 109
Juvenile water 140

Kankakee River 275
Kinetic energy 32
King Tube 171
Klamath River, CA 212
Koppen 79

La Porte, IN 275
Laboratory analysis of water quality 242
Laboratory models 256
Lacustrine Soils 163
    wetlands 135
Lag time, xx, 111, 201, 268
    and watershed shape 266
Lake Baikal 10
Lake Erie 55
Lake Hefner 275
Lake Michigan 275
Lake Tahoe 78
Lake-effect precipitation 55, 56
Lake/watershed system 244
Lakes 5, 8, 11, 76, 139, 140, 183, 240, 244, 273
    chemistry 198
    and wetlands 134
Lambert's Cosine Law 20
Laminar flow 204, 234
Land management xxi
    practices 270
Land managers 250, 281
Land Use 14, 104, 111, 115, 119, 203, 210, 216, 218, 227, 241, 251, 274, 285, 290, 311
    and albedo 19
    change 246

regulation 307
    studies 102
Landforms 250
Landscape esthetics 113
Latent heat 19, 25, 32, 64, 78
    of condensation of water 235
    of vaporization of water 235
Latitude 42, 73, 101, 104, 228
    equivalent 22
Leachate 239
Leaching 118, 154, 164
Leaf area indices, in model 125
    color, and albedo 19
    properties 88
    reflectivity, and albedo 19
    roughness 88
Leaf drip* 85
Length, watershed 252
Levees 10
Lightning 56, 238, 288
    suppression 275
Lime line 163
Limestones 151, 238
Limnology, science of 184, 233, 311
Linear hydrologic model 228
Litter* 58, 83, 84, 149, 150, 154, 163, 165, 176, 179
    water stored in 132
Livingston atmometer 76
Loam 150, 170
Loblolly Pine 86, 87
Local solar noon 27, 32, 131, 229
Lodgepole Pine 86, 296, 300
Lodgepole Pine/Alpine Fir 119
Loess 163
Logging 113, 280, 301
    reduction of infiltration by 177
    roads 177
Long Island, NY 140
Long-term trends runoff variations 196
    emission from the forest 26
    radiation 17, 23, 25, 38, 64, 78
    reradiation 98
Longleaf Pine 86

and grazing   288
Los Angeles   54, 84, 119
Low buffering capacity of Adirondack
    lakes   238
Low flow, in models   231
Low flows   241
Low pressure   50
Low pressure systems   35
Lysimeter*   93, 102, 175, 229
Lysimeters to measure percolation   180

Macropore flow   149
Magma   151
Magnesium   238
Maine   301
Maine   63, 73
Man, a factor of watershed formation   271
Manning's Formula   186, 236
Manometer, and soil moisture
    sampling   172
Map scales, and Appendix B   254
Marble   152
Marine soils   163
    wetlands   135
Marshall Creek watershed   270
Masking of runoff behavior   273
Mass-Transfer Method   78
Maximum Evapotranspiration, Season
    of   104, 105, 219, 222
Maximum Runoff, Season of   105, 219
    in model   125
Maximum Soil Moisture Utilization, Season
    of   104
Maximum Thermometer   40
Maximum to minimum flows, ratio of   255
Mean annual flood (See Flood, mean
    annual)
Mean annual precipitation   205
Mean daily flow   217
    temperature   42
Mean elevation of drainage basin   205
Mean precipitation calculation   67
Measures of central tendency   228
    of dispersion   228

Metabolic rate of the fish   241
Metamorphic rock   152
Meteor showers   30
Meteorologic drought   223
Meteorologist   xxi
Meteorology   246
Metric system   xxii
Mexico, gully control in   285
Mica   151
Michigan   64, 114
Mickey Mouse   234
Millswitch Creek Watershed   249, 270
Mineral soil   83
Minnesota   110
Misting   91
Mixing ratio   45
Model   2, 124, 241, 244, 300, 311
    defined   227
    length for similitude   256
    studies, laboratory   200
    and modeling   227
        the Watershed   255
    types of   228, 256
Modeling, and remote sensing   174
Mohawk River, NY   214
Moisture Index   105, 280
Molecular forces   95
    structure of water   234
Monitoring Water Quality   241
Monsoon   57
Montana   119
Mor humus type   165
Mount Rose Snow Sampler   62
Mount Wilson   54
Mountain pine beetle   119
Mountainous watersheds   213
Mull humus type   165
Mull layer   154
Multiple regression   229
Multiple Use and Sustained Yield Act   110
Municipal water supply watersheds   281
    watersheds   123, 248, 289, 301

National Forests   110, 122

linear programming resource allocation model   126
National Urban Runoff Program   306
National Weather Service   223
National Wetland Inventory   135
Natural resources management   311
Navigation   9, 276, 302
Nebraska   114
Net precipitation*   90, 229
Net radiation, in model   125
Neutron soil moisture meter   102, 130, 173, 180
New England   210
New Hampshire   110, 112, 212, 298
New Jersey   103
   in model   126
New Mexico   296
New York   109, 120, 121, 210, 214, 219, 222, 224, 238, 296, 307
   and wetlands   132
   flood study   219
New York City   302
New Zealand   108
Nitrates   238
   concentration in rainfall   238
   increase following conversion of chaparral to grass   115
   in model   126
   nitrogen   116
   and wetlands   136
Nitric acid, and acid rain   278
Nitrogen cycle   116
Nitrogen dioxide, and acid rain   278
   oxides   238
Nitrogen, fixed by lightning   238
Noncapillary pores   7, 149, 166
Nonpoint sources of pollution   248, 305
   control   306
Nonrecording Standard Rain and Snow Gage   58
North Carolina   85, 87, 280, 290
North Dakota   108
North Fork Infiltrometer   178
Nowmelt forecasting, in models   232

Nuclear Winter   25
Nutrient budgets   115, 233
   cycling   150, 153, 154, 166, 181
   complexity of chemical environment   117
   movement   97
   relations   112
   removal from a managed stand   116
   shock   116
   and wetlands   136
Nutrients   94, 240, 274, 280, 291
   and transmission in soil   181
   and wetlands   135

Occluded front*   53
Oceanographer   xxi
Oceans   9, 11, 15, 32, 31, 55, 73, 183, 228, 249
   and ozone depletion   278
   and wetlands   134
Oklahoma City   275
Oligotrophic lakes   184
Olympic Peninsula   53, 55, 57
Omnibus Flood Control Act of 1936   251
Onion Creek Experimental Forest   297
Open channels   256
Open space, and wetlands   133
Optimization, (*See* Models)   124
Oregon   55, 93, 113, 115, 118, 122
   and phreatophytes   127
Organic content   143, 165, 179
   material   165, 175, 83, 84, 150
Orientation, of watershed   252, 260, 270
Orographic precipitation   45, 53, 122
Orographic storm*   219
Orography   56, 69
Osmotic pressure   94, 130
Overgrazing   177, 274
Overland flow*   2, 115, 198, 200, 251
   effects of wildfire on   120
Oxygen   95, 127, 184, 238
   supersaturated   237
Ozone   278

P-RO 225
Paired watersheds 229, 290
   in models 124
Pakistan 230
Palmer Drought Index 223
Palustrine wetlands 135
Pan coefficient 76
Pan evaporation 108
Parent material 150, 153, 182
Parking lots 2, 249
Partial duration series 215
Partial pressure 43
Particulates, and the greenhouse
   effect 276
Patterns of flow 193
Peak runoff (flows, discharge) 13, 109, 111,
   119, 122, 137, 184, 196, 203, 215, 216,
   255, 263, 268, 274
   increase following conversion of
      chaparral to grass 115
Peak runoff, increase in by clear-
   cutting 110
Pedalfers 162
Pedestalling 283
Pedocals 162
Penman's E_t equation 227
   in model 100, 125
Pennsylvania 109, 212, 222, 275
Penstocks 256
Percent by volume, soil moisture 170
Percent by weight, soil moisture 170, 174
Percolating water 141
Percolation 4, 144, 165, 176, 179
   in models 230
Perennial stream 268
Perimeter, watershed 252
Permafrost 175
Permeability 4, 144, 148, 165, 177, 181
Pesticide 306
PF 143
PH 30, 165, 181, 241, 242, 244, 245
   of Adirondack lakes 238
   of rainfall 238
Phosphorous 116

Photogrammetry, and remote sensing 174
Photosynthesis xx, 27, 240, 244
Phreatic divide 252
Phreatophyte* 127
   control 128
   eradication 302
Physical characteristics of water 236
Physical models 229, 256
Phytometer* 98, 102
Piezometer 172
Piezometric surface 172
Pinyon-Juniper 113, 121
Plagioclase 151
Planck's Law 16
Planner xxii
Plant-soil-water relations, high degree of
   variability in 124
Plot, for measuring infiltration 178
Podzol 181
Podzolization, and acid rain 278
Polar front 48
Polarization of water molecule 30
Policy 311
Pollution 141, 233, 305
   and wetlands 132
Pollutograph 115, 236, 245
   differs for the different consituents 244
Ponderosa Pine 84, 114, 121, 300
   and grazing 288
Ponds 5, 6, 184, 240, 250
Population 228
Pore size 149
Pore space 143, 165, 170
Pores, soil 4
Porosity 143, 148, 165, 234
Porous medium 176, 181
   flow through 141
Portland, Oregon 301
Potable water 236
Potassium 120
Potential evapotranspiration 30
Poughkeepsie, NY 104
Prairie potholes 108

Precipitation* xix, 2, 4, 30, 32, 35, 50, 80, 87, 96, 100, 140, 143, 164, 176, 182, 184, 192, 209, 223, 238, 254
   amount of 57
   annual 228
      in the United States 162
   areal extent of 57
   average intensity per storm decreases as in channel 200
   drainage area increases 255
   duration of 57
   excess 180
   form of 57
   frequency of 57
   intensity-duration-frequency 205, 219, 310
   measurements, faith in 72
   in model 125
   purity of 30
   and runoff, relationship between 229
   sampling networks 63
   sampling rate 192
   variability of 66
   volume, total annual 9
Precipitation Effectiveness 155
Precipitation intensity* 57, 198, 204
Precipitation-runoff relations 224
   relationship, and vegetation conversion 291
Precision 242
Prescribed fire 289
Pressure 34
   atmospheric 35
Pressure gradient force 35, 50
Private water supply companies 302
   watershed 123
Probabilistic models 228
Probability of flood occurrence 216
Properties of Water 234
PROSPER 125
Protection, a phase of watershed management 279, 285
Public, and decisionmaking 311
Puddles 5, 6, 73, 250

Pumice 151

Quarter-flow interval 113, 210, 273
Quartz 151
Quickflow 12, 110, 200, 201, 225

Radar 64
Radiation 38, 42, 78, 98, 130, 164, 244, 254
Radiation fog 47
Rain gages 64
   design 58
Rain-on-snow peaks 196
Raindrops 176
   size of 57
Rainfall 4, 57, 110, 112, 175, 199
   amount of 65
   daily, in model 125
   frequency of 65
   simulator 254, 256
   and soil moisture sampling 172
Rainfall intensity 65, 176, 225, 245, 251
   (precipitation intensity) 205, 219, 310
   intensity-duration-frequency curves 66
Rainfall-runoff model 231
Rainstorms 12, 219
Random component, in stochastic models 231
   runoff variations 196
   sample, criterion for 228
Rangelands 123, 281, 287
   Best Management Pratices on 308
   infiltration on 177
Range management 248, 311
   pitting 287
Rating curve 191
Rational formula 204
   and models 227
Recession constant 12
Recession curve* 12, 180, 205, 225
   and models 227
Recession limb 200
Reciprocal Distance Squared Method 71

Recording Rain and Snow Gage   59
Records, long-term from forested
    watersheds   196
Recreation   9, 222, 276, 311
    and wetlands   132
Rectangular weirs   191
Recurrence interval   217, 220
Red alder   127
Red Fir-Lodgepole Pine   122
Red Pine   114
Reduction in runoff on reforested
    watersheds differs from that on
    abandoned lands   114
Reforestation   120, 240
Refuse Act of 1899   305
Regimen   183, 185, 250, 289
    and watershed shape   266
    improving the natural   294
Regional flood frequency curves   221
    flow duration curves, in models   230
    hydrology   231
Regression coefficients, high value of   229
Regular periodic runoff variations   196
Rehabilitation, a phase of watershed
    management   279, 280
Relative humidity   26, 44, 47, 50, 84, 95,
    166, 228, 260
    in Redwood stand   130
Remote sensing   63, 241
    Methods for measuring soil
        moisture   174
    and wetlands   135
Research   97, 103, 104, 109, 110, 115, 121,
    196, 203, 229, 233, 250, 255, 290
    and wetlands   134
Reservoirs   76, 202, 223, 276
    operations, in models   232
    water supply   63
Residence time   8, 184
    for ground water   140
Residual Soils   163
Respiration,   xx
Retention storage   7, 129, 139, 154, 166,
    170, 175, 179, 250, 273
    and models   227
Return period   xviii, xx, 13, 66, 204, 217
Reversible reactions in water   239
Revetments   283
Reynold's law   256
Rime   60, 91
Ring Infiltrometers   178
Rio Grande   296
Riparian vegetation   115, 279
    wetlands   135
    zone, removal of vegetation from, and
        conversion   291
    zones   285
Riprap outfalls   236
Rising limb*   12, 200, 201
    and water quality sampling   243
River basin hydrology, in models   232
    morphology, and phreatophytes   127
    regulation, and phreatophytes   127
Riverine wetlands   135, 136
    zones   200
Rivers   11, 140, 163, 183, 223, 244
Rivers and Harbors Act of 1899   305
Roads   286
    construction   274
Rock Composition   239
Rocky Mountain infiltrometer   178
    Region   122
    watersheds, in model   126
Rocky Mountains   53, 54, 295, 296
Root development   165
Root zone   104, 109, 114
    in model   126
Roots   179
Rotation grazing   287
Roughness   204
    coefficient, Manning's formula   187
Runoff   5, 10, 12, 30, 31, 57, 63, 66, 80, 84,
    91, 96, 100, 104, 109, 117, 123, 141,
    175, 183, 184, 185, 192, 223, 238, 239,
    243, 253, 260, 270, 303
    annual   196
    behavior   267, 270
    coefficient   204

Subject Index

increase in caused by forest vegetation cutting   109
increases following forest cutting   111
modeled   202
from mountain glaciers, in models   230
per unit of precipitation   224
seasonal distribution of   273
from snow   62
temperatures of   32
and wetlands   132
Runoff-causing event   12
Runoff-influencing factors   273
Rural areas   56, 246
Rural-urban interface, water quality changes at the   241

Sagebrush   300
Salt cedar, (*See* Tamarisk)
Salt Lake City   301
Salt River watershed   300
Salts, dissolved   31, 97
Saltwater wetlands   132
Sample   228
Sample representativeness   242
Sampling frequency, water quality   243
San Dimas Experimental Forest   84 102, 119
San Francisco, CA   181, 301
San Juan Mountains   275
Sands   141, 143, 170
Sandstones   151, 238
Santa Ana wind   54, 235, 281
Satellite imagery   50, 64, 230, 300
Saturation of soil   180
Saturation vapor pressure   42
over ice less than over water   235
Scaling   255
Sea of Cortez   53
Season   42, 205, 228
and albedo   19
Seattle, WA   301
Secchi disk transparency   241
Section 208   305

Sediment   31, 114, 122, 140, 191, 236, 246, 287
concentration of suspended   236
deposition, and phreatophytes   127
Management   306
increases upon installation of roads   115
no increase during logging   115
no significant increase in after yarding and felling harvestable timber   115
Sedimentary rock   151, 238
Sedimentation   109, 184, 249, 274, 276, 280, 304
Selection cutting   111, 113, 114, 281
Self-similarity   255
Sensible heat   25, 26, 78, 79, 124
Sensitivity analyses   124
Seventh Approximation   155
translation to Great Soil Groups   163
Shackham Brook reforestation and runoff   297
Shale   152
Shape, watershed   104, 212, 252, 257
Sharp-crested weirs   191
Shelterwood cutting   281
Short-wave radiation   17, 23, 25, 38, 64
and the greenhouse effect   276
radiation at night   26
Shortleaf pine   86
Sierra Nevada Mountains   114, 122, 196, 295
snowpack management in   297
Silica   237, 238, 241, 246
Silts   143
and wetlands   132
Silver iodide   43
Silvicultural activities, and albedo   19
Similitude (*See* Hydraulic, Hydrologic, or Watershed similitude)   256
Simulation models, types   124, 227
(*See* Models)
Size of watershed (*See* Watershed, size of)
Skew, of flood frequency curve   217
of flow duration curve   211
Skid trails   177, 280, 286

Skidding 118
Slash burning 113
Slash Pine 86
Slate 152
Sleet 60
Sling Psychrometer 45
Slope 22, 42, 104, 164, 198, 205, 250, 257
　of channel 187
Slopes, watershed 212
Small watershed 13, 139, 186, 210, 246,
　248, 250, 261, 270, 273, 301
　defined 251
　hydrology of 175
　water yield model 230
Small Watershed Protection and Flood
　Prevention Act 302
Smelter activity, and acid rain 278
Smithsonian Silver-Disk
　Pyroheliometer 27
Snow* 4, 57, 60, 62, 86, 131, 132, 196, 210,
　260
　accumulation 121
　　and snowmelt, effects of wildfire
　　　on 120
　albedo of 17
　course 62, 63
　cover, and remote sensing 174
　　in model 125
　depth, by remote sensing 63
　fences 64
　pillows 63
　water content 62, 64
　　by remote sensing 63
　　increases following timber
　　　harvesting 122
Snowfall 55, 270
　in mountainous areas 62
Snowflake 62, 235
Snowmelt 12, 13, 64, 110, 112, 164, 196,
　199, 202, 213, 216, 219, 255, 257, 260,
　281
　flood peaks delays 301
　in model 126
　modelling 231

runoff 200, 236, 301
runoff, and acid rain 278
Snowmelt Runoff Model 64
Snowpack 5, 43, 62, 64, 86, 91, 123, 224,
　235, 275, 290
　energy budget of 27
　color of 64
　improving runoff from 294
　Management 296
　management 64, 248
　water equivalent 112
Sodbuster 307
Sodium 116
Soft water lakes, and acid rain 278
Soil 2, 4, 14, 27, 58, 73, 91, 94, 130, 137,
　139, 154, 198, 271, 273, 311
Soil and water conservation district 307
Soil Conservation Service 63, 97, 170, 178,
　251, 296, 302
Soil depth 104, 250, 257, 273
　depth at high elevations 260
　development 164
　dryness index 175
　fauna 165
　fertility 150, 165
　fertility, and albedo 19
　mantle 5
　material 152, 153, 164
　moisture Recharge, Season of 105
　moisture 9, 64, 82, 98, 150, 179, 224
Soil moisture content 206, 230
　increased by clear-cutting 110
　by remote sensing 63
Soil moisture deficit 105, 112
　in models 230
Soil moisture determinations, and remote
　sensing 174
　measurements 102
　movement, in models 230
　sampling 130
　storage 166, 182, 222
　　effects of wildfire on 120
　and drought 223
　in model 125

Subject Index 403

in Redwood stand   130
   monitoring   170
Soil Moisture Recharge, Season of   219
   particle sizes   150
      gradient, importance of   179
      surface area   147
Soil profile   163
Soil sample analysis (gravimetric)   171
Soil scientist   xxi
Soil separates, size limits   147
Soil storage   30, 57, 100
Soil structure   143, 176, 179
Soil texture   143, 170, 176, 179, 273
   triangle   143
Soil water   7, 10, 143, 154
   depletion   122
Soil wettability   149
Soil, "age" of   152
   protection of the   82
Soil-forming factors   150, 151
Soil-vegetation survey system   155
Soil/Vegetation/Water Classification   181
Solar beam radiation   27, 112
Solar Constant   16, 27
Solar energy   240
   heating   35
   radiation   15
      and ozone depletion   278
Solum   154, 164
Solution channels   152
Solvent, water as a   234
Specific conductance   237
Specific heat, of water   16, 31, 235
Specific humidity   45
Spectral analysis, in models   232
Sphere of influence of soil moisture
   measurement   174
Spontaneous condensation   43
Spring snowmelt   218
Sprinkling Infiltrometers   178
Spruce bog   133
St. Lawrence River   55
St. Louis Creek watershed   296
   snowpack management research in   297

Stalactite frost   (See Frost, Stalactite)
Standard deviation   228
Standard Methods   242, 245
Standard pressure   35
Standard Project Flood (See Flood,
   Standard Project)
Standard Rain and Snow Gage   72
Standard Thermometer   40
Stanford Watershed Model   232
State, change of   16
Static models   229
Station-angle method   67, 69
Steam fog   47
Steamboat Springs, Colorado   91
Stefan-Boltzmann Constant   16
Stemflow*   xx, 5, 86, 274
   measurement of   87
Stilling well, in stream gage   191
Stochastic models   228, 230
Stomata*   95
   and phreatophytes   128
Storage   xv, xvi, 5, 6, 10, 30, 77, 82, 89, 129,
   137, 165, 183, 192, 224, 248, 250, 276
   formula   8
Storm   12
   classification   46
   duration   56
Storm flow   (runoff, discharge) 5, 12, 14,
   110, 139, 142, 198, 201, 203, 209, 225,
   246
   distribution of   57
   increase in volume of by clear-
      cutting   110
   hydrograph   273
      effect of forest cutting on   110
   movement, direction of and runoff   263
   volume lowered by timber
      harvesting   111
Stream (See also streamflow)   11, 273
   behavior   57, 66, 80, 111, 198, 210, 261
      effect of forest cutting on   110
   channel   283
   density   267
   gage   12, 203

improvement 32, 302
management 216
network 32
order* 266
regimen xx
rises, classification of 225
size 216
slope, alteration of 303
system 295
Stream/watershed system 210, 216-218, 221, 222, 229, 310
in models 231
Streambank erosion control 302
Streambed roughness, alteration of 303
alteration of 303
Streamflow 5, 11, 98, 104, 114, 122, 142, 191, 193, 215, 224
changes following farmland abandonment 120
changes following reforestation 120
declines following vegetation conversion 113
effect of forest cutting on 300
hydrograph separation, in models 230
increased in model 125
increase in 101
increases following outbreak of mountain pine beetle 119
increases following outbreak of bark beetles 119
in model 125
velocity, and suspended particle size and amount 236
water quality 109
in models 232
Streamflow or channel routing 227
Streams 252
Strip cropping 303
Structure, of soil (See Soil structure)
Sublimation* 4, 25
Subalpine Fir 289, 300
Subalpine zone 122, 260
Sublimation 299

Subsurface flow (runoff)* 5, 7, 110, 141, 142, 175, 185, 200, 205, 225, 227, 250
in models 230
and wetlands 133, 134
Suburban land use 56, 214, 246
Sulfate 118
Sulfides, and acid rain 278
Sulfuric acid, and acid rain 278, 280
Sulphur dioxide 280
Sunspot cycles 223, 228
Surface erosion, effects of wildfire on 120
Surface runoff (flow)* 2, 5, 7, 14, 110, 115, 140, 175, 177, 185, 198, 200, 225, 227, 236, 250, 283
and wetlands 134
Surface tension of water 234
and similitude 256
Suspended sediment, in model 126
Suspended solids 236
and wetlands 135
Susquehanna River, NY 219
Sustained yield forest management 122
Swampbuster 307
Synthetic streamflows, in models 232

Tamarisk, 127, 128
Taste of water 236
Technical Bulletin No. 40 66
Tehachipi Mountains 54
Temperate Zone, flood occurrence in 216
Temperature 4, 34, 36, 42, 74, 95, 164, 175, 205, 222, 228, 236, 241, 246, 260
average 25
inversion, in Redwood stand 130
as a geomorphic agent 250
in model 126
in Redwood stand 130
and soil moisture sampling relations 172
Temperature Efficiency 100, 155
Tennessee 280
Tennessee Valley Authority 97
Tensiometer, for variable source area dtermination 200

Subject Index 405

Tensiometric Methods for measuring soil moisture   172
Tension   143, 168, 172, 179
   barriers   149
   gradients   95, 139
Terminal velocity, of raindrops   58, 83
Terracing   287, 303
Terrasphere   xvii, 139, 183, 184, 237, 238
   and phreatophytes   128
Texas   57, 73, 115
   and phreatophytes   127
Texture, soil (*See* Soil, texture)
Thermodynamics, first law of   32
Thermometers   40
   minimum   40
   Standard,   40
Thinning   114, 240
Thornthwaite and Mather water budget   100, 104, 229
   and models   227
Thornthwaite climatic classification   79
Thornthwaite, C. W.   85, 155
Thornthwaite-Holzman equation   74, 78
Three Bar watersheds   291
Throughfall*   xx, 5, 85, 89, 110, 229
   gage   85
Thunderstorm   53, 55, 56, 65, 66, 85, 218, 238, 245, 275
Tidal influences of the moon and sun   35
   marshes   136
   wetlands   133, 134
Timber harvesting, impacts of on water quality   118
   impacts on storm flow   of   118
Timber management   281, 311
Time   271
Time (a soil-forming factor)   151, 152, 182
Time of Concentration   xix, 13, 204, 266
   and watershed shape   261
Time of day   42
Time of rise   xx
Time series analysis, in models   231
Tioga River, NY   222
Tipping bucket gage   60

Topographic divide   252
   map   257
Topographic sampler   262
Topography   53, 151, 163, 200, 249
   modifying radiation   22
Topsoil   199
Tornados   53, 56
Torrents   283
Total dissolved solids (TDS)   31, 234, 237
Total suspended solids (TSS)   31
Towing icebergs   10
Transmissibility   4
Transmission   4, 176, 180
Transpiration*   xx, 5, 77, 80, 94, 96, 112, 122, 130, 137, 168, 273, 291
   in model   125
   in models   230
   measuring   96
Trapezoids, used for streamflow measurement   188
Tree-ring analysis   223
Trenched plots   117
Tributaries, and stream order, number and class   266
Trincheras   285
Trivia   273
Tulip Poplar   87
Turbidity   118, 236
Turbulent flow   235
Turgor pressure   88
TV weather forecaster   231

Ultraviolet radiation, and ozone depletion   278
Understory vegetation   83
Undisturbed soils   83
Ungaged watersheds, estimating floods peaks from   222
Unit hydrograph   14, 198
   and models   227
   prediction of peak flows with   203
   theory   201
      assumptions underlying   202

United States   53, 64, 65, 66, 84, 100, 105, 110, 119, 121, 187
   agricultural land in   306
   and acid rain   278
   average annual runoff in   194, 197
   diversion in western   275
   first watershed management research in   296
   forest types in   292
   ground water withdrawals   140
   gully control in   285
   monthly distribution of runoff   195
   monthly patterns of flow   210
   normal distribution of runoff by months   197
   patterns of flow in   193
   population change in   311
   raingages in   58
   research in   296
   soil distribution in   155
   streamflow measurement in   186
   use of DDT in   305
   vegetation differences due to aspect   261
   wildland management activities in   244
Units   xxii
Universal Recording Rain and Snow Gage   60
Upstream land treatment measures   303
   watershed   311
Urban environment   311
   hydrology   248
   lands, BMPs on   308
   stormwater runoff   306
Urbanization   14, 56, 214, 246
   effect on flood peak magnitude   220
Utah   102
V-notch weirs   191

Validation, of models   228
Vapor pressure   42, 74, 229
   deficit   26, 32, 44, 100
   gradient   44, 66, 73, 78, 84, 90, 95, 124, 235, 260

Variability   4
Variable Source Area*   xviii, xix, 11, 13, 198, 199, 263
Vegetation   xvi, 14, 91, 137, 164
   conversion   113, 248
   cutting   102, 248
   growth   276
   management   xxi, 294
   succession   166
   surfaces, water stored on   132
   type   42
   type or cover   205
   types, and wetlands   133
Vegetative cover   4, 10, 19, 29, 97, 254, 270, 279
   and wetlands   134
Velocity head rod   187
Velocity of water   186
Vermont   177, 200
Virginia   307
Viscosity of water   234, 235
   and similitude   30, 151, 238, 256
Volcanic activity, and acid rain   278
   and the greenhouse effect   277

Wading rod   190
Wagon Wheel Gap   109, 296, 311
Wall-and-step pattern of forest cutting   299
Wappingers Creek watershed, NY   210
Warm front*   52
Warm rains on snowpack   224
Washington   55, 57, 122
Washington State University   149
Washout   238
Waste Management   306
Waste water treatment   233
WATBAL   126
Water   90, 249, 250
   in atmosphere, movement of   47
   in circulation   29
   film on soil   179
   issues   311
   molecule   234
Water balance   (*See* Water budget)